# Human Tooth Crown and Root Morphology

## The Arizona State University Dental Anthropology System

This guide to scoring crown and root traits in human dentitions substantially builds on a seminal 1991 work by Turner, Nichol, and Scott. It provides detailed descriptions and multiple illustrations of each crown and root trait to help guide researchers to make consistent observations on trait expression, greatly reducing observer error.

The book also reflects exciting new developments driven by technology that have significant ramifications for dental anthropology, particularly the recent development of a web-based application that computes the probability that an individual belongs to a particular genogeographic grouping based on combinations of crown and root traits; as such, the utility of these variables is expanded to forensic anthropology.

It is ideal for researchers and graduate students in the fields of dental, physical, and forensic anthropology and will serve as a methodological guide for many years to come.

**G. Richard Scott** is Foundation Professor of Anthropology at the University of Nevada, Reno. He focuses on native populations of the American Southwest, Alaskan Eskimos, Norse in the North Atlantic, and Spanish Basques.

**Joel D. Irish** is a Professor of Biological Anthropology at Liverpool John Moores University. He has traversed the length and breadth of Africa studying teeth from Plio-Pleistocene hominins to recent Arabs in the north and Zulu in the south.

Both Richard Scott and Joel Irish are past presidents of the Dental Anthropology Association.

# Human Tooth Crown and Root Morphology

## The Arizona State University Dental Anthropology System

**G. Richard Scott**
University of Nevada, Reno

**Joel D. Irish**
Liverpool John Moores University

Shaftesbury Road, Cambridge CB2 8EA, United Kingdom

One Liberty Plaza, 20th Floor, New York, NY 10006, USA

477 Williamstown Road, Port Melbourne, VIC 3207, Australia

314–321, 3rd Floor, Plot 3, Splendor Forum, Jasola District Centre, New Delhi – 110025, India

103 Penang Road, #05–06/07, Visioncrest Commercial, Singapore 238467

Cambridge University Press is part of Cambridge University Press & Assessment, a department of the University of Cambridge.

We share the University's mission to contribute to society through the pursuit of education, learning and research at the highest international levels of excellence.

www.cambridge.org
Information on this title: www.cambridge.org/9781107480735

DOI: 10.1017/9781316156629



First published 2017

*A catalogue record for this publication is available from the British Library*

*Library of Congress Cataloging-in-Publication data*
Names: Scott, George Richard, author. | Irish, Joel D., author.
Title: Human tooth crown and root morphology : the Arizona State University dental anthropology system / G. Richard Scott, Joel D. Irish.
Description: Cambridge, United Kingdom : New York, NY : Cambridge University Press, 2017. | Includes bibliographical references and index.
Identifiers: LCCN 2017003669 | ISBN 9781107480735 (pbk. : alk. paper)
Subjects: | MESH: Tooth Crown—anatomy & histology | Tooth Root—anatomy & histology | Anthropology, Physical—methods | Paleodontology—methods
Classification: LCC RK305 | NLM WU 101 | DDC 617.6/3—dc23 LC record available at https://lccn.loc.gov/2017003669

ISBN 978-1-107-48073-5 Paperback

This guidebook is dedicated to the memory of
Christy G. Turner II, the primary architect
of the ASUDAS and our mentor

# Contents

# Acknowledgments

This volume is, first and foremost, a testimony to the legacy of Regent's Professor Christy G. Turner II, whose prescience and decades of hard work made dental morphology a significant area of research in physical anthropology. After Christy's untimely passing in 2013, GRS initiated the Christy Turner legacy project, which involved scanning a mountain of data sheets, slides, and computer printouts so they would not be lost to the silverfish in the back bedroom/library. Without the support of Christy's youngest daughter, Korri, and his second wife, Olga Pavlova, this task would not have been possible. They always make me feel like part of the family, and I can only hope our efforts now and in the future do justice to Christy's memory, his family, and the broader field of dental anthropology.

The authors would like to recognize several individuals who played important roles in the production of this guidebook. First, we thank Cambridge University Press for their unstinting support of physical anthropology in general and dental anthropology in particular. Editors at the Press, especially Caroline Mowatt and Dominic Lewis, are acknowledged for their many services that led to the production of this volume.

In addition, GRS thanks colleagues Scott Burnett and Christopher Stojanowski for providing specific trait illustrations we could not otherwise locate, and the following graduate students who provided a field test for trait descriptions in a seminar on dental anthropology: Donovan Adams, Rebecca George, Callie Greenhaw, Kelly Heim, Amelia Hubbard, Shannon Klainer, Amanda O'Neill, Kirk Schmitz, and Victoria Swenson. They were encouraged to be blunt and provide input on the traits that were easy to score and those that proved the most difficult.

JDI thanks the many individuals who played a role in nurturing my interest in all things dental, including Christy Turner, Don Morris, and Albert A. Dahlberg. As well, thank you to the dozens of individuals at museums and other institutions around the world for access to collections. But without cash, of course, nothing would have been possible, so the following granting agencies are acknowledged for allowing me to study thousands of African and other dentitions: National Science Foundation (BCS-0840674, BNS-0104731, BNS-9013942), Wenner-Gren Foundation (#7557), National Geographic Committee for Research & Exploration (#8116-06), among others. My parents provided me with lifelong support, and my wife Carol has filled this role for 21+ years now.

# Part I
# Introduction

During the first half of the twentieth century, scoring dental morphology was idiosyncratic. That is, researchers who wanted to study morphology would decide which traits they wanted to score and the system of recordation for scoring them. Except for basic observations such as molar cusp number, there was relatively little agreement on how to score these traits. Hrdlička (1920) was among the first to realize that observations had to be fine-tuned for shovel-shaped incisors, which showed extensive variation in presence expressions, with a dramatic difference between American whites and Native Americans. Toward this end, he developed a four-grade scale that included no shovel, trace, semi-shovel, and full shovel. Hrdlička (1924) also used a rank scale to score the precuspidal fossa (i.e., anterior fovea). He recognized that tooth crown traits were not simply discrete, presence/absence variables, but felt the need to recognize the variability of presence forms, which often ranged from slight to pronounced. Because of its wide range of variation in American whites and Europeans, Carabelli's trait was often scored on a ranked scale (e.g., Dietz 1944, Kraus 1951), but these scales varied by observer. Beyond these examples, many traits were scored as present or absent even when presence forms varied from slight to pronounced (or small to large). The problem was that individuals had different ideas as to what constituted trait presence, and the result was widely contrasting trait frequencies for closely related groups.

Albert A. Dahlberg, a dentist by training and trade but one blessed with an inquisitive anthropological and biological mind when it came to teeth, recognized the pitfalls and inconsistencies in scoring dental morphology. To rectify the situation, he set himself the task of developing ranked standards for 16 traits (Dahlberg 1956). After setting up ranked scales, he duplicated them in plaster and distributed them to dental researchers throughout the world. This was the first necessary step to reducing inter-observer error in scoring tooth crown traits (he did not include roots in this effort). Although observer error was resolved to some extent, it was not eliminated altogether.

Utilizing the Dahlberg standards in his dissertation on Arctic populations, Christy G. Turner II (1967a) wanted to take the study of dental morphology to another level. Al Dahlberg and Bertram Kraus both played big roles in this development as they sought to unravel the genetics of morphological trait expression, with some hope these efforts would eventually mirror the rapid rise in serological genetics in the early 1950s. The first efforts by Turner (1967b, 1969) to use crown and root traits in microevolutionary studies set the stage for addressing a broad range of anthropological problems through dental morphology. Even so, if dental morphology was to play a more significant role in the study of human variation, more traits had to be defined, more standards had to be established, and general principles had to be developed.

In 1970, Turner developed the first two plaques (lower molar cusp 6 and cusp 7) in what was to become known as the Arizona State University Dental Anthropology System. Going beyond crowns, Turner (1971) studied roots and laid out the three-wave model for the peopling of the Americas based on the distribution of three-rooted lower first molars (3RM1). At the same time, Scott (1973) developed a number of standards for his dissertation on dental morphology that focused on a genetic analysis of families and variation among native populations of the American Southwest. From roughly 1970 to 1990, Turner worked with a number of other graduate students to develop standard plaques for a variety of crown and root traits.

Prior to Turner's work, dental morphology was mostly descriptive and had very little impact on the broader world of physical anthropology. That changed substantially in the 1980s when teeth played a key role in discussions of the peopling of the New World, the dental dichotomy in Asia, and the settlement of Australia and the Pacific. Even geneticists allow that their findings are often in concert with those of dental morphology (cf. Reich *et al.* 2012). Now dental morphology is being used in a wide variety of contexts, from intra-cemetery analysis and regional microdifferentiation to ancestry estimation in forensic anthropology.

## Why a Guidebook?

Although serendipitous, the production of this guidebook marks the 25th anniversary of "Scoring procedures for key morphological traits of the permanent dentition: the Arizona State University dental anthropology system" (Turner *et al.* 1991). This article has received a great deal of attention since its publication, being cited almost 400 times and viewed over 1600 times (academia.edu). Many students and researchers have followed the methodological guidelines set forth in that article, used along with the standard plaques distributed by Arizona State University (ASU) to over 400 researchers throughout the world.

Given the success of the 1991 article, some might say "let sleeping dogs lie." How can you improve on something so widely read and followed? GRS raised this issue with Christy Turner in 2013, just months before his passing, and Turner agreed that the methodological guidelines set forth in the original article could be expanded and improved upon. For example, the 1991 article only had four photos. Most variables were briefly described but not illustrated. That was due, in large part, to the limitations in the production of an edited volume (Kelley and Larsen 1991). An authored book has fewer limitations regarding illustrations, and Cambridge University Press has allowed us to take advantage of that in this volume.

The goal of this guidebook is to facilitate more research on crown and root trait variation throughout the world. Over several decades, Turner made observations on hundreds of samples and over 30,000 individuals, with an intensive focus on the New World, Asia, Australia, and the Pacific. His observations on Europeans were limited and on Africans even more limited. JDI has remedied the paucity of dental morphological observations on African populations (Irish 1993, etc.), but there is much more work to be done in other parts of the world. Tens of thousands of human skeletons/dental casts/loose teeth and hundreds of anthropological questions are waiting to be pursued through the assessment of crown and root morphology.

The old bugaboo of inter- and intra-observer error during the first half of the twentieth century remains a minor issue in the age following standardization. Making dental observations on crown and root traits still requires training and experience, but this is true in all scientific endeavors. The primary goal of this volume is to help students and researchers alike make systematic observations on dental morphology with a minimum of error. Before proceeding with individual trait descriptions, we provide basic terms required for research in dental anthropology and morphology.

## Terminology

### Teeth and Fields

Mammals typically have four types of teeth: cutting teeth at the front of each jaw (incisors), piercing teeth immediately behind the cutting teeth (canines), grinding teeth at the back of the jaws (molars), and all-purpose teeth that can either slice/dice (carnivores) or grind (herbivores) between the canines and molars (premolars) (Hillson 2005). In developing the concept of dental fields, P.M. Butler (1939) only included incisors, canines, and molars, given that premolars could develop either in line with the grinding molars or with the cutting, slicing anterior teeth. In a classic paper adapting the concept of morphogenetic fields to the human dentition, Dahlberg (1945) included premolars as a separate field along with incisors, canines, and molars.

Although dental clinicians use numbers to denote specific teeth, this system is less useful in anthropology. Many researchers use a letter to describe jaw location (U = upper; L = lower), another letter to describe tooth type (I = incisor; C = canine; P = premolar; M = molar), and a number to note a tooth's position within a morphogenetic field (e.g., 1 = central incisor, first molar). The ancestral mammalian dental formula of 3–1–4–3 involved some tooth reduction during primate evolution. Catarrhine primates (Old World monkeys and apes) lost their first two premolars (P1, P2), so researchers who specialize on primate and fossil hominin dentitions refer to the two premolars in hominoids and hominins as P3 and P4. Although there is no argument that the first two premolars were lost during the course of evolution, many anthropologists who focus on recent human populations refer to the premolars as P1 and P2. That is the convention we adopt in this guidebook. Although antimeric asymmetry does pose minor issues in studies of dental morphology, we are not overly concerned with designations

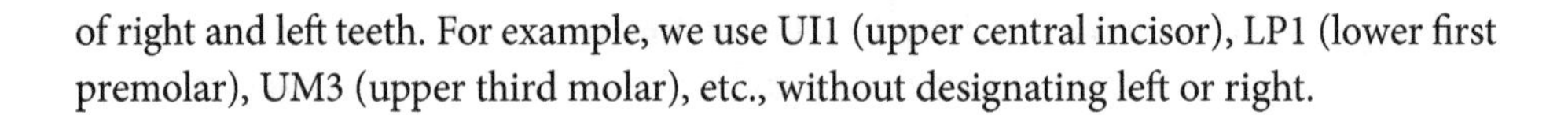

of right and left teeth. For example, we use UI1 (upper central incisor), LP1 (lower first premolar), UM3 (upper third molar), etc., without designating left or right.

## Orientation

Crowns and roots are often described in terms of their relationship to the midline of the upper or lower jaw and direction toward the lips/cheeks or tongue (Figure i). The midline runs between the central incisors of both jaws. When a surface, cusp, or trait runs in the direction of or toward the midline, the term mesial is used to indicate direction. When these same features run away from the midline (toward the back of each quadrant), the term of orientation is distal. The surface of all teeth in both jaws on the inside of the mouth and toward the tongue is lingual. For the cheek teeth (premolars and molars), the surface in contact with the cheek is buccal. For anterior teeth (incisors and

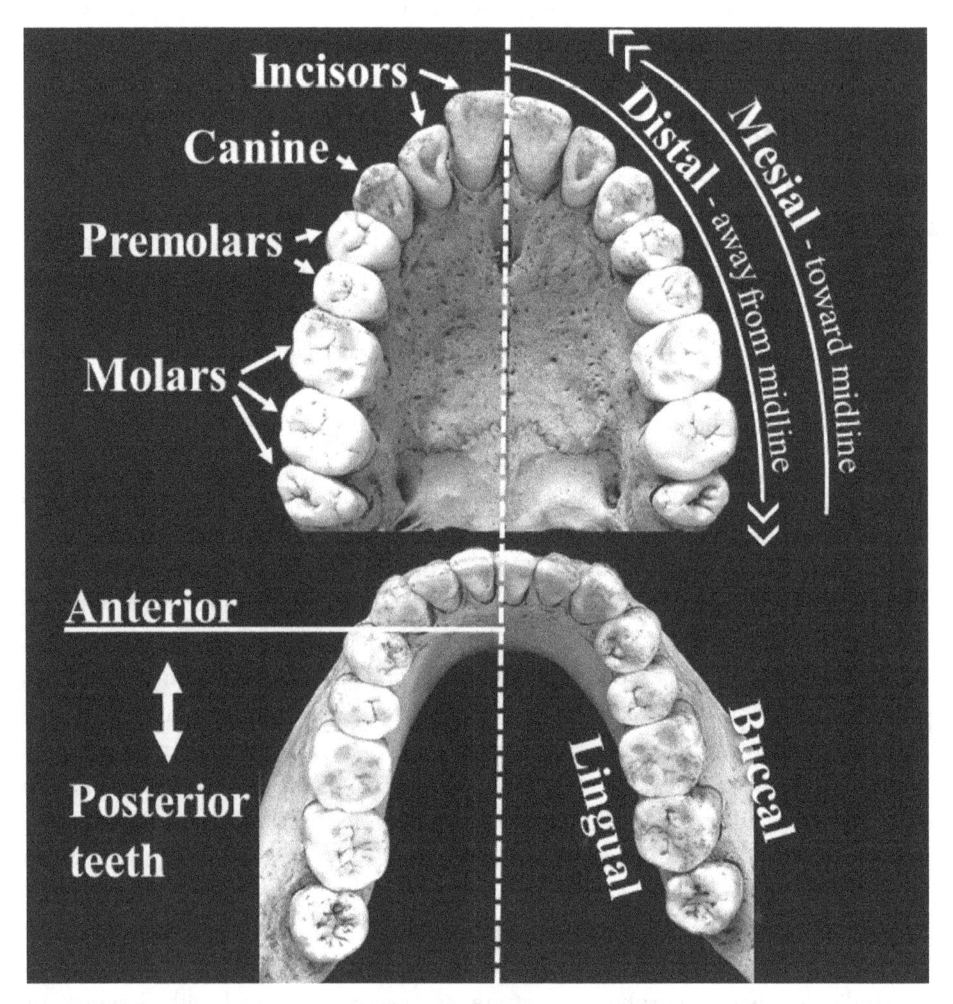


**Figure i** Basic terms of orientation for upper and lower teeth.

canines), the surface in contact with the lips is labial. To avoid the use of two different terms, dental anatomists combine labial and buccal under the term facial (cf. Carlsen 1987). These terms are often used in conjunction to pinpoint location on a crown (e.g., mesiolingual cusp) or root (e.g., distobuccal root).

The left and right quadrants of either jaw have teeth that are, for the most part, mirror images of one another. The corresponding right and left teeth, e.g., right upper first molar and left upper first molar, are antimeres. Corresponding teeth in the two jaws, or isomeres, are not mirror images (e.g., upper left central incisor, lower left central incisor). The teeth in the two jaws evolved to enhance masticatory efficiency so they fit together (ideally), but they differ in size and morphology. The surfaces of the crowns that come in contact when the upper and lower jaws occlude are incisal for the anterior teeth (incisors and canines) and occlusal for the posterior teeth (premolars and molars).

## Lobes and Cusps

In *Dental Morphology*, Carlsen (1987) lays out general dental anatomical principles that are of utility in the study of nonmetric crown and root trait variation. Early dental anthropologists used the term cusp to describe the major units of each tooth. Carlson also discusses cusps but in the context of another macromorphological unit, the lobe (Figures ii, iii, iv). Each tooth, for example, is made up of one to five lobes. The anterior teeth have one lobe, the premolars basically two lobes, and the molars four to five lobes, with often fewer in the second and third molars. For the anterior teeth, there is a centrally located essential lobe segment and two accessory lobe segments, one mesial and one distal. Lobe segments have both facial and lingual components, each of which is called a lobe section. On the essential lobe section of canines and posterior teeth, there is usually an essential ridge that extends up to the cusp tip. The accessory lobe sections end at a point lower than the cusp tip of the essential lobe. Trichotomous lobes are not as evident in the incisors, although the divisions are reflected at an early age in incisal mamelons. For the upper canine, the essential ridge is a prominent feature of the crown and is distinctly set off from the mesial and distal lobe sections. Although the lower canine is spatulate and does not usually exhibit a distinct essential ridge, it still follows this basic form. For both canines, it is not unusual to find a distinct ridge on the lingual aspect of the distal accessory lobe section, referred to as the distal accessory ridge. This has two connotations. It specifies the location on the tooth and also notes that it is accessory, meaning that it may or may not be present (i.e., a nonmetric trait).

The cusps of the upper molars follow the Cope–Osborn nomenclature (Gregory 1916) (Figure v). The trigon, which has deep roots in the mammalian fossil record, has three major cusps: the protocone (lingual), the paracone (mesiobuccal), and the metacone (distobuccal). The hypocone is an additional cusp that was added in many mammalian lineages on the distolingual corner of the trigon (Hunter and Jernvall 1995). The cusps are numbered relative to their presumed appearance in the mammalian fossil record (protocone = 1, paracone = 2, metacone = 3, and hypocone = 4). Each of

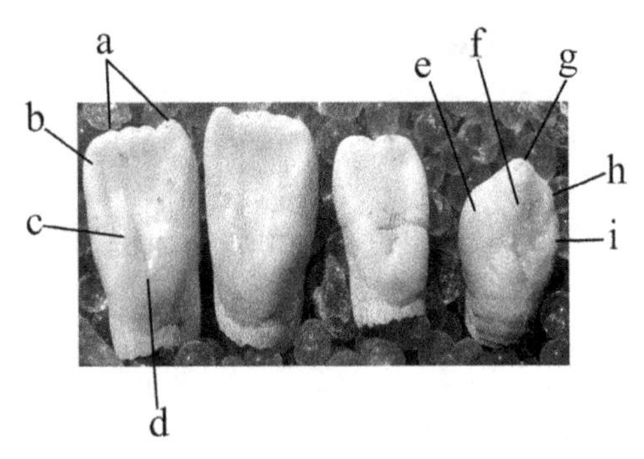


**Figure ii** Unworn upper central incisors, a left upper lateral incisor, and a left upper canine used to illustrate basic dental terms: (a) mamelons; (b) distal marginal ridge; (c) tuberculum projection; (d) basal cingulum; (e) mesial marginal ridge; (f) essential ridge of lingual lobe section; (g) cusp; (h) distal accessory ridge; (i) distal marginal ridge.

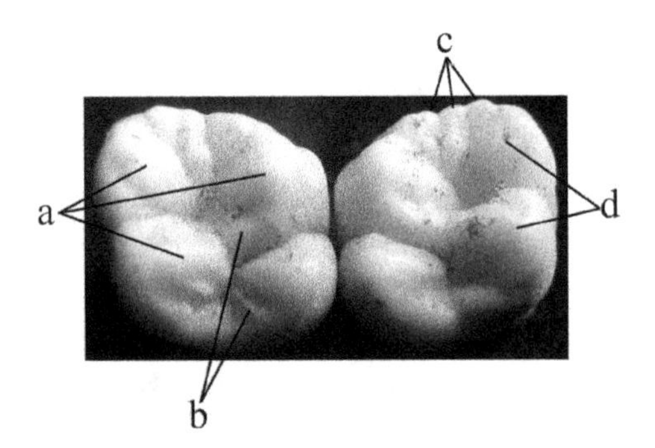


**Figure iii** Unerupted upper first molars used to illustrate: (a) essential ridges of the three major cusps of the trigon; (b) accessory ridges; (c) marginal ridge complex with three mesial marginal tubercles (trait 17); (d) cusp tips of the paracone and metacone.

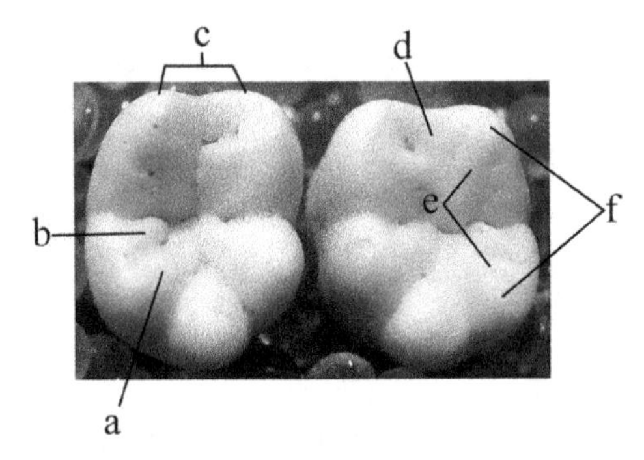


**Figure iv** Unerupted lower first molars used to illustrate: (a) essential ridge of hypoconid; (b) mesial accessory ridge of hypoconid; (c) marginal ridge complex; (d) anterior fovea; (e) essential ridges of protoconid and hypoconid; (f) cusps of protoconid and hypoconid.

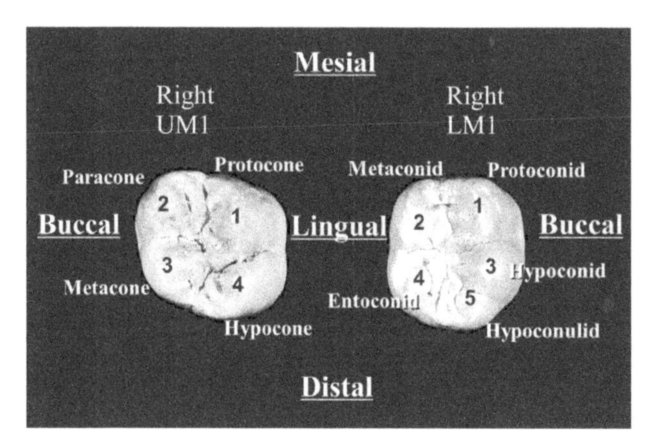


**Figure v** Cusp names and numbers for upper and lower right first molars.

the major cusps of the trigon have essential and accessory lobe segments which in turn have facial and occlusal lobe sections. For each cusp, the essential ridge ends in an elevation that is the highest point of that lobe section, which in the Carlson system is the cusp of the lobe segment. As with the upper canine, the essential ridge is flanked by accessory lobe sections that may or may not exhibit accessory ridges.

The evolution of the lower molars was more complicated than that of the upper molars because lower molars have two major components rather than one. The mesial component is the trigonid, made up of the protoconid (cusp 1, mesiobuccal) and the metaconid (cusp 2, mesiolingual). There was a third cusp (paraconid) that was situated between and anterior to the protoconid and metaconid, but this cusp was lost in primate evolution during the Oligocene. Distal to the trigonid is the talonid. Opposite of the trigonid, the talonid started out with two cusps, the hypoconid (cusp 3, distobuccal) and the entoconid (cusp 4, distolingual), but eventually added the hypoconulid (cusp 5), situated between and posterior to the hypoconid and entoconid.

By convention, large cusps of the upper dentition are referred to as cones while those in lower jaw are conids. The same distinction is maintained for minor features (e.g., styles or conules in the upper jaw, stylids or conulids in the lower jaw). These terms recur frequently in trait descriptions. Other terms, such as accessory ridges and inter-segmental grooves, apply to both upper and lower teeth.

While crowns are divided into lobes, roots are divided into units called cones. The number of cones and roots of each tooth is dictated by the presence of root grooves and inter-radicular projections (Figure vi). A groove, for example, divides the cones but they remain coalesced. An inter-radicular projection is a bifurcation of root cones into two or more separate roots (sometimes called separation structures). An upper incisor typically has two root cones, but these rarely show a bifurcation. The lower canine, on the other hand, usually shows two root cones but in some groups there is an inter-radicular projection that separates the cones, producing two-rooted lower canines (another nonmetric trait).

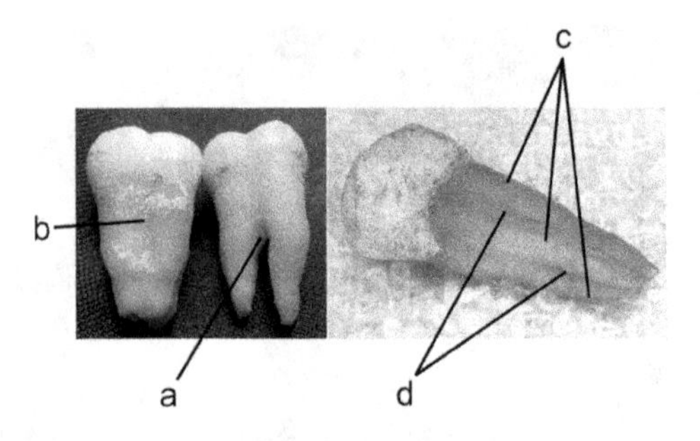


**Figure vi** Upper premolars used to illustrate: (a) inter-radicular projection; (b) root groove separating two root cones; (c) three root cones; (d) two root grooves.

Cones separated from other cones by root grooves but without separation structures are called radicals. On the original Arizona State University Dental Anthropology System (ASUDAS) scoring sheet (Turner *et al.* 1991), there was a row for the number of radicals exhibited by each tooth. To our knowledge, Turner amassed a great deal of information on radical number in world populations but never tabulated this information, so its significance is unknown. Researchers can score this variable, but we do not include it in the following trait descriptions.

## References

Butler, P.M. (1939). Studies of the mammalian dentition. Differentiation of the post-canine dentition. *Proceedings of the Zoological Society of London* B 109, 1–36.

Carlsen, O. (1987). *Dental Morphology*. Copenhagen: Munksgaard.

Dahlberg, A.A. (1945). The changing dentition of man. *Journal of the American Dental Association* 32, 676–690.

(1956). Materials for the establishment of standards for classification of tooth characters, attributes, and techniques in morphological studies of the dentition. Zollar Laboratory of Dental Anthropology, University of Chicago (mimeo).

Dietz, V.H. (1944). A common dental morphotropic factor: the Carabelli cusp. *Journal of the American Dental Association* 31, 78–89.

Gregory, W.K. (1916). Studies on the evolution of the primates. I. The Cope–Osborn "theory of trituberculy" and the ancestral molar patterns of the Primates. *Bulletin of the American Museum of Natural History* 35, 239–257.

Hillson, S. (2005). *Teeth*, 2nd edn. Cambridge: Cambridge University Press.

Hrdlička, A. (1920). Shovel-shaped teeth. *American Journal of Physical Anthropology* 3, 429–465.

(1924). New data on the teeth of early man and certain fossil European apes. *American Journal of Physical Anthropology* 7, 109–132.

Hunter, J.P., and Jernvall, J. (1995). The hypocone as a key innovation in mammalian evolution. *Proceedings of the National Academy of Sciences of the USA* 92, 10718–10722.

Irish, J.D. (1993). *Biological Affinities of Late Pleistocene through Modern African Aboriginal Populations: The Dental Evidence.* PhD dissertation, Department of Anthropology, Arizona State University, Tempe.

Kelley, M.A., and Larsen, C.S., eds. (1991). *Advances in Dental Anthropology.* New York: Wiley-Liss.

Kraus, B. S. (1951). Carabelli's anomaly of the maxillary molar teeth. *American Journal of Human Genetics* 3, 348–355.

Reich, D., Patterson, N., Campbell, D., *et al.* (2012). Reconstructing Native American population history. *Nature* 488, 370–375.

Scott, G.R. (1973). *Dental Morphology: A Genetic Study of American White Families and Variation in Living Southwest Indians.* PhD dissertation, Department of Anthropology, Arizona State University, Tempe.

Turner, C.G., II (1967a). *The Dentition of Arctic Peoples.* PhD dissertation, Department of Anthropology, University of Wisconsin, Madison.

(1967b). Dental genetics and microevolution in prehistoric and living Koniag Eskimo. *Journal of Dental Research* 46 (suppl. to no. 5), 911–917.

(1969). Microevolutionary interpretations from the dentition. *American Journal of Physical Anthropology* 30, 421–426.

(1970). New classifications of non-metrical dental variation: cusps 6 and 7. Paper presented at 39th Annual Meeting of the American Association of Physical Anthropologists, Washington, DC.

(1971). Three-rooted mandibular first permanent molars and the question of American Indian origins. *American Journal of Physical Anthropology* 34, 229–241.

Turner, C.G., II, Nichol, C.R., and Scott, G.R. (1991). Scoring procedures for key morphological traits of the permanent dentition: the Arizona State University dental anthropology system. In *Advances in Dental Anthropology*, ed. M.A. Kelley and C.S. Larsen. New York: Wiley-Liss, pp. 13–31.

## NOTES

# Part II

# Crown and Root Trait Descriptions

The organization of this book was inspired by one of GRS's favorite reads during graduate school. Michael H. Day's *Guide to Fossil Man* (editions in 1965, 1967, 1977, 1986) systematically described fossils throughout the world in one standard format. For each fossil (or fossil locale), he would cover synonyms and other names, site, found by, geology, associated finds, dating, morphology, dimensions, affinities, and references. For the neophyte physical anthropologist, this was poolside reading in Arizona in the late 1960s.

Tooth morphology presents different descriptive parameters than fossil hominins, but we can parallel the effort of Day by providing the same ten categories for each crown and root trait. The categories are:

(1) Name
(2) Observed on
(3) Key tooth
(4) Synonyms
(5) Description
(6) Classification
(7) Breakpoints
(8) Potential problems in scoring
(9) Geographic variation
(10) Select bibliography

Each trait description is accompanied by a photo of the associated standard plaque, photos of traits for which there are no plaques, and photos that help illustrate some aspect of trait recordation. Scoring dental traits is a visual process, so illustrations are a significant aid in making accurate and consistent observations.

In this part of the guidebook, 42 traits are defined and described for each of these ten categories. The majority of descriptions focus on crown and root traits, although, in line with the ASUDAS, we include two oral tori (palatine and mandibular torus) and one mandibular variant (rocker jaw). The trait numbering system is, with some exceptions (e.g., UM bifurcated hypocone, UM marginal ridge tubercles, LM mid-trigonid crest), in direct accord with the trait order presented on the ASUDAS scoring sheet (Part III). If a researcher chooses to use the sheet they can follow the order set forth here.

## Reference

Day, M.H. (1986). *Guide to Fossil Man*, 4th edn. Chicago: University of Chicago Press.

# 1

# Winging

## Observed on

UI1, LI1

## Key Tooth

UI1

## Synonyms

Rotated maxillary central incisors (Enoki and Dahlberg 1958), mesiopalatal torsion (Enoki and Nakamura 1959), wing teeth (Escobar 1979), bilateral mesiopalatal rotation (Iizuka 1976), mesial–palatal version (Rothhammer *et al.* 1968)

## Description

Distinct from other morphological traits, winging does not involve a variable structure on the crowns or roots. Rather, it is defined by patterned rotation of the upper central incisors. In the original ASUDAS description, following the classification by Dahlberg (1963), winging could be bilateral or unilateral with the marginal borders of the central incisors rotated outward (winging) or inward (counter-winging). As the key patterned variable is bilateral winging, the classification recommended here is a modification of the 1991 description. Although not as common as UI1 winging, upper deciduous central incisors and lower central incisors can also exhibit bilateral winging.

## Classification

Absence and three degrees of trait presence (present study)

One can use angles to establish ranked expressions for winging. From an occlusal view, place a straight object (ruler, pencil, etc.) on the labial surface of the upper incisors (see white lines in Figure 1.1a–d). If the straight object does not come in contact with the distal margins of the central incisors and the angle along the labial axis is 180° or greater, there is no winging. If the distal margins are rotated in such a way that the line falls above the mesial margins, bilateral winging is present; in such instances, the angle formed from a point at the midline of the central incisors to the distal margins is less than 180°.

Grade 0 (absence): if the line is parallel to the labial surfaces or if the distal margins fall below the line, winging is absent. Angle ≥180°.

Grade 1 (trace winging): the mesial margins of the upper incisors fall slightly below the line. Angle 160–180°.

Grade 2 (moderate winging): the mesial margins are more removed from the line. Angle 135–159°.

Grade 3 (pronounced winging): there is a distinct distance between the line and the mesial margins. Angle <135°.

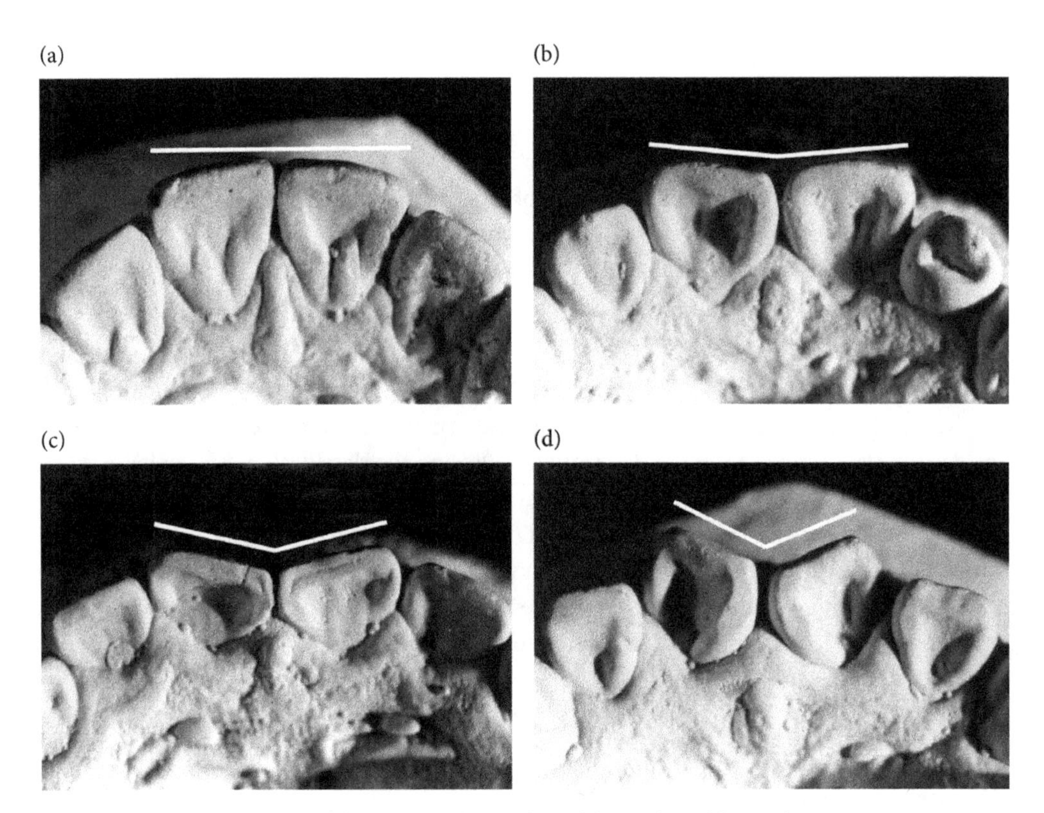


**Figure 1.1** Winging: (a) grade 0; (b) grade 1; (c) grade 2; (d) grade 3.

## Breakpoint

1 (any form of bilateral winging is scored as present)

## Potential Problems in Scoring

As winging involves tooth rotation, environmentally induced rotation can be confused with genetically based rotation. However, rotation associated with crowding of the anterior teeth can usually be inferred because expression is typically asymmetrical.

## Geographic Variation

Bilateral incisor winging is strongly associated with Asian populations. It is common in north and east Asians (20–35%), and reaches its maximum expression in Native American populations (ca. 50%). It is also common (20–30%) in Southeast Asian and derived populations (Polynesians, Micronesians, Melanesians). It is least common (<10%) in Europeans, Australians, New Guineans, and Africans. The tables in the Appendix follow the classification in Turner *et al.* (1991): grade 1 equals winging while grades 2, 3, and 4 represent different forms of non-winging.

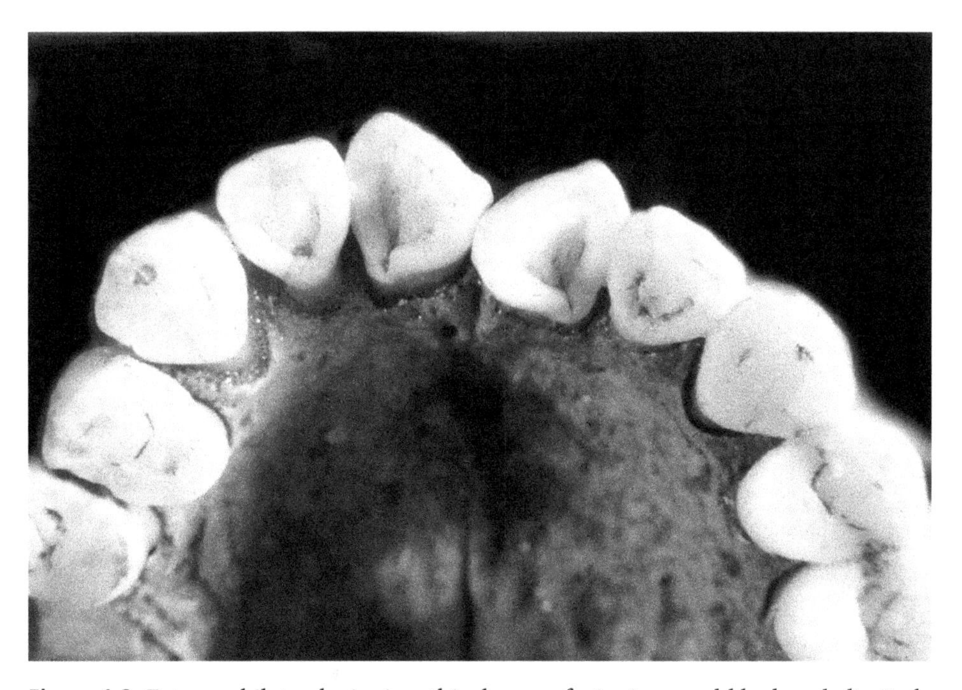

**Figure 1.2** Extreme bilateral winging; this degree of winging would be largely limited to Native Americans.

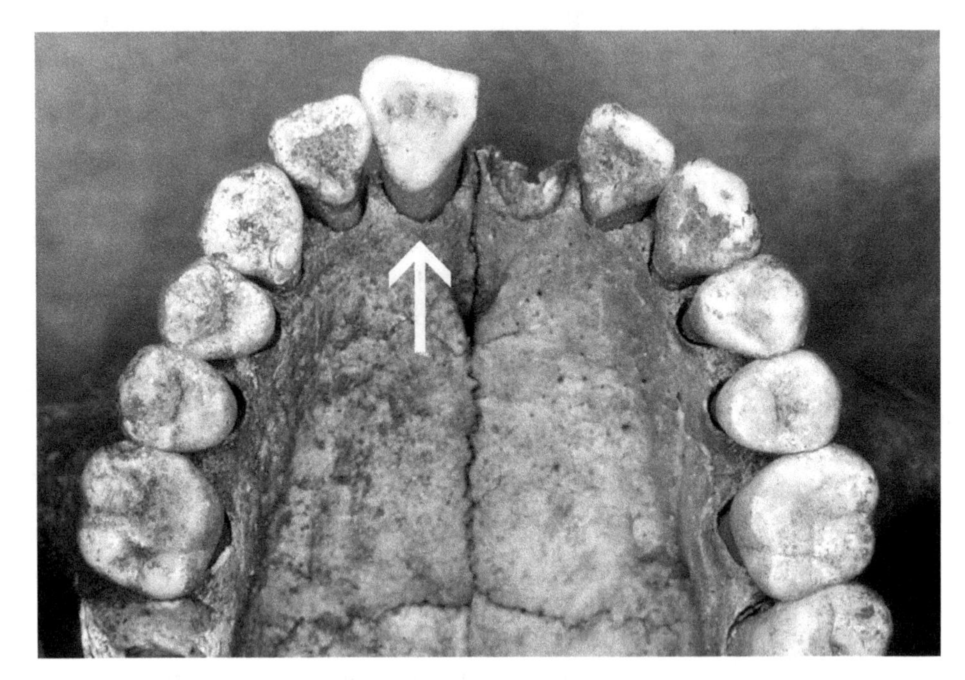

**Figure 1.3** Despite the absence of one UI1, winging is clearly indicated in this individual.

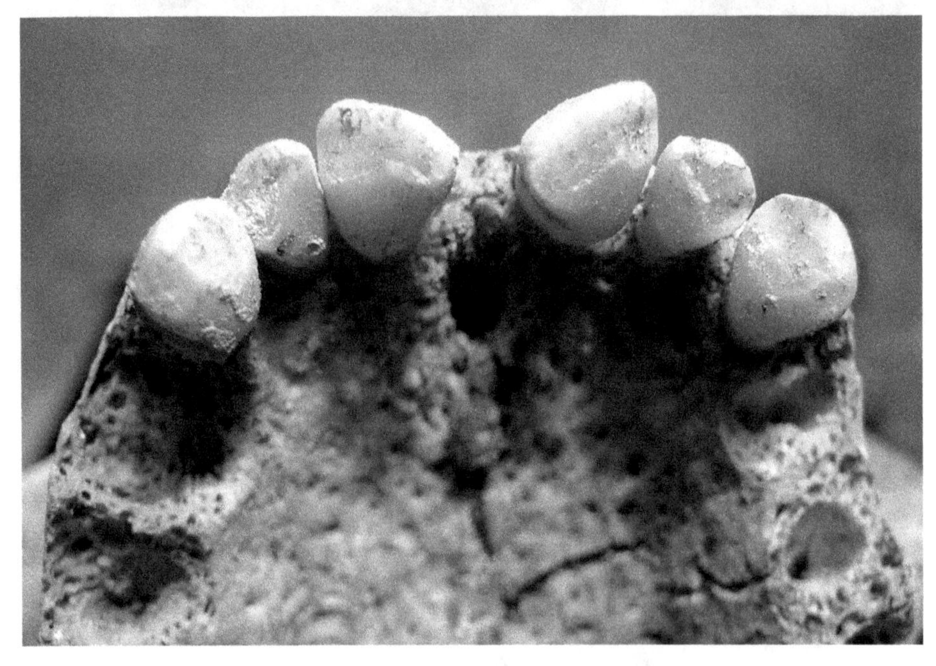

**Figure 1.4** Individual in a Basque skeletal collection exhibiting winging (may not, however, be Basque).

## Select Bibliography


Dahlberg, A.A. (1959). A wing-like appearance of upper central incisors among American Indians. *Journal of Dental Research* 38, 203–204.

(1963). Analysis of the American Indian dentition. In *Dental Anthropology*, ed. D.R. Brothwell. New York: Pergamon Press, pp. 149–178.

Enoki, K., and Dahlberg, A.A. (1958). Rotated maxillary central incisors. *Orthodontic Journal of Japan* 17, 157–169.

Enoki, K., and Nakamura, E. (1959). Bilateral rotation (mesiopalatal torsion) of maxillary central incisors. *Journal of Dental Research* 38, 204.

Escobar, V.H. (1979). *A Genetic Study of Upper Central Incisor Rotation (Wing Teeth) in the Pima Indians*. PhD dissertation, Indiana University, Bloomington.

Escobar, V., Melnick, M., and Conneally, P.M. (1976). The inheritance of bilateral rotation of maxillary central incisors. *American Journal of Physical Anthropology* 45, 109–116.

Iizuka, A. (1976). The bilateral mesiopalatal rotation of upper central incisors. *Journal of Anthropological Society of Nippon* 84, 31–47.

Ling, J.Y.K., and Wong, R.W.K. (2010). Incisor winging in Chinese. *The Open Anthropology Journal* 3, 8–11.

Rothhammer, F, Laserre, E, Blanco, R, Covarrubias, E., and Dixon, M. (1968). Microevolution in human populations. IV. Shovel shape, mesial-palatal version and other dental traits in Pewenche Indians. *Zeitschrift für Morphologie und Anthropologie* 60, 162–169.

Turner, C.G., II, Nichol, C.R., and Scott, G.R. (1991). Scoring procedures for key morphological traits of the permanent dentition: the Arizona State University dental anthropology system. In *Advances in Dental Anthropology*, ed. M.A. Kelley and C.S. Larsen. New York: Wiley-Liss, pp. 13–31.

# NOTES

# 2

# Labial Convexity

## Observed on

Labial surface of UI1

## Key Tooth

UI1

## Synonyms

N/A

## Description

In modern humans, the labial surface of the upper incisors is almost flat. In earlier hominins, especially *Homo heidelbergensis* and Neanderthals, the labial surface is often markedly convex. This level of convexity is rarely seen in recent humans, but there is some variation in the curvature of the labial surface of the upper incisors.

## Classification

Nichol *et al.* (1984)

Grade 0: labial surface is flat
Grade 1: labial surface exhibits trace convexity
Grade 2: labial surface exhibits weak convexity
Grade 3: labial surface exhibits moderate convexity
Grade 4: labial surface exhibits pronounced convexity
Grade 5: pronounced convexity not observed in modern humans but present in earlier hominins (not shown on ASUDAS plaque)

**Figure 2.1** ASUDAS plaque for labial curvature, with absence and four degrees of trait presence.

## Breakpoint

Grade 2+

## Potential Problems in Scoring

Labial convexity can be scored when wear is moderate on UI1. Focus should be on the central one-third of the labial surface. The presence of labial convexity precludes double-shoveling.

## Geographic Variation

Asian and Asian-derived groups typically have little or no labial convexity. Africans show the highest frequency of labial convexity, although it may be as common in New Guinea. European and Pacific populations fall between the African and Sino-American extremes. It is far more common and pronounced in Neanderthals and *Homo heidelbergensis* than in any modern human population (Bailey 2006, Martinón-Torres *et al.* 2012), so much so that the scale established by Nichol *et al.* (1984) does not have a rank that matches earlier hominin extremes.

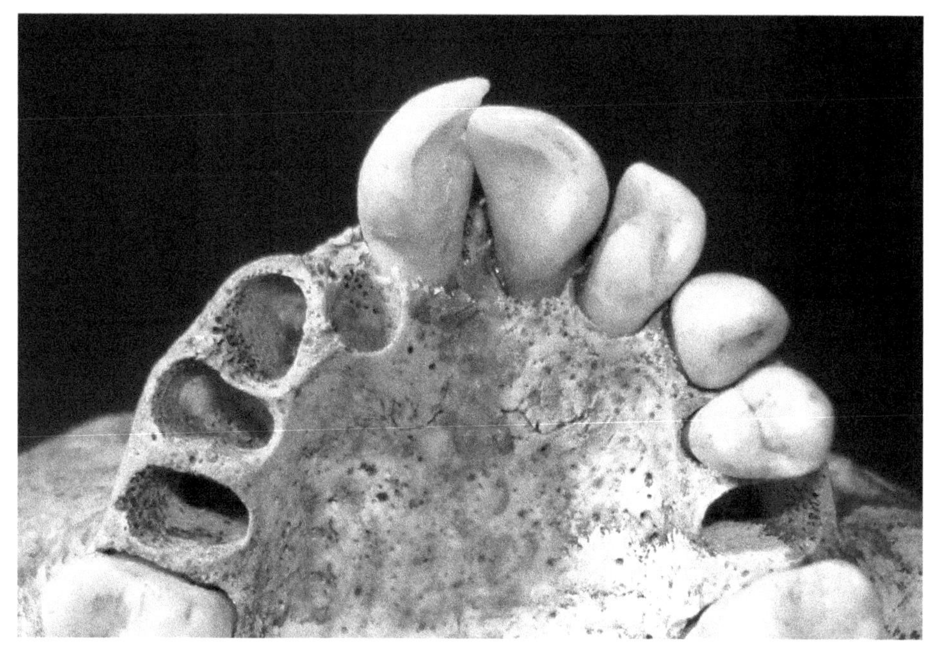

**Figure 2.2** Pronounced labial convexity on UI1 (and talon cusp on UI2).

## Select Bibliography

Bailey, S. (2006). Beyond shovel-shaped incisors: Neanderthal dental morphology in a comparative context. *Periodicum Biologorum* 108, 253–267.

Martinón-Torres, M., Bermúdez de Castro, J.M., Gomez-Robles, A., Prado-Simon, L., and Arsuaga, J.L. (2012). Morphological description and comparison of the dental remains from Atapuerca-Sima de los Huesos site (Spain). *Journal of Human Evolution* 62, 7–58.

Nichol, C.R., Turner, C.G., II, and Dahlberg, A.A. (1984). Variation in the convexity of the human maxillary incisor labial surface. *American Journal of Physical Anthropology* 63, 361–370.

# NOTES

# 3

# Palatine Torus

## Observed on

Palatine bones and palatal process of maxilla (i.e., roof of hard palate)

## Key Tooth

N/A (oral torus)

## Synonyms

*Torus palatinus* (Thoma 1937, Woo 1950), palatal exostosis, maxillary torus, maxillary hyperostoses (Hrdlička 1940)

## Description

Palatine torus is a bony exostosis that is expressed on both sides of the midline on the hard palate. Its point of origin is on the palatine bones. From a point of initial constriction, it expands in breadth onto the hard palate and generally narrows moving in the direction of the incisive foramen.

## Classification

The tori comes in myriad of forms and sizes, but Turner *et al.* (1991) followed a basic size scheme that included:

Grade 0: absence
Grade 1: small (elevated 1–2 mm)
Grade 2: moderate (elevated 2–5 mm)
Grade 3: marked (covers more of palate, 5–10 mm relief)
Grade 4: very marked (>10 mm high and broad)

## Breakpoint

In skeletons, assuming researchers agree on what constitutes a low grade of expression, palatine torus can be characterized by total trait frequency. Some researchers score this torus in living individuals, where low grades of expression are difficult to observe. To compare data on skeletal and living samples, medium torus expression would serve as a better breakpoint.

## Potential Problems in Scoring

There are some unusual bony exostoses that run along the midline of the palate that may not represent palatine torus (e.g., the ridge of bone that runs along both sides of the intermaxillary suture sometimes grows vertically but not laterally). Figures 3.1 to 3.8 show varying degrees of palatine torus expression and unusual exostoses that are not part of the trait complex.

## Geographic Variation

Palatine torus is most common in East Asian and Native American populations, with frequencies that are slightly higher than those of Europeans. The lowest frequencies are found in Pacific populations. Sub-Saharan Africans fall between these extremes for palatine torus frequencies.

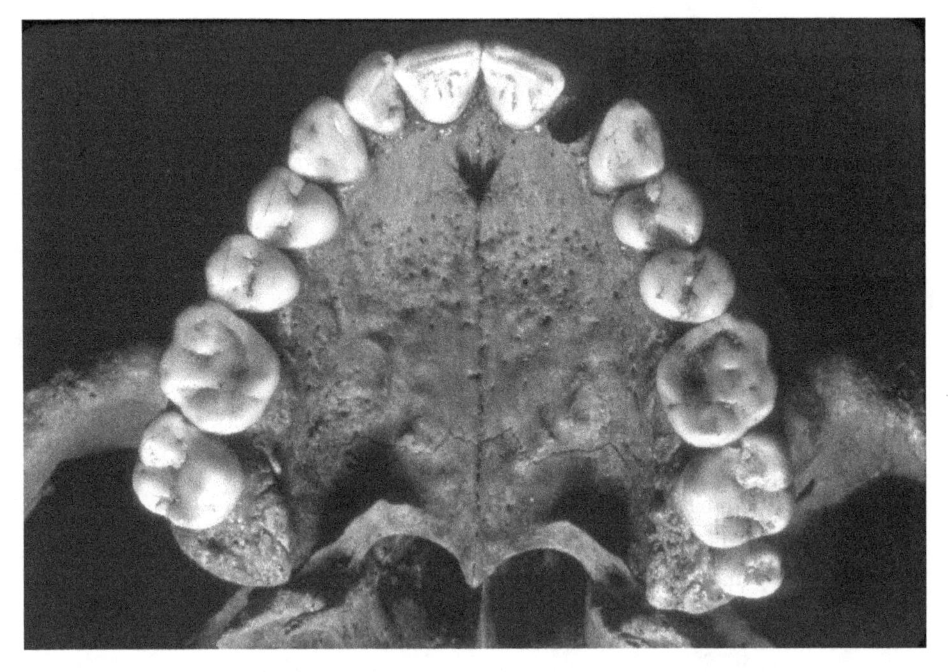

**Figure 3.1** Grade 0. No palatine torus present along intermaxillary suture.

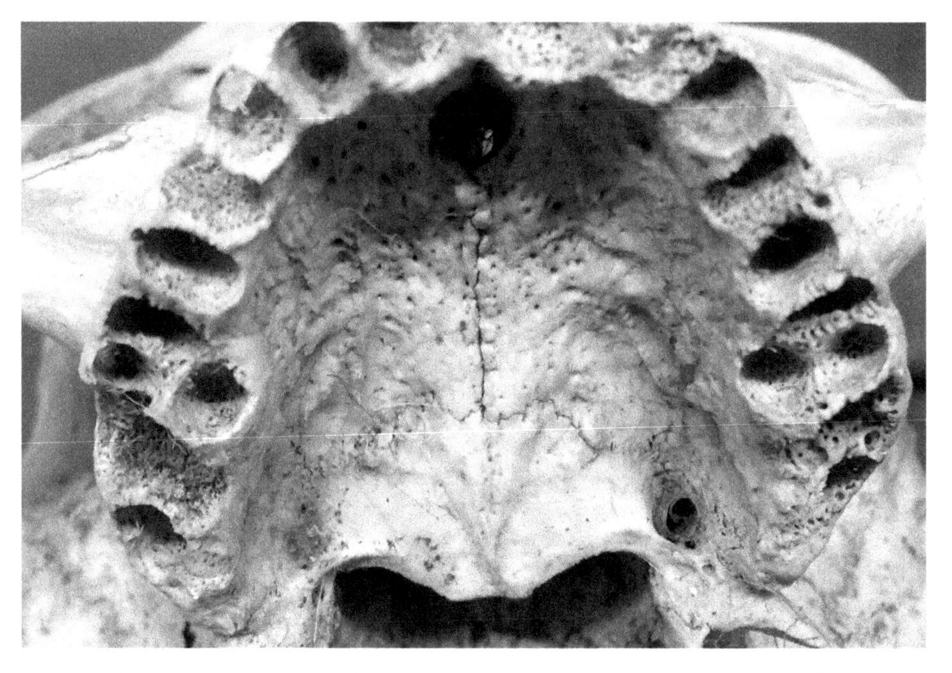

**Figure 3.2** Grade 1. Small palatine torus, constricted on palatine bones but flares out toward hard palate; does not extend to incisive foramen.

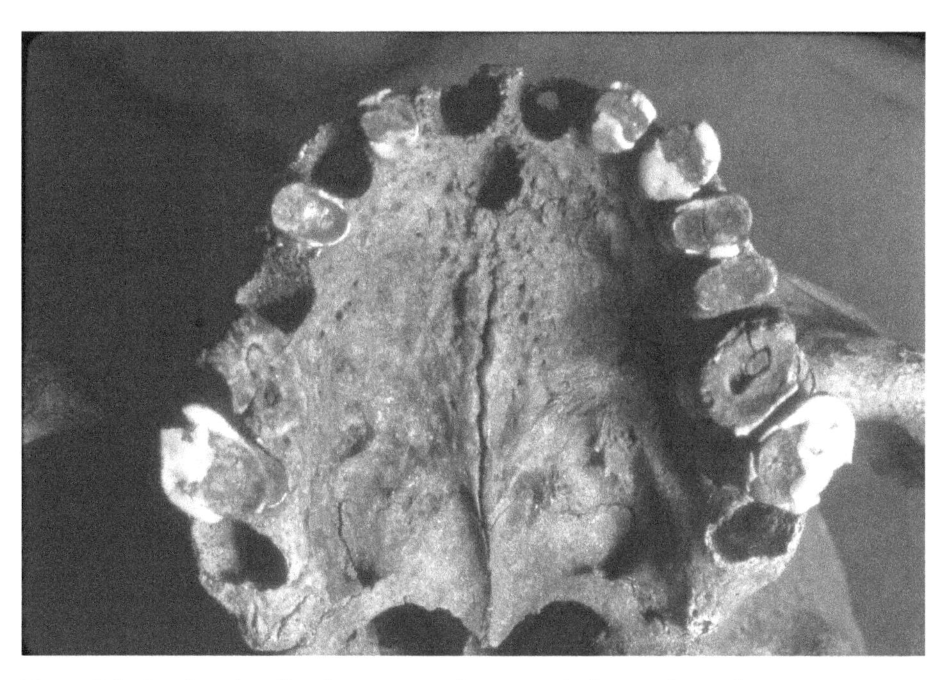

**Figure 3.3** Grade 1. Small palatine torus that extends from palatine bones to incisive foramen, but overall configuration narrow and low.

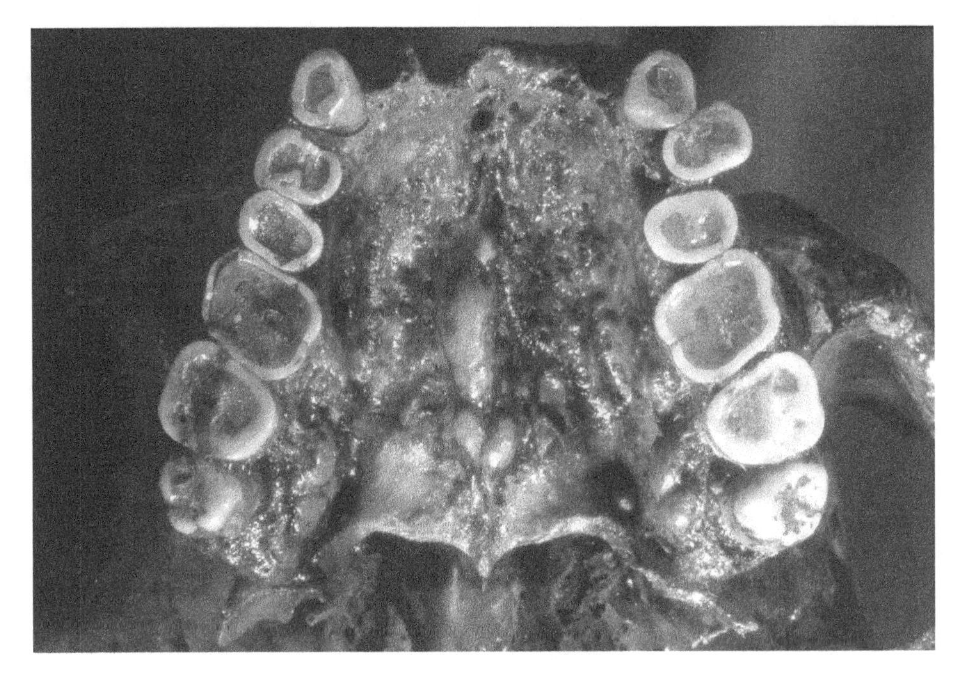

**Figure 3.4** Grade 2. Moderate palatine torus, expressed more distinctly than grade 1 forms.

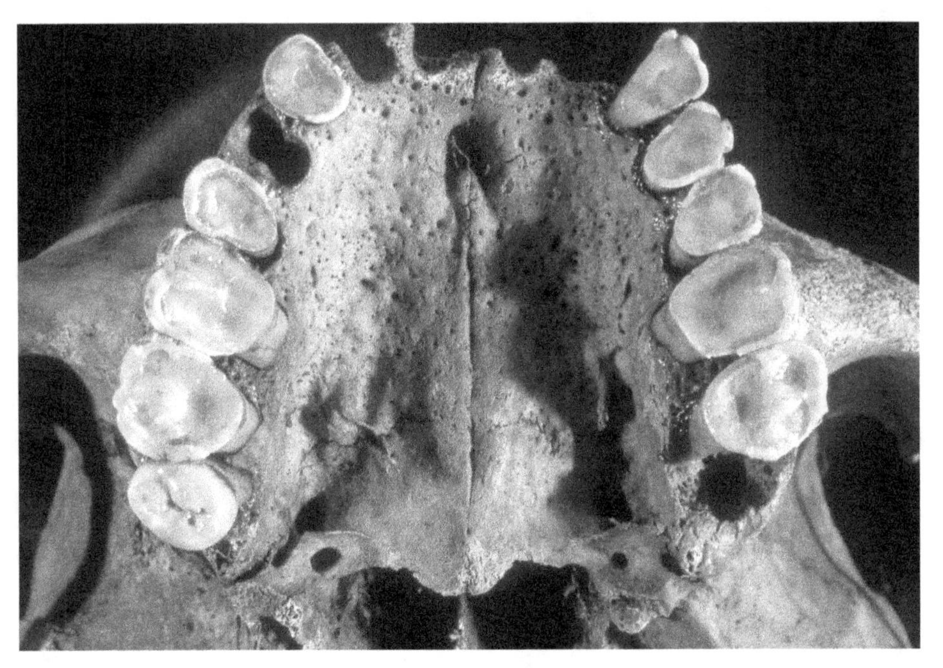

**Figure 3.5** Grade 2. Another form of moderate palatine torus that extends from palatine bones to incisive foramen; broader and higher than grade 1 expression.

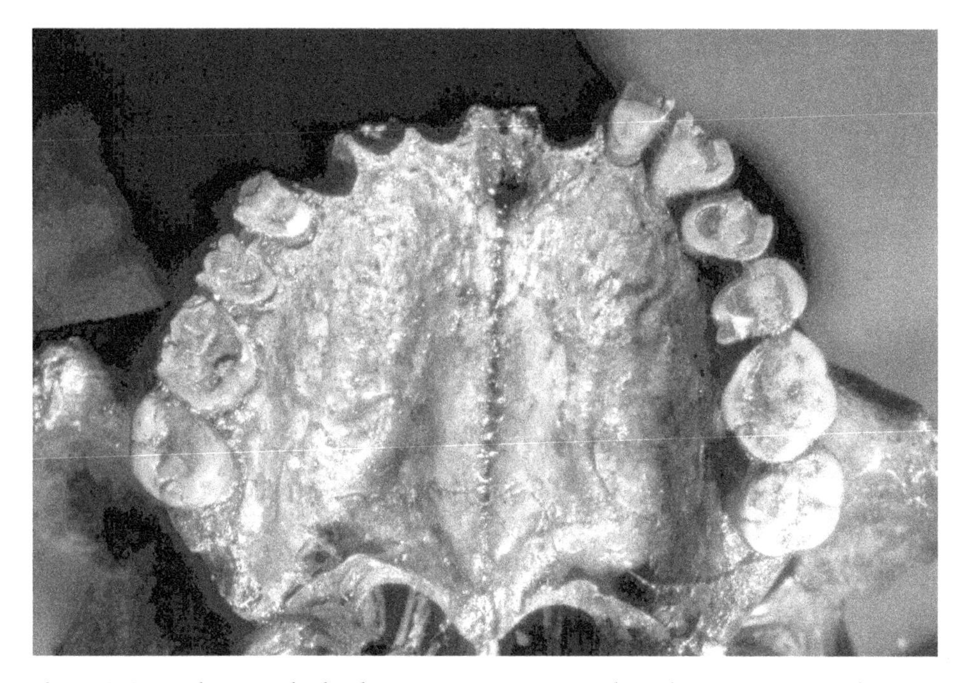

**Figure 3.6** Grade 3. Marked palatine torus assuming relatively common cigar form; broader and higher than moderate forms.

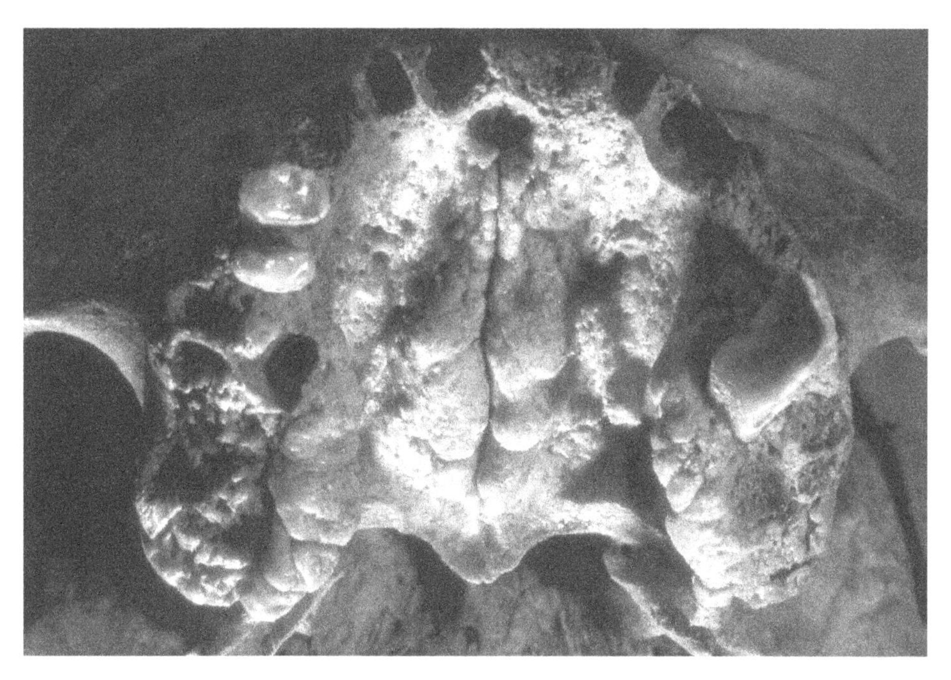

**Figure 3.7** Grade 4. Very marked form of palatine torus expressed along midline as large irregular structure.

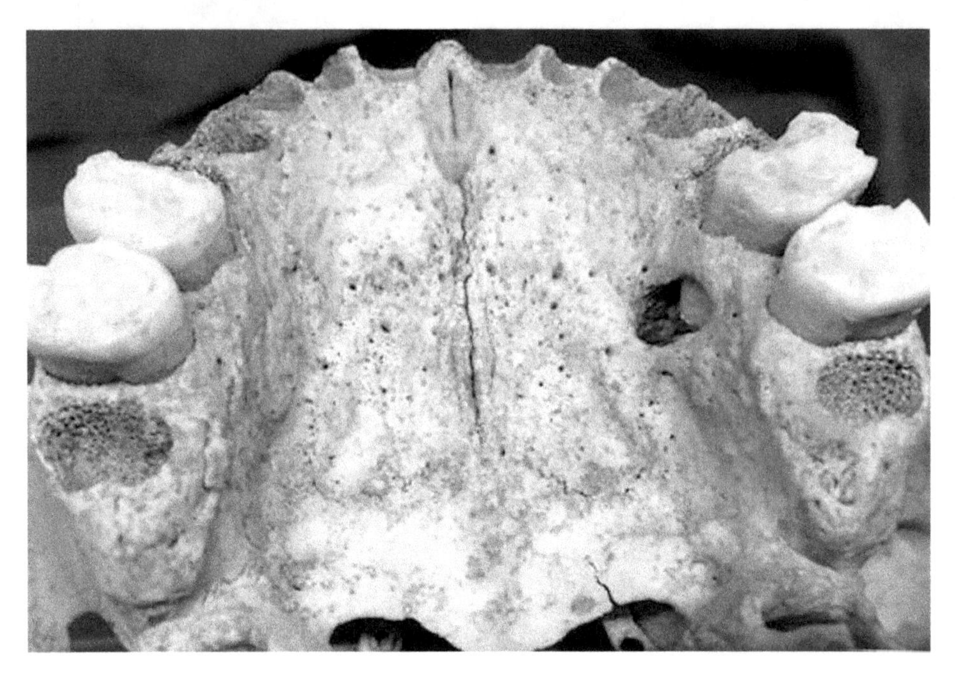

**Figure 3.8** Elevated ridges along both sides of the intermaxillary suture are not considered part of the palatine torus complex.

## Select Bibliography

Halffman, C.M., and Irish, J.D. (2004). Palatine torus in the pre-conquest inhabitants of the Canary Islands. *Homo – Journal of Comparative Human Biology* 55, 101–111.

Halffman, C.M., Scott, G.R., and Pedersen, P.O. (1992). Palatine torus in the Greenlandic Norse. *American Journal of Physical Anthropology* 88, 145–161.

Hooton, E.A. (1918). On certain Eskimoid characters in Icelandic skulls. *American Journal of Physical Anthropology* 1, 58–62.

Hrdlička, A. (1940). Mandibular and maxillary hypertostoses. *American Journal of Physical Anthropology* 27, 1–67.

Miller, H.C., and Roth, H. (1940). Torus palatinus: a statistical study. *Journal of the American Dental Association* 27, 1950–1957.

Ohno, N., Sakai, T., and Mizutani, T. (1988). Prevalence of torus palatinus and torus mandibularis in five Asian populations. *Aichi-Gakuin Dental Science* 1, 1–8.

Seah, Y.H. (2009). Torus palatinus and torus mandibularis: a review of the literature. *Australian Dental Journal* 40, 318–321.

Suzuki, M., and Sakai, T. (1960). A familial study of torus palatinus and torus mandibularis. *American Journal of Physical Anthropology* 18, 262–273.

Thoma, K.H. (1937). Torus palatinus. *International Journal of Orthodontics and Oral Surgery* 23, 194–202.

Turner, C.G., II, Nichol, C.R., and Scott, G.R. (1991). Scoring procedures for key morphological traits of the permanent dentition: the Arizona State University dental anthropology system. In *Advances in Dental Anthropology*, ed. M.A. Kelley and C.S. Larsen. New York: Wiley-Liss, pp. 13–31.

Woo, J.-K. (1950). Torus palatinus. *American Journal of Physical Anthropology* 8, 81–100.

# NOTES

# 4

# Shoveling

## Observed on

Lingual surface of UI1, UI2, UC, LI1, LI2, LC

## Key Tooth

UI1

## Synonyms

Shovel-shaped incisors, ridge and fossa formation (i.e., keilo-koilodonty) (Hrdlička 1920), mesial and distal marginal ridges (Mizoguchi 1985)

## Description

Shoveling is part of the marginal ridge complex of the upper and lower anterior teeth. It involves the development of mesial and distal marginal ridges on the lingual surface. Mizoguchi (1985) scored the mesial and distal ridges separately, but they are strongly correlated. For this reason, most researchers consider the two marginal ridges as a single trait. Shoveling is most developed on the upper incisors but can also be expressed on the lower incisors and both upper and lower canines. The prominence of the essential lingual lobe on the upper canines precludes the formation of a fossa, one of the hallmarks of shoveling. This is the major reason most workers do not score shoveling on the upper canine. This is less of a problem on the lower canine, but shoveling on the lower anterior teeth in general is much less pronounced than on the upper anterior teeth. For these reasons, most researchers focus on UI1 and UI2 shoveling expression.

## Classification

Scott (1973)

UI1 ranked scale:

Grade 0 (absence): it is rare for UI1 to express the complete absence of marginal ridges (see Figure 4.1a for example). For this reason, grade 0 on the UI1 shoveling plaque actually shows very slight marginal ridge expression.

Grade 1 (trace): marginal ridges can be discerned, but expression is slight, with mesial marginal ridge not extending to the basal eminence.

Grade 2 (low moderate): ridges more pronounced, with mesial marginal ridge extending further down on basal eminence.

Grade 3 (high moderate): ridges more pronounced, almost coalescing at basal eminence.

Grade 4 (low pronounced): well-developed ridges that converge at basal eminence.

Grade 5 (medium pronounced): more pronounced marginal ridges meeting at basal eminence.

Grade 6 (high pronounced): pronounced ridges that meet at basal eminence, almost folding around on themselves.

Grade 7 (extreme pronounced): any expression that exceeds grade 6 can be placed in grade 7. It is a rare expression, so much so that a good example for the plaque was never found. This grade would involve marginal ridges that folded around on themselves, similar to grade 6 on the UI2 shoveling plaque.

(a)

(b)

**Figure 4.1** Shoveling plaques: (a) UI1; (b) UI2.

UI2 ranked scale:

- Grade 0 (absence): as for UI1, complete absence of marginal ridges is rare on UI2. There are slight marginal ridges on the plaque for grade 0.
- Grade 1 (trace): characterized by the presence of faint mesial and distal marginal ridges.
- Grade 2 (low moderate): moderate marginal ridges with little fossa formation.
- Grade 3 (high moderate): distinct marginal ridges but only moderate lingual fossa.
- Grade 4 (low pronounced): well-developed marginal ridges that come in contact at the lingual base of the crown.
- Grade 5 (medium pronounced): well-developed marginal ridges, forming a distinct lingual fossa.
- Grade 6 (semi-barreled): the marginal ridges wrap around and contact at a low point on the basal eminence.
- Grade 7 (barreled): the marginal ridges are so pronounced that they contact at almost the incisal surface of the basal eminence, assuming a full barrel shape.

## Breakpoint

Although some researchers use total frequency to characterize a sample for shoveling (i.e., 1–6/0–6), others use either grade 2 or grade 3 as a breakpoint. Any breakpoint should provide the same inter-group contrasts, but a problem arises when data are reported only by breakpoint and not by degree of expression, as not all researchers use the same breakpoints. For groups with modest shoveling a breakpoint of 2 is often used, while studies of groups with pronounced shoveling often use grade 3 as a breakpoint.

## Potential Problems in Scoring

Although the mesial and distal marginal ridges are strongly correlated, the relationship is not perfect. In some instances, a mesial marginal ridge may be present when the distal marginal ridge is completely absent. Such a phenotype would usually be scored as grade 0, as a moderate marginal ridge would not be found in association with distal ridge absence. Because anterior teeth are often used as a third hand, abrasive crown wear is a serious issue in scoring shoveling. For the lower grades of shoveling, the marginal ridges do not extend to the basal eminence. If the crown is worn down by ¼ to ⅓, there is no remnant of the trait. Shoveling should only be scored when there is slight wear on the upper incisors.

## Geographic Variation

Hrdlička (1920) noted that shoveling characterized North Asian and New World populations. In some Native American populations, shoveling is almost invariant, with ranked scales approximating a normal distribution, with grades 3 or 4 as the modal

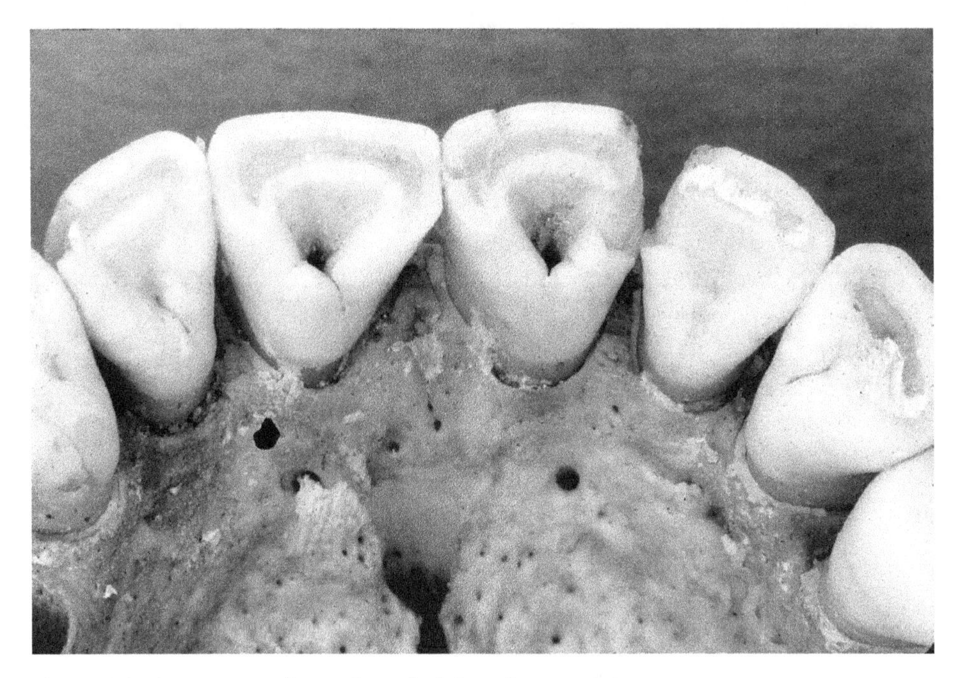

**Figure 4.2** A rare case of semi-barreled shoveling on UI1.

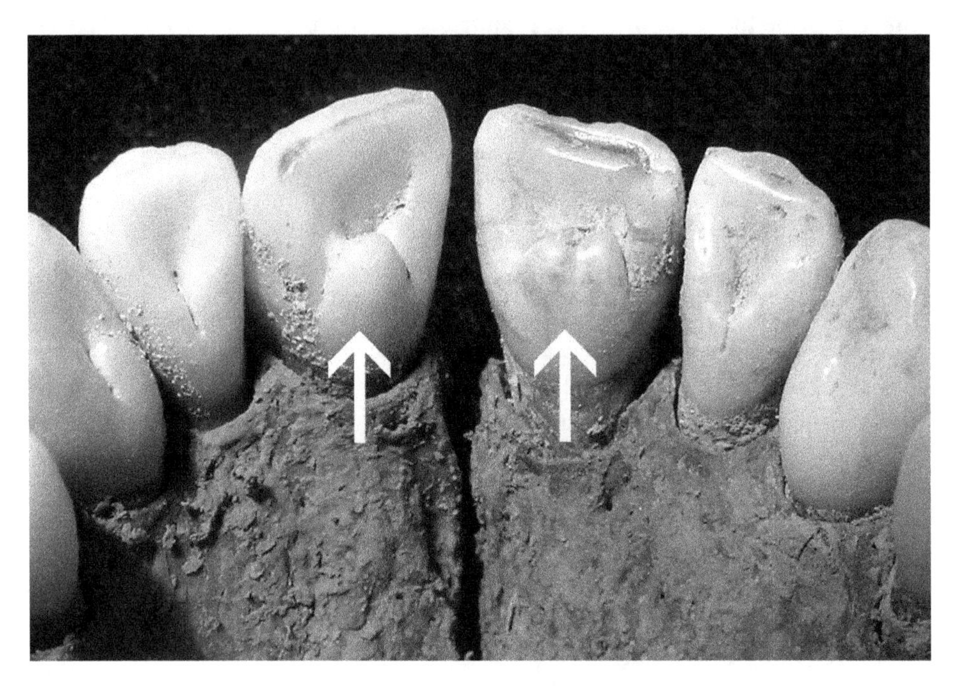

**Figure 4.3** The presence of *tuberculum dentale* can obscure marginal ridge formation.

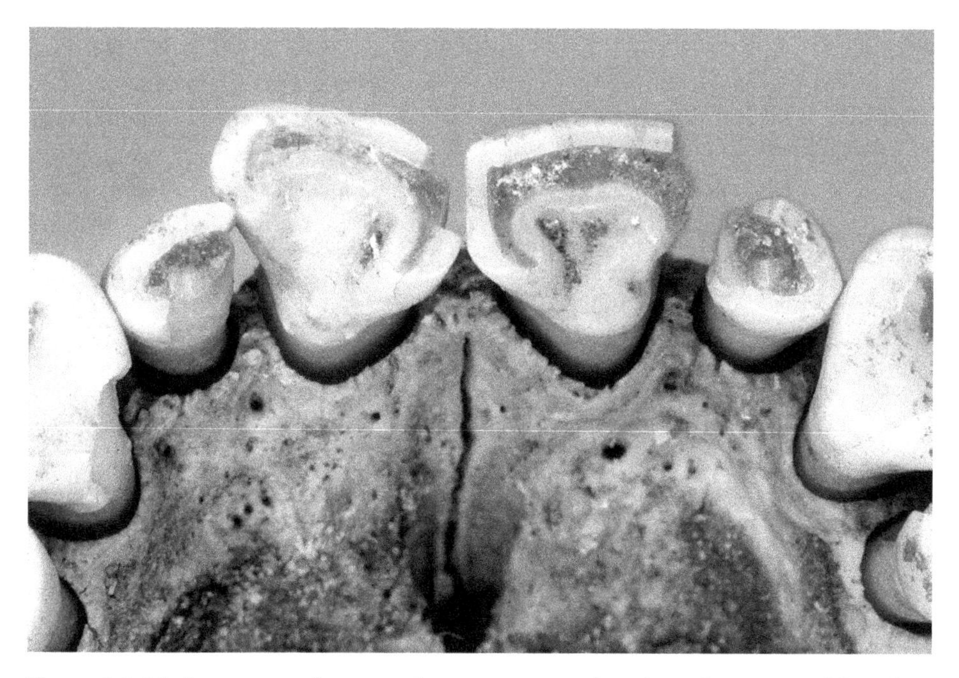

**Figure 4.4** Moderate wear does not obscure a strong shoveling phenotype, although it does make it difficult to assign to a specific grade.

expression. Southeast Asian and derived populations (e.g., Polynesians, Micronesians) exhibit less shoveling than North Asians, but they rank second on the world scale. At the opposite end of the shoveling scale are Western Eurasians, where grades 0 or 1 are the modal expressions; expression at or above grade 3 is uncommon. Sub-Saharan African, Australian, and Melanesian populations exhibit slightly more shoveling than Western Eurasians but fall below Southeast Asians.

## Select Bibliography

Aas, I.H.M., and Risnes, S. (1979). The depth of the lingual fossa in permanent incisors of Norwegians. I. Method of measurement, statistical distribution and sex dimorphism. *American Journal of Physical Anthropology* 50, 335–340.

Blanco, R., and Chakraborty, R. (1977). The genetics of shovel shape in maxillary central incisors in man. *American Journal of Physical Anthropology* 44, 233–236.

Carbonell, V.M. (1963). Variations in the frequency of shovel-shaped incisors in different populations. In *Dental Anthropology*, ed. D.R. Brothwell. New York: Pergamon Press, pp. 211–234.

Dahlberg, A.A., and Mikkelsen, O. (1947). The shovel-shaped character in the teeth of the Pima Indians. *American Journal of Physical Anthropology* 5, 234–235.

Harris, E.F. (1980). Sex differences in lingual marginal ridging on the human maxillary central incisor. *American Journal of Physical Anthropology* 52, 541–548.

Hrdlička, A. (1920). Shovel-shaped teeth. *American Journal of Physical Anthropology* 3, 429–465.

Kimura, R., Yamaguchi, T., Takeda, M., *et al.* (2009). A common variation in EDAR is a genetic determinant of shovel-shaped incisors. *American Journal of Human Genetics* 85, 528–535.

Mizoguchi, Y. (1985). *Shovelling: A Statistical Analysis of its Morphology*. Tokyo: University of Tokyo Press.

Portin, P., and Alvesalo, L. (1974). The inheritance of shovel shape in maxillary central incisors. *American Journal of Physical Anthropology* 41, 59–62.

Scott, G.R. (1973). *Dental Morphology: A Genetic Study of American White Families and Variation in Living Southwest Indians*. PhD dissertation, Department of Anthropology, Arizona State University, Tempe.

(1977). Interaction between shoveling of the maxillary and mandibular incisors. *Journal of Dental Research* 56, 1423.

Suzuki, M., and Sakai, T. (1966). Morphological analysis of the shovel-shaped teeth. *Journal of the Anthropological Society of Nippon* 74, 202–218.

# 5

# Double-Shoveling

## Observed on

UI1, UI2

## Key Tooth

UI1

## Synonyms

Labial-ridged mesial (Snyder 1960), phenomenal labial raised marginal ridges (Rabkin 1943), ¾ double-shoveling

## Description

Shoveling, without qualification, refers to mesial and distal marginal ridges on the lingual surface of the incisors. Marginal ridges can also manifest themselves on the labial surface of the incisors, especially UI1. When the trait is present, it is typically more pronounced on the mesial than the distal margin (sometimes called ¾ double-shoveling, as ridges are on both lingual marginal ridges and one labial marginal ridge).

## Classification

Turner *et al.* (1991)

Grade 0 (absence): no labial marginal ridges present; surface is smooth
Grade 1 (faint): very faint labial ridging, more evident on mesial than distal margin
Grade 2 (trace): ridge more distinct than faint expression of grade 1 but still slight
Grade 3 (slight): ridges distinct enough to be palpated
Grade 4 (moderate): ridging clearly evident along at least one half of crown height
Grade 5 (pronounced): very distinct ridging expressed from incisal edge to crown root junction
Grade 6: (very pronounced): extreme double-shoveling with well-developed ridges along both the mesial and distal labial margins

**Figure 5.1** UI1 double-shoveling plaque.

## Breakpoint

Grade 2

## Potential Problems in Scoring

Upper central incisors have one lobe, with a centrally positioned essential lobe segment and mesial and distal accessory lobe segments. A mesial lobe set off from the rest of the labial surface by a groove can be confused with a low grade of double-shoveling.

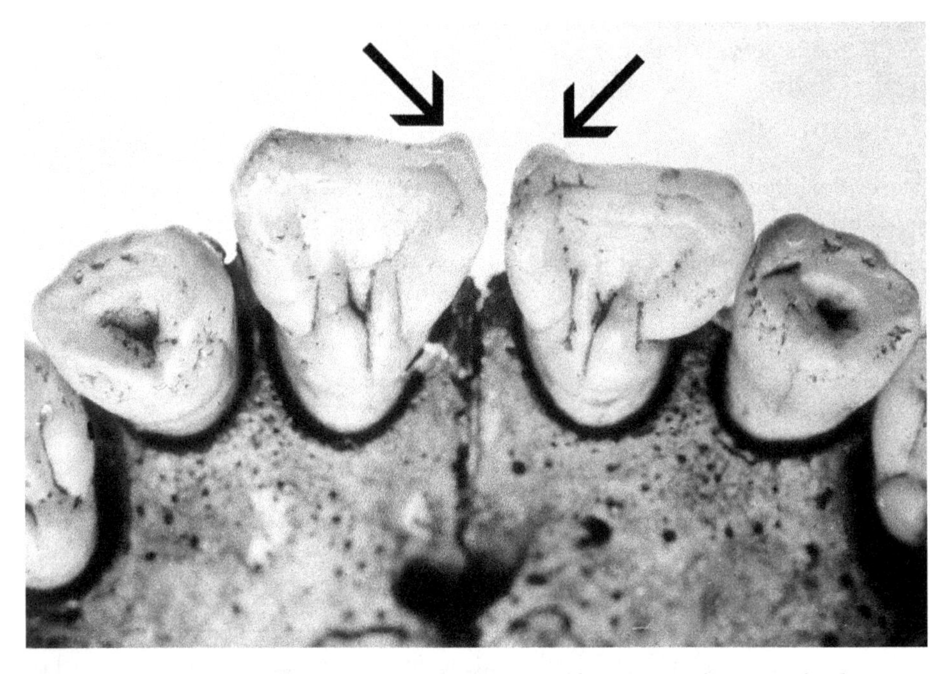

**Figure 5.2** An occlusal view shows a well-developed labial mesial marginal ridge but no corresponding distal ridge (this is the form sometimes referred to as ¾ double-shoveling).

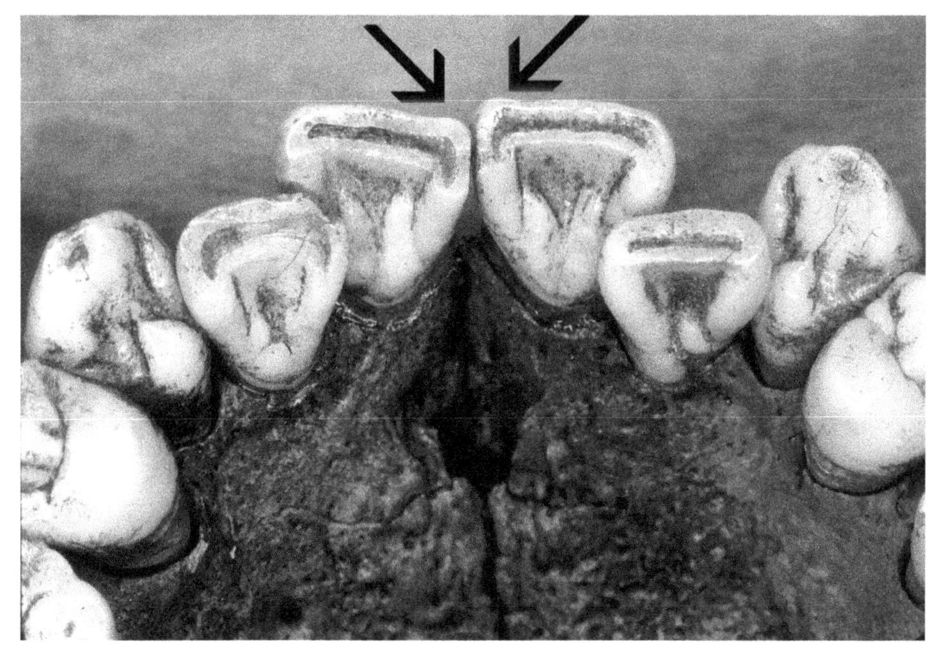

**Figure 5.3** Less pronounced ¾ double-shoveling.

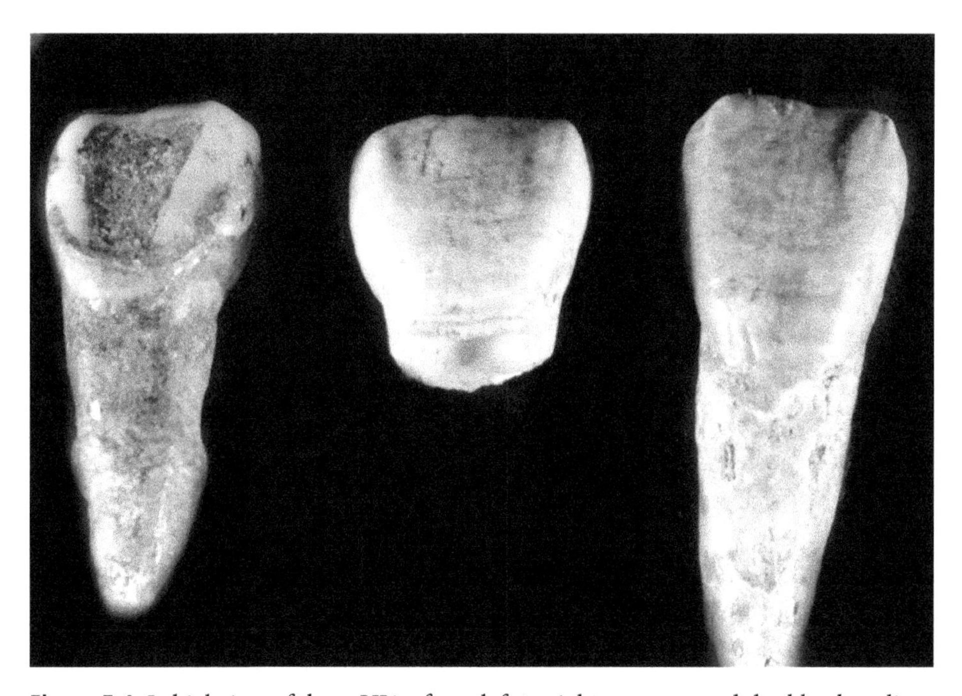

**Figure 5.4** Labial view of three UI1s: from left to right, pronounced double-shoveling, moderate double-shoveling, moderate ¾ double-shoveling.

In such instances, the ridge is typically broader than the defined ridges associated with double-shoveling. In general, labial marginal ridges are never as pronounced as lingual marginal ridges, though the differences between grades can be subtle. The key is to make certain the ridges can be palpated with a finger or fingernail, a requirement for a trait to fall above the breakpoint.

## Geographic Variation

Double-shoveling parallels the variation shown by the more commonly scored lingual shoveling. It is in highest frequency in Asian and Asian-derived populations, especially North and East Asians and all populations of the Americas. It is intermediate in Southeast Asian and derived Pacific populations. It is less common in Europeans, Africans, and Australians.

### Select Bibliography

Dahlberg, A.A., and Mikkelsen, O. (1947). The shovel-shaped character in the teeth of the Pima Indians. *American Journal of Physical Anthropology* 5, 234–235.

Moorrees, C.F.A. (1957). *The Aleut Dentition: A Correlative Study of Dental Characteristics in an Eskimoid People.* Cambridge, MA: Harvard University Press.

Rabkin, S.B. (1943). Dental conditions among prehistoric Indians of Kentucky. *Journal of Dental Research* 22, 355–366.

Snyder, R.G. (1959). *The Dental Morphology of the Point of Pines Indians.* PhD dissertation, Department of Anthropology, University of Arizona, Tucson.

(1960). Mesial margin ridging of incisor labial surfaces. *Journal of Dental Research* 39, 361–364.

Turner, C.G., II (1969). Microevolutionary interpretations from the dentition. *American Journal of Physical Anthropology* 30, 421–426.

Turner, C.G., II, Nichol, C.R., and Scott, G.R. (1991). Scoring procedures for key morphological traits of the permanent dentition: the Arizona State University dental anthropology system. In *Advances in Dental Anthropology*, ed. M.A. Kelley and C.S. Larsen. New York: Wiley-Liss, pp. 13–31.

# 6

# Interruption Grooves

## Observed on

UI1, UI2

## Key Tooth

UI2

## Synonyms

Coronal-radicular grooves (Brabant 1971), palatoradicular groove, distolingual groove, radiculolingual groove, palatogingival groove (Sharma *et al.* 2015)

## Description

Interruption grooves are referred to as such because they are manifest as distinct depressions or grooves that interrupt the normal course of the mesial or distal marginal ridges or even the basal cingulum. They differ in a number of significant ways from most nonmetric dental traits. In some instances, the groove is restricted to one or more marginal ridges and does not extend apically toward the root. In other instances, however, the groove extends vertically along the lingual margin of the basal cingulum and extends without a break onto the root. Because these grooves can be evident on different parts of the tooth crown and/or root, their expression has yet to be classified in the same manner as other traits where presence varies from slight to pronounced.

## Classification

Turner (1967) (no plaque)

The classification adopted by Turner (1967) notes the presence and location of grooves by letters rather than numbers. That is:

0 = absence of grooves on lingual marginal ridges and basal cingula
M = groove on mesiolingual marginal ridge

D = groove on distolingual marginal ridge
MD = grooves on both mesiolingual and distolingual marginal ridges
med = groove on medial aspect of basal cingulum, sometimes extending onto root

## Breakpoint

When there is a groove present on any lingual marginal ridge or on the basal cingulum, the trait is scored as present.

## Potential Problems in Scoring

It is easier to score interruption grooves on real teeth as opposed to dental casts. If grooves are evident on the uppermost cervical third of the crown, they can be scored consistently. When grooves are manifest on or around the basal cingulum, it is difficult to distinguish them from *tuberculum dentale.*

## Geographic Variation

UI2 interruption grooves are most common in East Asian and derived populations, where they are in a frequency of 40–60%. They are slightly less common in Europeans and Southeast Asian and derived populations, where frequencies fall between 35% and 40%. African and Austral-Melanesian populations have the world's lowest frequencies of around 10–20%.

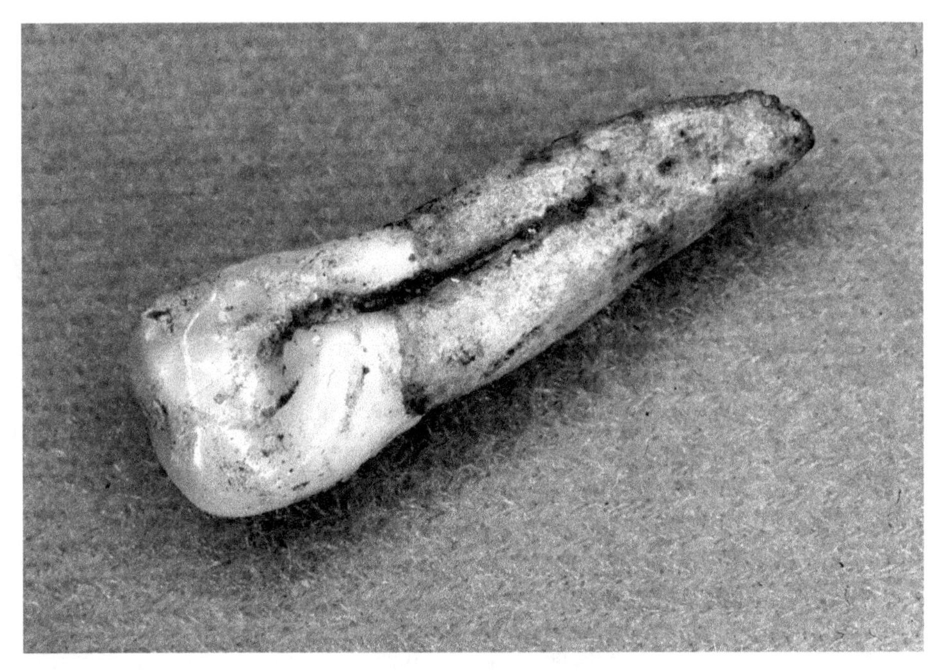

**Figure 6.1** Pronounced interruption groove that extends from the crown to the root.

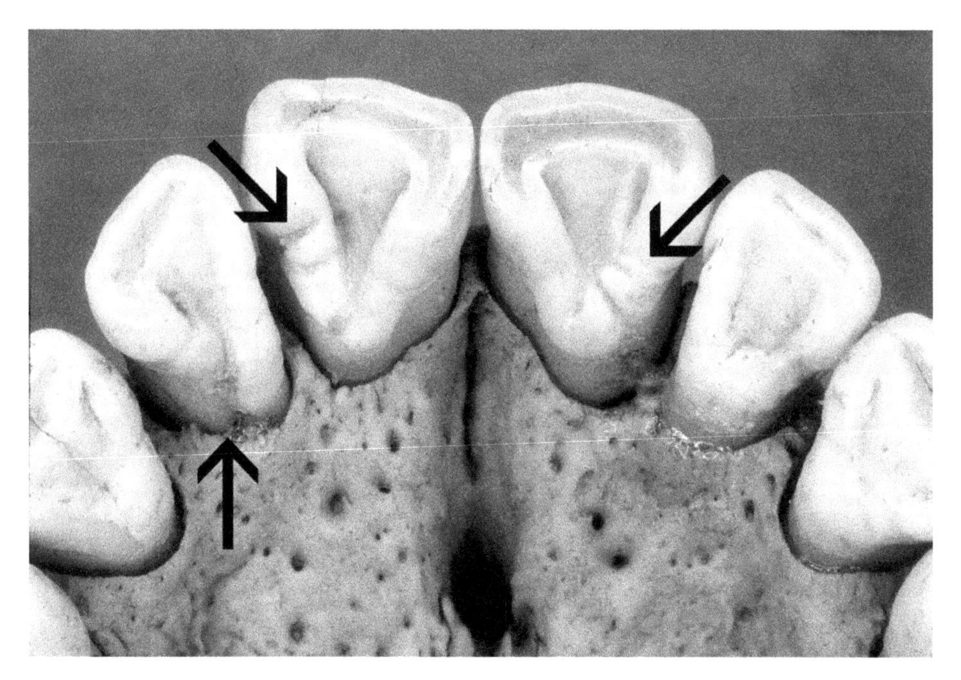

**Figure 6.2** Right UI2 shows interruption groove that extends onto the root, but the interruption grooves on UI1 involve only the marginal ridges.

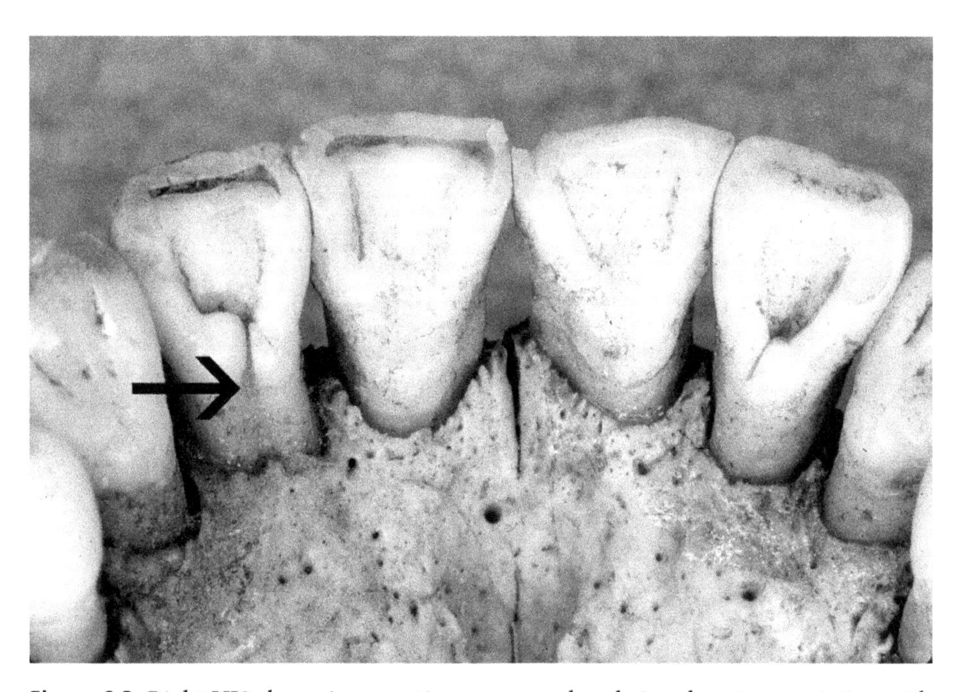

**Figure 6.3** Right UI2 shows interruption groove at basal cingulum in association with some form of *tuberculum dentale*; left UI2 shows interruption groove at basal cingulum that does not extend onto the root.

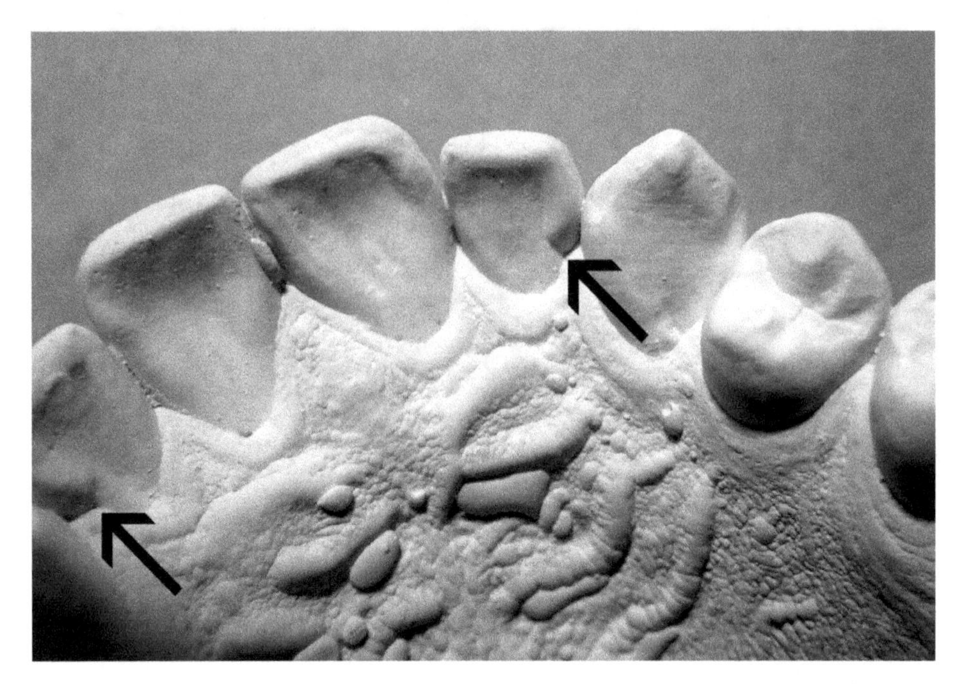

**Figure 6.4** In this cast, interruption grooves are symmetrical on UI2, with the groove on the distal marginal ridge at cervical third of crown.

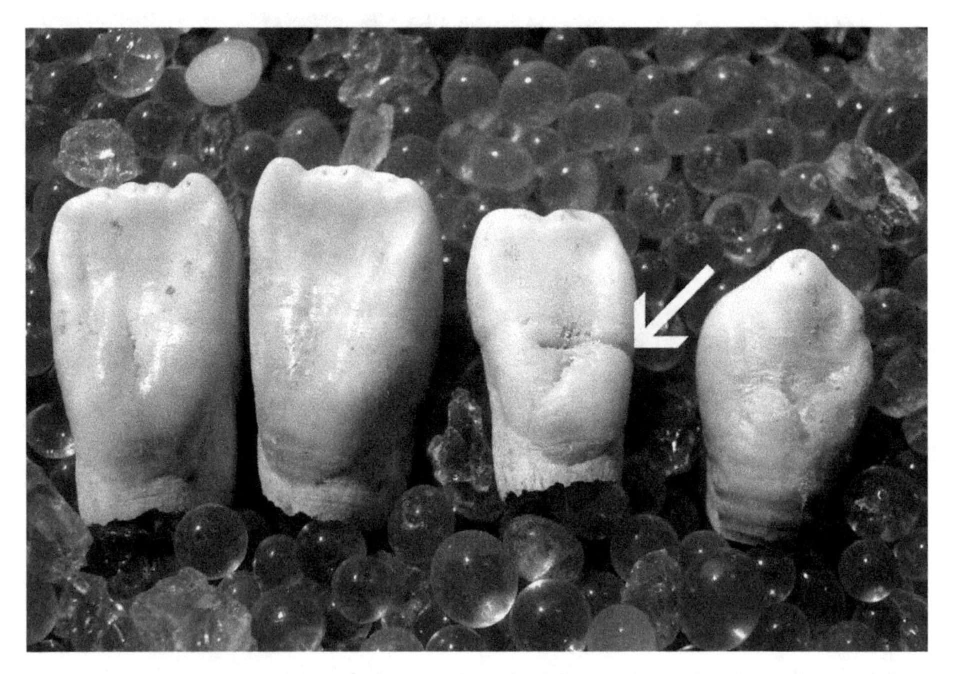

**Figure 6.5** The interruption groove on UI2 is evident on the mesial margin and does not extend onto the root.

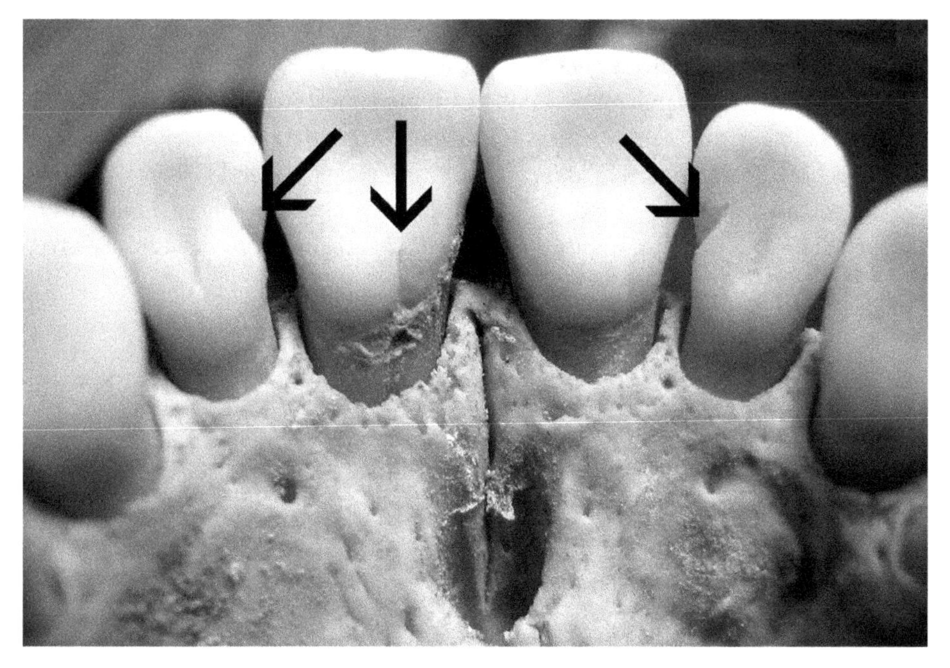

**Figure 6.6** The interruption grooves on UI2 are symmetrical and do not extend onto the root; right UI1 has an interruption groove on the basal cingulum.

## Select Bibliography

Alsoleihat, F., and Khraisat, A. (2013). The phenetic distances of the living Druze from other human populations suggest a major genetic drift from the Western Eurasian ancestral category. *Homo – Journal of Comparative Human Biology* 64, 377–390.

Brabant, H.E. (1971). The human dentition during the Megalithic era. In *Dental Morphology and Evolution*, ed. A.A. Dahlberg. Chicago: University of Chicago Press, pp. 283–297.

Gu, Y. (2011). A micro-computed tomographic analysis of maxillary lateral incisors with radicular grooves. *Journal of Endodontics* 37, 789–792.

Hrdlička, A. (1921). Further studies of tooth morphology. *American Journal of Physical Anthropology* 4, 141–176.

Lukacs, J.R. (1987). Biological relationships derived from morphology of permanent teeth: recent evidence from prehistoric India. *Anthropologischer Anzeiger* 45, 97–116.

Nichol, C.R. (1989). Complex segregation analysis of dental morphological variants. *American Journal of Physical Anthropology* 78, 37–59.

Sharma, S., Deepak, P., Vivek, S., and Dutta, S.R. (2015). Palatogingival groove: recognizing and managing the hidden tract in a maxillary incisor: a case report. *Journal of International Oral Health* 7, 110–114.

Turner, C.G., II (1967). *The Dentition of Arctic Peoples*. PhD dissertation, Department of Anthropology, University of Wisconsin, Madison

Turner, C.G., II, and Hanihara, K. (1977). Additional features of the Ainu dentition. V. Peopling of the Pacific. *American Journal of Physical Anthropology* 46, 13–24.

Ullinger, J.M., Sheridan, S.G., Hawkey, D.E., Turner, C.G., II, and Cooley, R. (2005). Bioarchaeological analysis of cultural transition in the Southern Levant using dental nonmetric traits. *American Journal of Physical Anthropology* 128, 466–476.

# 7

# Tuberculum Dentale

## Observed on

UI1, UI2, UC

## Key Tooth

UI2

## Synonyms

TD; canine tubercle (Scott 1971); dental tubercle, basal cingulum, tuberculum projection (Carlsen 1987)

## Description

Cingular projections on the lingual surface of the upper anterior teeth are relatively common in modern human populations and have a deep history in hominin and hominoid evolution. They typically take the form of ridges and/or tubercles. As expression varies among the three anterior teeth, their classifications are described separately.

## Classification

Scott (1973)

### UI1

On the upper central incisor, *tuberculum dentale* (TD) is typically expressed in the form of ridges. These ridges vary in size and number. The minimal expression is a single faint ridge. The maximal expression is two and sometimes three pronounced ridges. Scott (1973) set up a classification that included both ridge count and degree of expression (e.g., 1–3 = one moderately developed ridge, 2–1 = two slightly developed ridges). Such two-part rank numbers make analysis difficult, so this classification has

**Figure 7.1** In the plaque for UI1 tuberculum dentale, grade 1 shows slight ridges, grade 3 shows moderate ridges, and grade 4 shows a pronounced ridge.

been modified to take both ridge number and size into account. The goal is to rank expression somewhere along a broad scale from slight to pronounced.

Grade 0: no ridges on basal eminence
Grade 1: one slight ridge
Grade 2: two slight ridges, or one moderate ridge
Grade 3: one moderate ridge and one slight ridge, or one pronounced ridge
Grade 4: two moderate ridges, or one slight and one pronounced ridge
Grade 5: one moderate ridge and one pronounced ridge
Grade 6: two pronounced ridges, or three slight and/or moderate ridges

## UI2

Upper lateral incisors are notoriously variable in size, morphology, and form. *Tuberculum dentale* on this tooth is in keeping with this variability. The lowest manifestations of TD on UI2 take the form of ridges, much like those expressed on UI1. Unlike UI1, the next level of expression is in the form of tubercles. The dividing line between a ridge and tubercle is arbitrary, but the following definitions might help. A ridge is an elevation where the two sides are roughly parallel and the terminus blends into the lingual surface of the tooth as it extends toward the incisal margin. For a tubercle, the terminus is typically rounded and distinct (with or without a free apex) and does not gradually blend into the lingual surface. Figure 7.2 illustrates the scale for scoring TD on UI2.

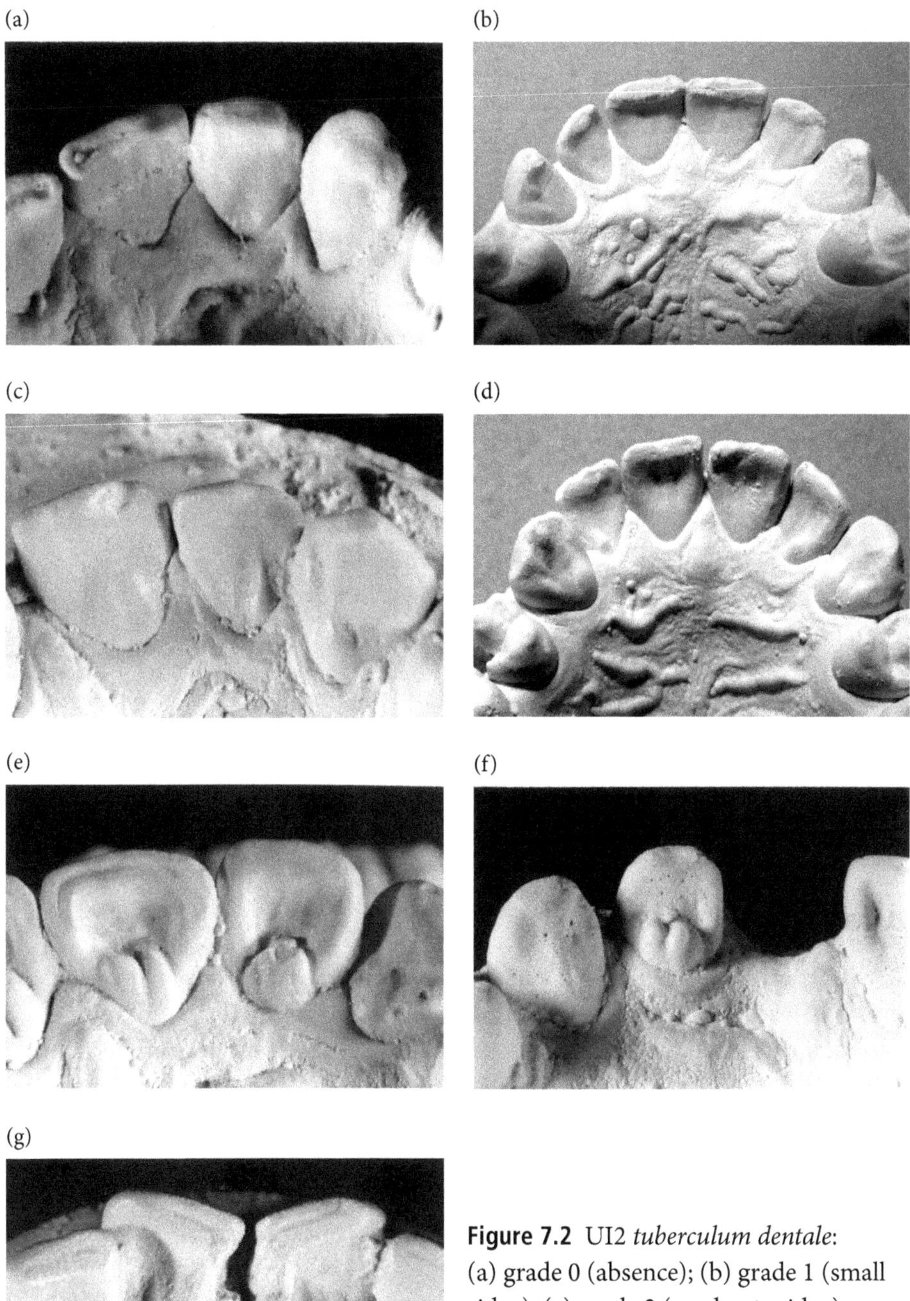


**Figure 7.2** UI2 *tuberculum dentale*: (a) grade 0 (absence); (b) grade 1 (small ridge); (c) grade 2 (moderate ridge); (d) grade 3 (small tubercle); (e) grade 5 (large tubercle); (f) grade 5 (double tubercle); (g) grade 6 (welt at base of crown).

Grade 0: absence of ridge or tubercle formation
Grade 1: slight ridge
Grade 2: moderate ridge
Grade 3: small tubercle
Grade 4: moderate tubercle
Grade 5: large or double tubercle
Grade 6: welt on basal cingulum (distinct from talon cusp; see Trait 23)

## UC

The TD gradient in the upper anterior teeth goes from mostly ridges (UI1) to ridges and tubercles (UI2) to mostly tubercles (UC). In some cases, TD on the upper canine does take the form of a ridge, but it typically takes the form of a tubercle ranging from slight to pronounced.

Grade 0: absence of any ridge or tubercle formation
Grade 1: very slight tubercle, characterized by a single groove
Grade 2: slight tubercle, outlined by two grooves
Grade 3: moderate tubercle
Grade 4: medium tubercle with no free apex
Grade 5: large tubercle with free apex
Grade 6: pronounced tubercle with free apex
Grade 7: hyper-pronounced tubercle with free apex

## Breakpoint

To characterize a population with a single frequency, a breakpoint of grade 2 and above should be used to characterize *tuberculum dentale* variation. This would include all distinct manifestations of the trait, as grade 1 represents faint variation for all three upper anterior teeth.

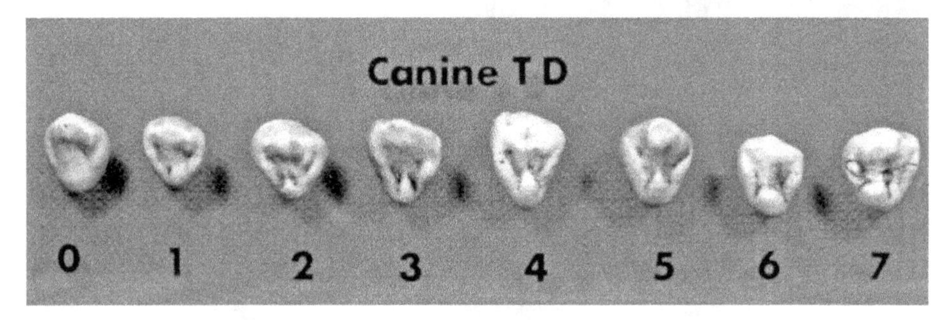


**Figure 7.3** Canine tubercle plaque (never distributed with other plaques).

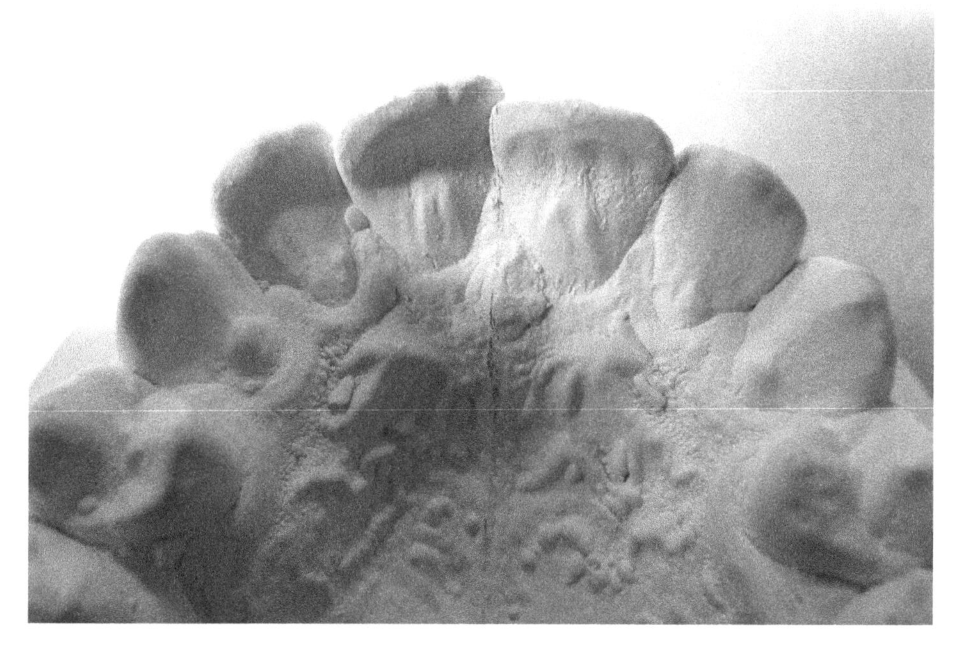

**Figure 7.4** *Tuberculum dentale* on the three upper anterior teeth co-vary, but the relationship is not perfect. In this individual, TD is grade 4 for UI1, grade 4 for UI2, grade 5 for right UC, and grade 4 for left UC.

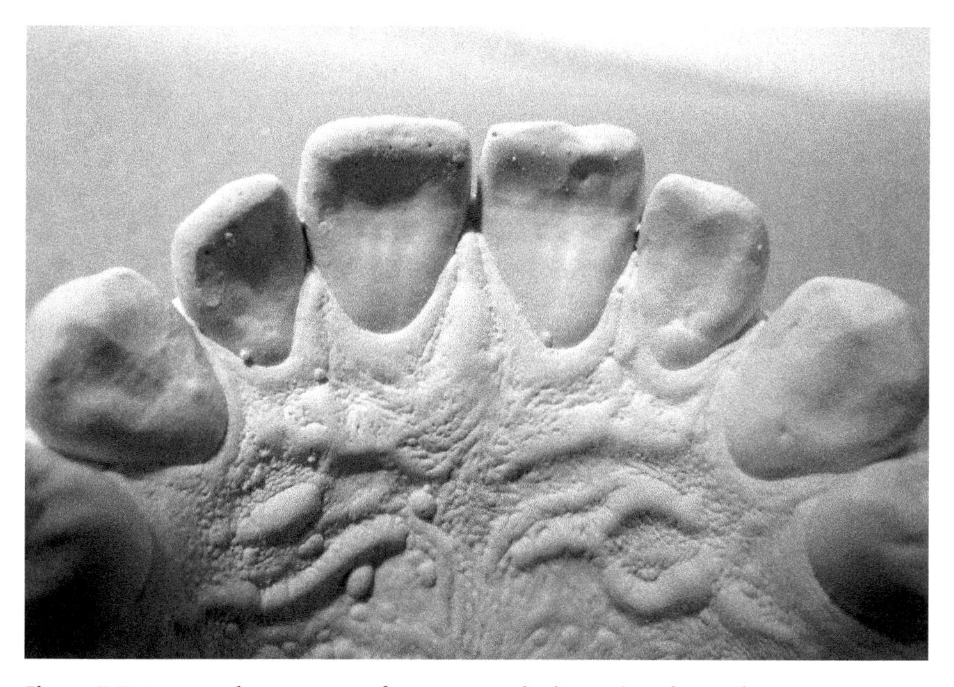

**Figure 7.5** Despite the presence of grooves on the lingual surfaces of UI1 and UI2, this individual would be scored as zero, as features reflect part of lobe segments, not TD (left UI1 shows possible interruption groove).

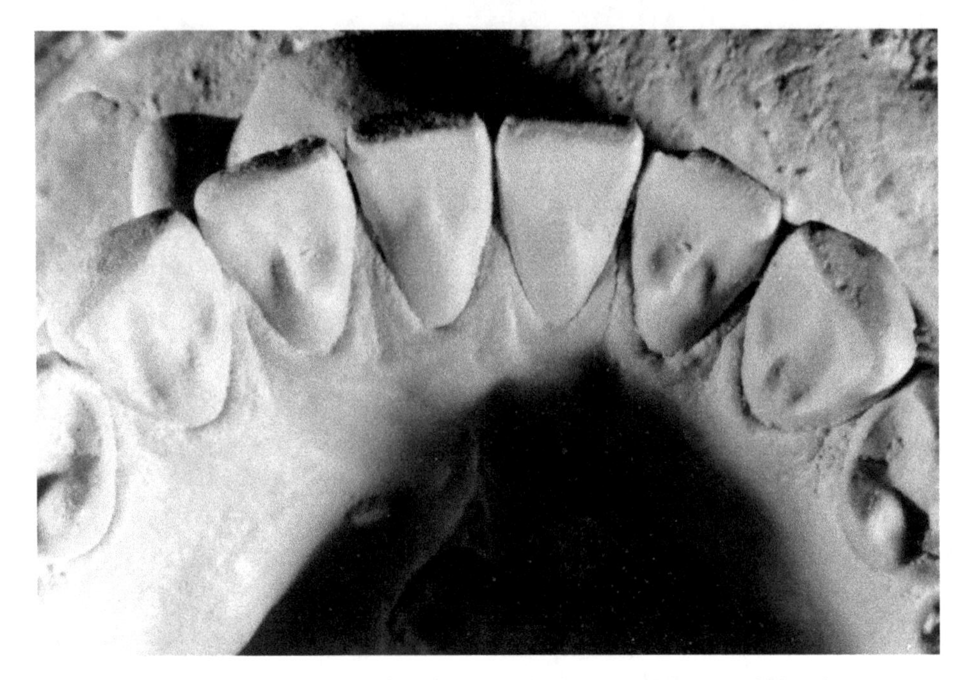

**Figure 7.6** *Tuberculum dentale* is very rare on the lower anterior teeth, but here is an individual that expresses a ridge form of TD on all lower incisors and canines.

## Potential Problems in Scoring

If bite is edge to edge, these cingular variants can be observed even with moderate incisal wear. When there is overbite, the incisal surface of the lower anterior teeth occludes with the basal eminence of the upper anterior teeth. This produces direct wear on *tuberculum dentale*, making scoring difficult.

## Geographic Variation

Various manifestations of *tuberculum dentale* can be seen throughout the hominin fossil record, with some of the most pronounced expressions observed on Neanderthal upper central incisors. In recent human populations, expressions on UI2 are relatively common, varying generally between 20% and 50%. It is more common in Asian and Asian-derived populations than in Europeans.

### Select Bibliography

Carlsen, O. (1987). *Dental Morphology*. Copenhagen: Munksgaard.

Hrdlička, A. (1921). Further studies of tooth morphology. *American Journal of Physical Anthropology* 4, 141–176.

Lasker, G.W. (1950). Genetic analysis of racial traits of the teeth. *Cold Spring Harbor Symposia on Quantitative Biology* 15, 191–203.

Nichol, C.R., and Turner, C.G., II (1986). Intra- and interobserver concordance in classifying dental morphology. *American Journal of Physical Anthropology* 69, 299–315.

Scott, G.R. (1971). Canine tuberculum dentale. *American Journal of Physical Anthropology* 35, 294.

(1973). *Dental Morphology: a Genetic Study of American White Families and Variation in Living Southwest Indians*. PhD dissertation, Department of Anthropology, Arizona State University, Tempe.

(1977). Lingual tubercles and the maxillary incisor-canine field. *Journal of Dental Research* 56, 1192.

Turner, C.G., II, and Hanihara, K. (1977). Additional features of the Ainu dentition. V. Peopling of the Pacific. *American Journal of Physical Anthropology* 46, 13–24.

# NOTES

# 8

# Bushman Canine

## Observed on

UC

## Key Tooth

UC

## Synonyms

Canine mesial ridge (Turner *et al.* 1991)

## Description

As early as the 1930s, the upper canines of some South African San were said to resemble premolars by Oranje (1934), and again by Galloway (1937) in his study of burials from the sites of Mapungubwe and Bambandyanalo in the Northern Province. In 1975, Morris confirmed these characterizations, noting that upper canines resembled lower first premolars in >43% of a sample of Botswana San. Consequently, he named the trait "Bushman canine," and noted it in South African Sotho as well. Turner standardized scoring of the trait by incorporating it into the ASUDAS, changing the name to canine mesial ridge.

Here we use the original term, Bushman canine, in consideration of Morris's original definition and because of its virtually exclusive Sub-Saharan occurrence, particularly in the San and their close relatives, the Khoekhoe. Morris's original definition states that the UC lingual surface expresses a hypertrophied mesial ridge and tubercle, i.e., *tuberculum dentale*, which are coalesced to the point where neither can be identified separately. Canines may have large mesial ridges or tubercles, but to be scored as a Bushman canine, the two must coalesce – and reach a point where the lingual sulcus is distal to the midline of the tooth.

## Classification

Morris (1975), Sakuma *et al.* (1991), Irish and Morris (1996a, 1996b)

The ASUDAS classifies the trait by dividing it into four rank-scale grades of expression:

Grade 0: mesial and distal lingual ridges are the same size. Neither is attached to the *tuberculum dentale,* if present.

Grade 1: mesiolingual ridge is larger than distolingual and is weakly attached to the *tuberculum dentale.*

Grade 2: mesiolingual ridge is larger than the distolingual and is moderately attached to the *tuberculum dentale.*

Grade 3: Morris's type form. Mesiolingual ridge is much larger than the distolingual and is fully incorporated into the *tuberculum dentale.*

## Breakpoint

Grade 1+

## Potential Problems in Scoring

Although the four ASUDAS grades adequately describe the continuum from absence to full Bushman canine expression, the original plaque can be misleading if it is not used in conjunction with the above definitions. Specifically, the original grade 2 example does not sufficiently demonstrate the morphology described above. This incongruity can lead, and has led, to cases of misidentification. In Figure 8.2, the grade 2 example has been "photo-shopped" to enhance the size of the mesial ridge, along with moving the tubercle distally so that the feature does extend at a point beyond (i.e., distal to) the lingual sulcus. The remaining grades are representative of the corresponding definitions.

## Geographic Variation

The second author (JDI) substantiated Morris's claim for a pan-African distribution. Of 15 Sub-Saharan samples studied ($n = 824$), 14 possess the trait (grades 1–3). Trait presence ranges from 0% in a Congo sample to 40.6% in South African Khoekhoe. Others with high frequencies are: southern African San (35.1%), southern African Nguni (33.3%), and pooled samples from Togo and Benin (35.3%), and Nigeria and Cameroon (22.2%). The overall frequency across Sub-Saharan Africa is 18.1%. It is also found, presumably due to gene flow, in North African peoples, at a rate of 6.1%. Turner recorded the trait in samples of Europeans, Southeast Asians, Northeast Asians,

Aleuts/Eskimos, Native North and South Americans, Australians/Tasmanians, and Melanesians. He found either complete absence or presence in significantly lower frequencies than in Africans (i.e., <5%, and presumably at grades 1 and 2). According to Morris, the type form (grade 3) is not found in non-Africans.



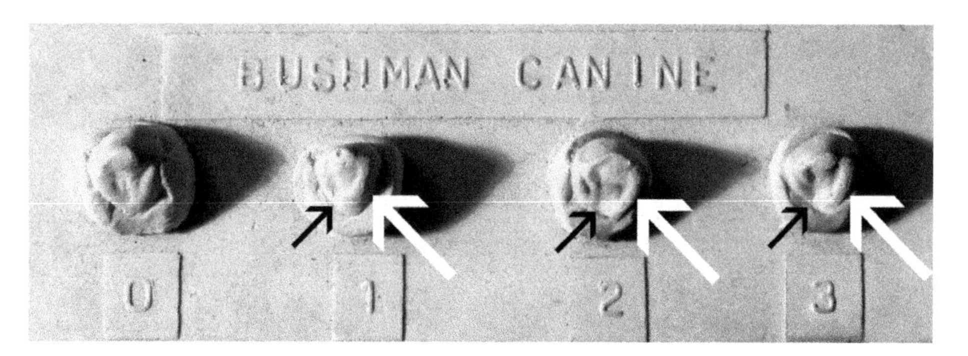


**Figure 8.1** Bushman canine plaque. The white arrows point toward the hypertrophied mesial ridge, as it coalesces with an increasingly large *tuberculum dentale* in grades 1, 2, and 3. The black arrow identifies the distal position of the lingual sulcus – which is a key characteristic of the trait.

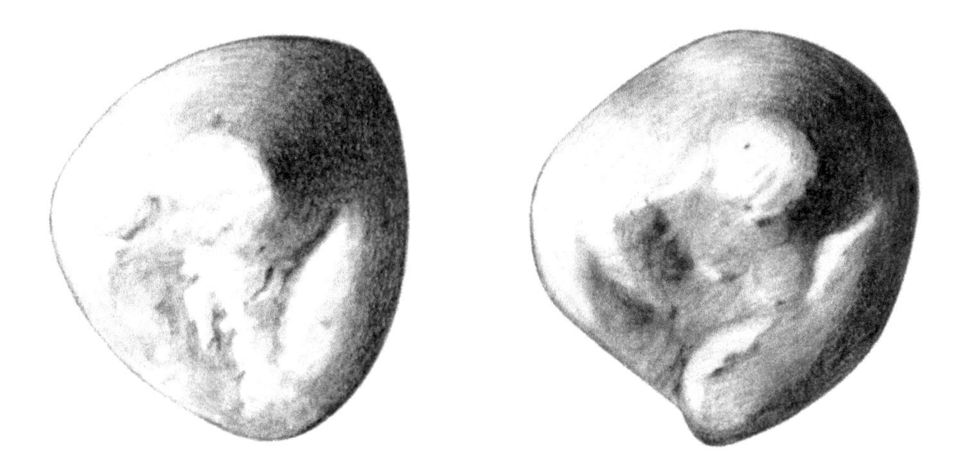

**Figure 8.2** Occlusal-view line drawing of grade 0 (left) and grade 3 (right) Bushman canines. For grade 3, note the large mesial ridge and *tuberculum dentale* that coalesce and, importantly, reach a point where the lingual sulcus is distal to the midline of the tooth (drawing by Jeffry M. Irish).

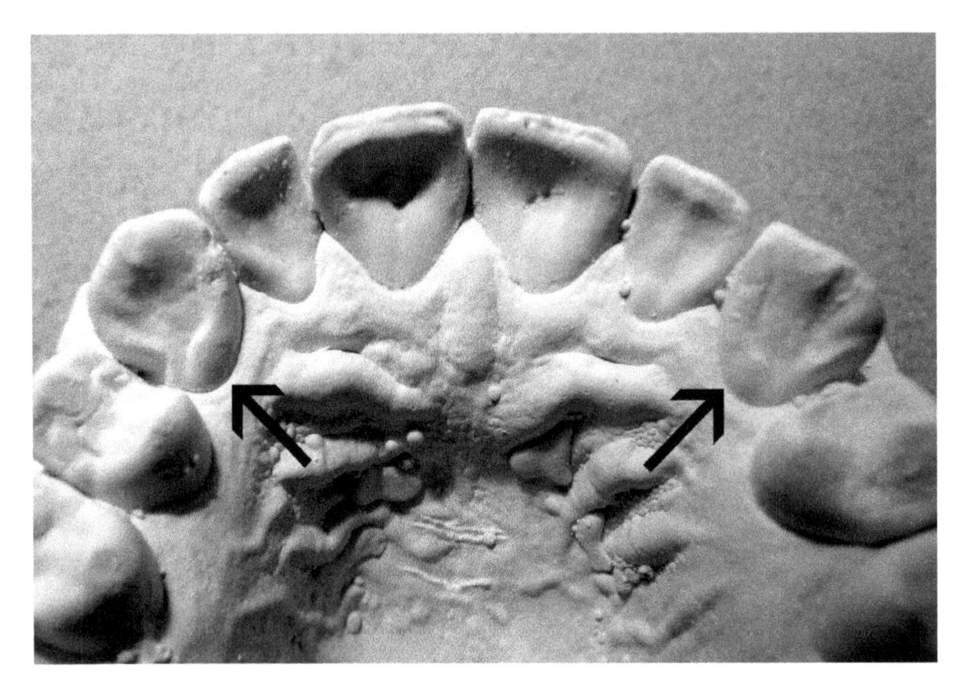

Figure 8.3 Bushman canine present on right UC (grade 2) but absent on left UC.

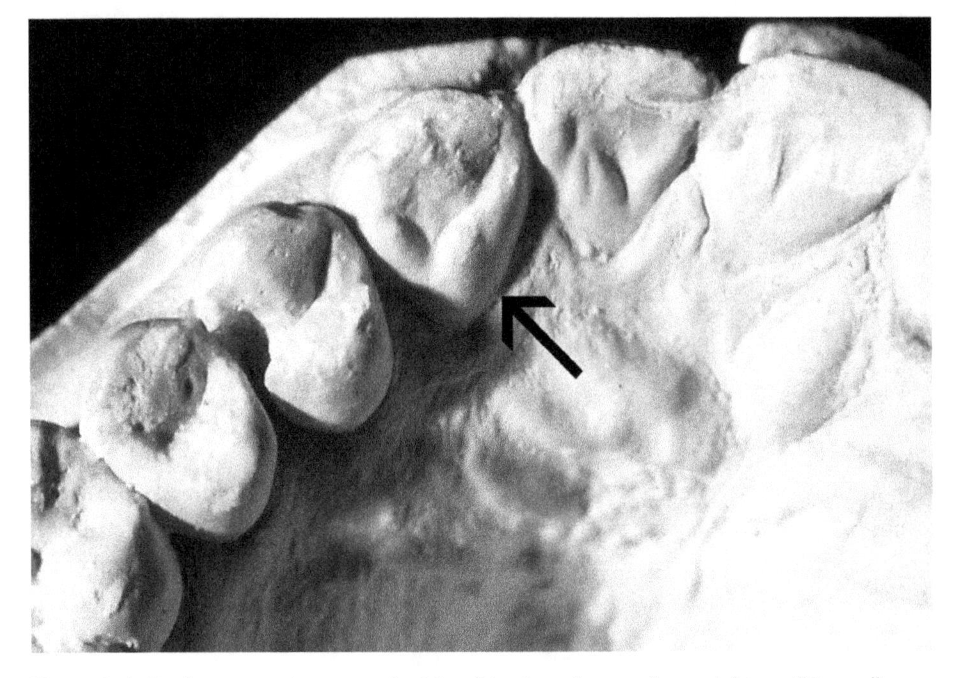

Figure 8.4 Bushman canine at grade 3 level in dental cast of east African "Bantu."

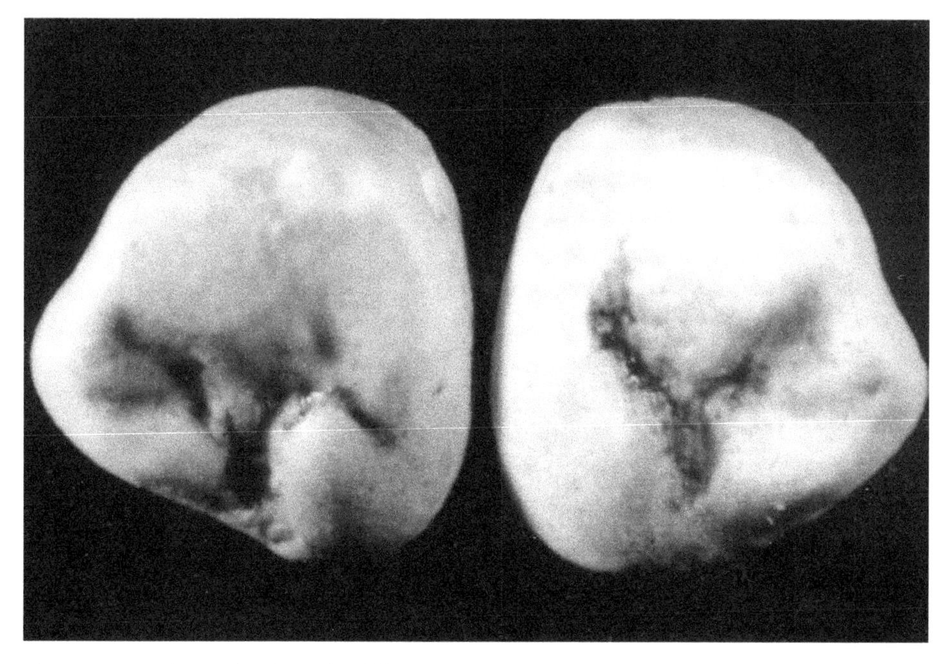

**Figure 8.5** Left UC (right side of figure) shows clear Bushman canine, while right antimere shows distinct *tuberculum dentale* that is not incorporated into mesial ridge.

## Select Bibliography

Galloway, A. (1937). The skeletal remains of Mapungubwe. In *Mapungubwe: Ancient Bantu Civilization on the Limpopo*, ed. L. Fouche. Cambridge: Cambridge University Press, pp. 127–174.

(1959). *The Skeletal Remains of Bambandyanalo*. Johannesburg: Witwatersrand University Press.

Haeussler, A.M., Irish, J.D., Morris, D.H., and Turner, C.G., II (1989). Morphological and metrical comparison of San and Central Sotho dentitions from southern Africa. *American Journal of Physical Anthropology* 78, 115–122.

Irish, J.D. (1998). Dental morphological affinities of Late Pleistocene through recent sub-Saharan and North African peoples. *Bulletins et Mémoires de la Société d'anthropologie de Paris* 10, 237–272.

Irish, J.D., and Morris, D.H. (1996a). Technical note: canine mesial ridge (Bushmen canine) dental trait definition. *American Journal of Physical Anthropology* 99, 357–359.

(1996b). A supplemental description of the Bushman maxillary canine polymorphism. *South African Journal of Science* 92, 351–353.

Manabe, Y., Rokutanda, A., Kitagawa, Y., and Oyamada, J. (1991). Genealogical position of native Taiwanese (Bunun tribe) in East Asian populations based on tooth crown morphology. *Journal of the Anthropological Society of Nippon* 99, 33–47.

Morris, D.H. (1975). Bushmen maxillary canine polymorphism. *South African Journal of Science* 71, 333–335.

Oranje, P. (1934). The dentition of the Bush race. *South African Journal of Science* 31, 576.

Sakuma, M., Irish, J.D., and Morris, D.H. (1991). The Bushman maxillary canine of the Chewa tribe in east-central Africa. *Journal of the Anthropological Society of Nippon* 99, 411–417.

# 9

# Canine Distal Accessory Ridge

## Observed on

UC, LC

## Key Tooth

UC/LC

## Synonyms

DAR

## Description

The lingual lobe segment of the upper and lower canines typically expresses a medially positioned essential ridge, a mesial marginal ridge, and a distal marginal ridge. The mesial marginal ridge is primarily vertical in orientation and often extends almost to the level of the cusp tip. The distal marginal ridge is shorter and courses away at an angle from the essential ridge. Between the essential ridge and distal marginal ridge, an additional ridge can be manifest on the lingual aspect of the distal lobe segment, hence the name distal accessory ridge (DAR). Although the upper and lower canines are morphologically distinct, with the upper canine more conical and the lower canine more incisiform, both can exhibit distal accessory ridges. The ridge is more common and pronounced on the upper canine, and, given the different forms of the two canines, separate classifications were established for each.

## Classification

Scott (1973, 1977)

UC and LC distal accessory ridge (two plaques):

Grade 0: trait absence
Grade 1: faint expression (not shown on UC DAR plaque)

(a)

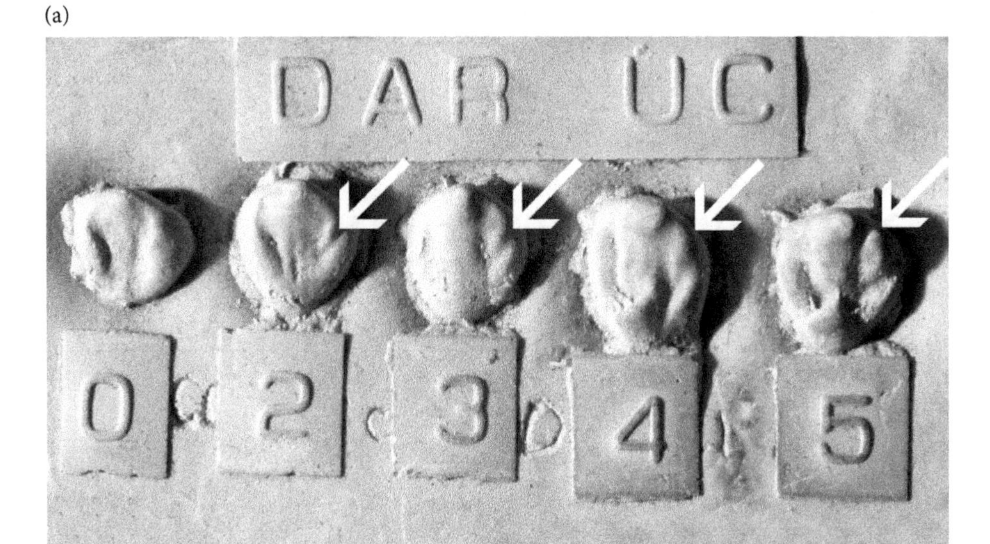


(b)

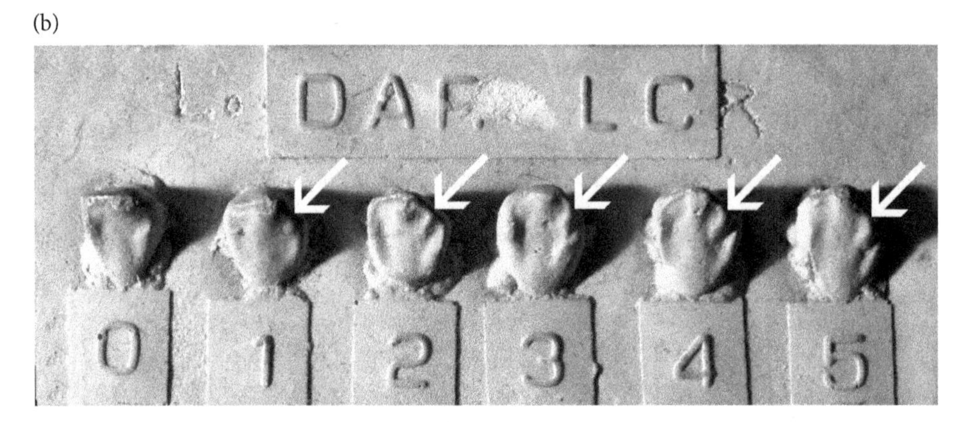


**Figure 9.1** Canine distal accessory ridge plaques: (a) UC; (b) LC.

Grade 2: slight expression
Grade 3: moderate development
Grade 4: strongly developed
Grade 5: pronounced expression

## Breakpoint

Turner adopted UC DAR as one of his 29 key morphological crown and root traits. For analysis, he used grade 2 as a breakpoint. Grade 2 would be a good breakpoint for either UC or LC DAR, as grade 1 expressions are subtle and can only be scored on unworn teeth.

## Potential Problems in Scoring

The canine distal accessory ridge provides one of the first contact surfaces in occlusion and it manifests wear at an early age. There is no dentine component so one has to exercise caution in scoring trait absence. Not only is lingual wear an issue, the incisal edge that contacts the DAR also wears at an early age. Because of the scoring problems introduced by wear, sample size is often reduced in groups with a normal age profile. This is less of an issue if a sample is composed primarily of children and teens.

## Geographic Variation

World variation has yet to be summarized, but DAR is in higher frequency in Native Americans than in Europeans.

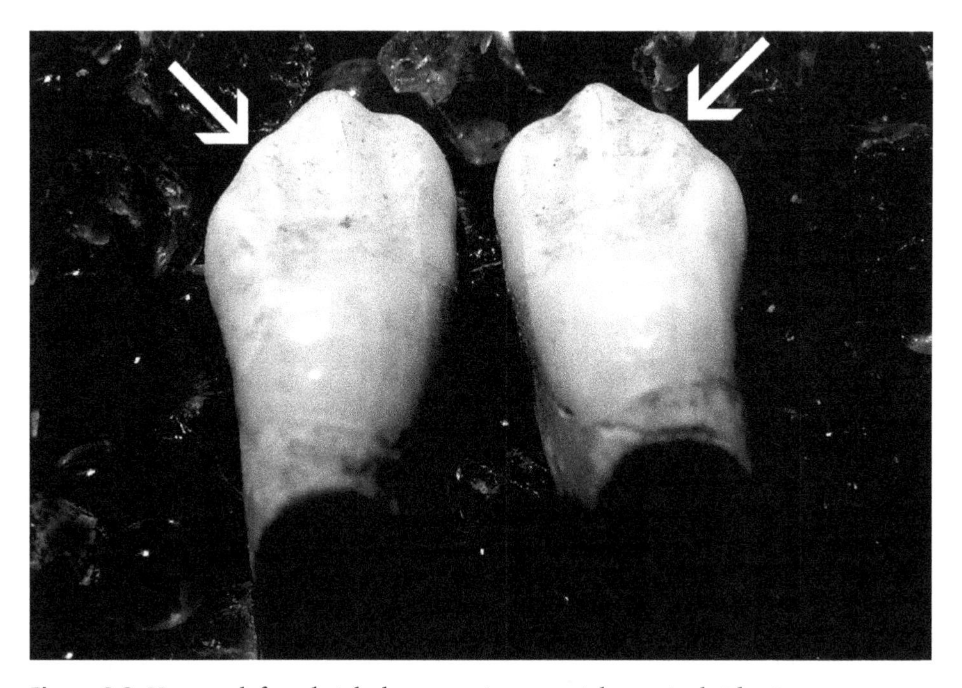

**Figure 9.2** Unworn left and right lower canines; mesial marginal ridge is more vertical than distal marginal ridge. Distal accessory ridges are between essential ridge that ends in cusp tip and distal marginal ridge.

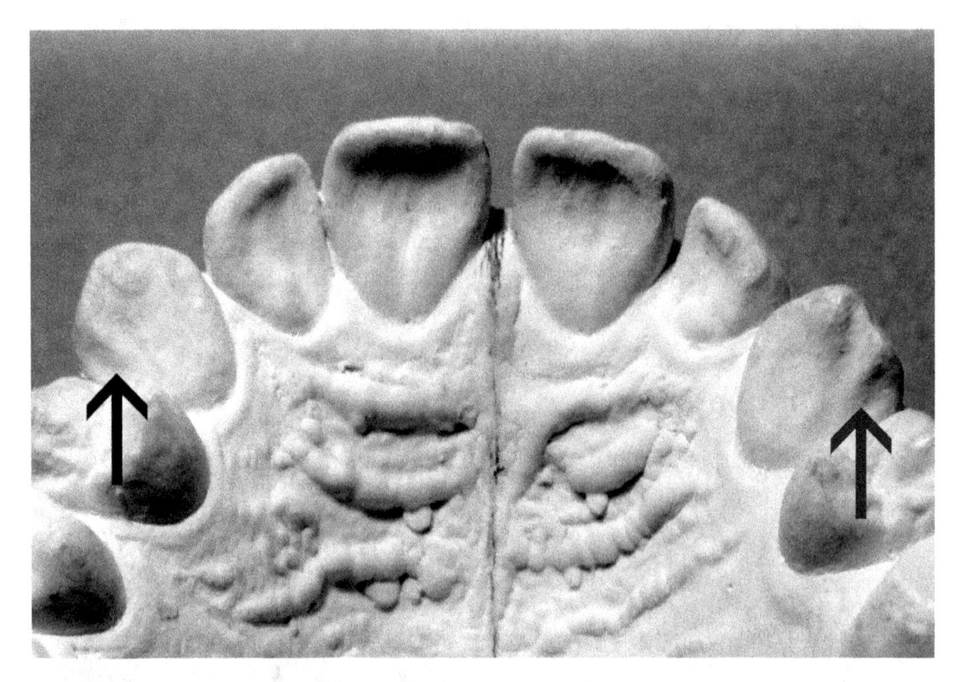

**Figure 9.3** Large symmetrical UC distal accessory ridges.

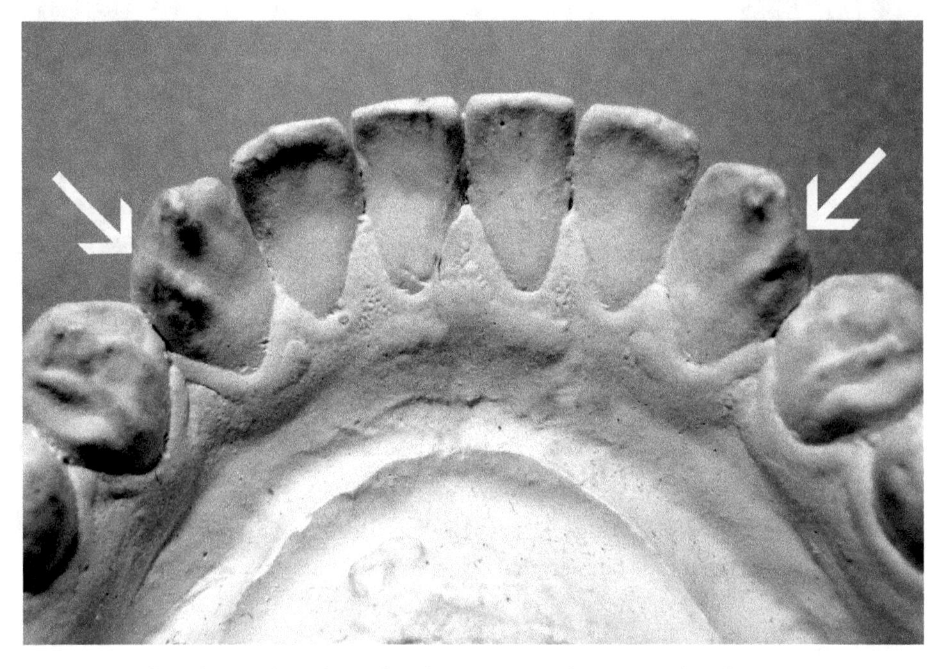

**Figure 9.4** Sharply developed LC distal accessory ridges; LP1 also shows very distinctive distal accessory ridges.

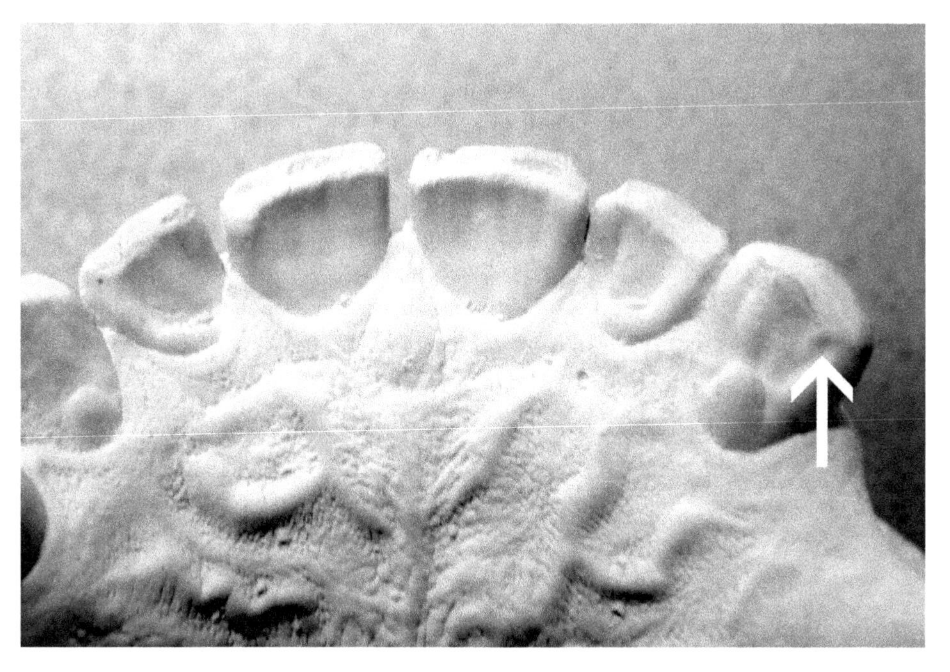

**Figure 9.5** Even with some incisal wear, the UC distal accessory ridge is still evident.

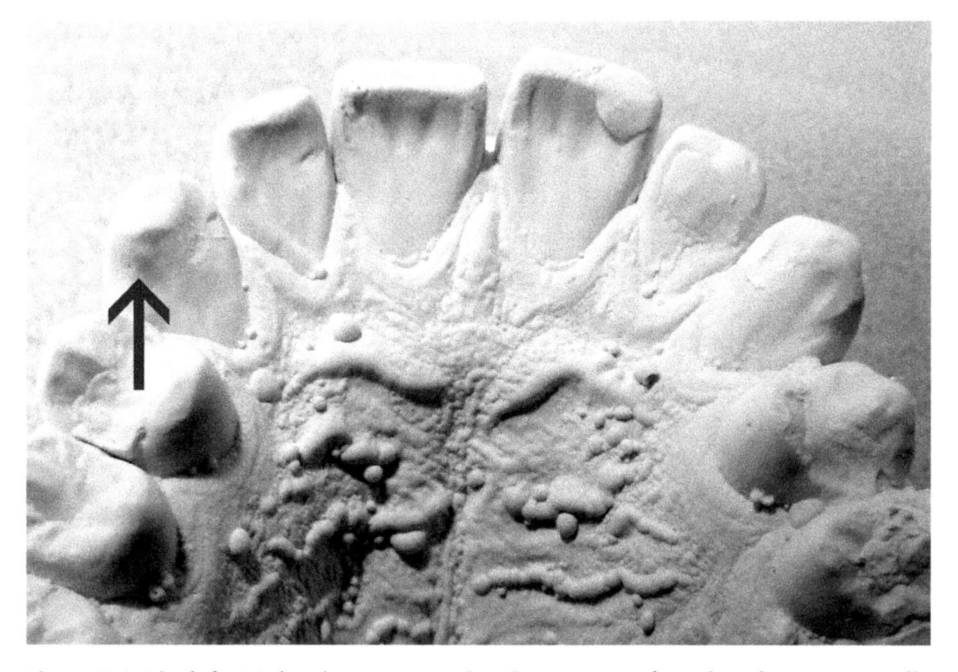

**Figure 9.6** The left UC distal accessory ridge shows a wear facet, but the trait can still be scored. Only the cusp tip is worn on the right UC and the distal accessory ridge is still present.

## Select Bibliography

Noss, J.F., Scott, G.R., Potter, R.H.Y., Dahlberg, A.A., and Dahlberg, T. (1983). The influence of crown size dimorphism on sex differences in the Carabelli trait and the canine distal accessory ridge in man. *Archives of Oral Biology* 28, 527–530.

Scott, G.R. (1973). *Dental Morphology: A Genetic Study of American White Families and Variation in Living Southwest Indians*. PhD dissertation, Department of Anthropology, Arizona State University, Tempe.

(1977). Classification, sex dimorphism, association, and population variation of the canine distal accessory ridge. *Human Biology* 49, 453–469.

Scott, G.R., Potter, R.H.Y., Noss, J.F., Dahlberg, A.A., and Dahlberg, T. (1983). The dental morphology of Pima Indians. *American Journal of Physical Anthropology*, 61, 13–31.

# 10

# Premolar Accessory Ridges

## Observed on

UP1, UP2, LP1, LP2

## Key Tooth

UP2

## Synonyms

Accessory occlusal ridges (Gilmore 1968), premolar accessory ridges (Scott 1973), MxPAR (Burnett 1998)

## Description

The lingual lobe segment of upper and lower premolar buccal cusps is divided into three lobe sections: the central section exhibits the essential ridge. This ridge is flanked on two sides by the mesial and distal lobe sections. Accessory ridges may or may not be expressed on the occlusal surfaces of these lobe sections, hence the terms mesial and distal accessory ridges. Of the four possible loci for ridge expression (P1 mesial and distal, P2 mesial and distal), distal ridges are more common than mesial ridges and ridge frequencies are higher on P2 than on P1. Because of wear, oral pathologies, and other factors that preclude observation on a particular lobe section, each locus should be scored separately.

## Classification

Burnett (1998)

For either upper or lower premolars, the ridges should be scored in the following manner (adapted from Burnett *et al.* 2010):

Grade 0: absence of ridge
Grade T: truncated ridge (is not continuous from buccal cusp to sagittal sulcus)

Grade 1: trace (a slight, continuous ridge from buccal cusp to sagittal sulcus)
Grade 2: small (thin continuous ridge)
Grade 3: medium (moderately thick continuous ridge)
Grade 4: pronounced (thick continuous ridge)

## Breakpoint

Grade 2+

## Potential Problems in Scoring

These accessory ridges are on the occlusal surface and have no dentine component. As with other such traits, they wear down quickly, with wear facets appearing shortly after tooth eruption. The ridges are often obscured on dental casts by occlusal caries and casting error, even when the teeth show little attrition.

## Geographic Variation

Maxillary premolar ridges were not key traits in C.G. Turner's analysis of worldwide dental variation, so geographic information on UP MxPAR is limited. Burnett *et al.* (2010) found these ridges were more common in Northeast Asian and Native American populations (ca. 60–80%) than in Western Eurasians, who had frequencies of 25–35%. Solomon Island and "Bantu" samples fell between these extremes. Scott (1973) reported consistently higher frequencies of upper premolar ridges in ten Southwest Native American samples compared to two American white samples. Mihailidis *et al.* (2013) found Australian aborigines and Southeast Asians had lower frequencies than Europeans and East Asians.

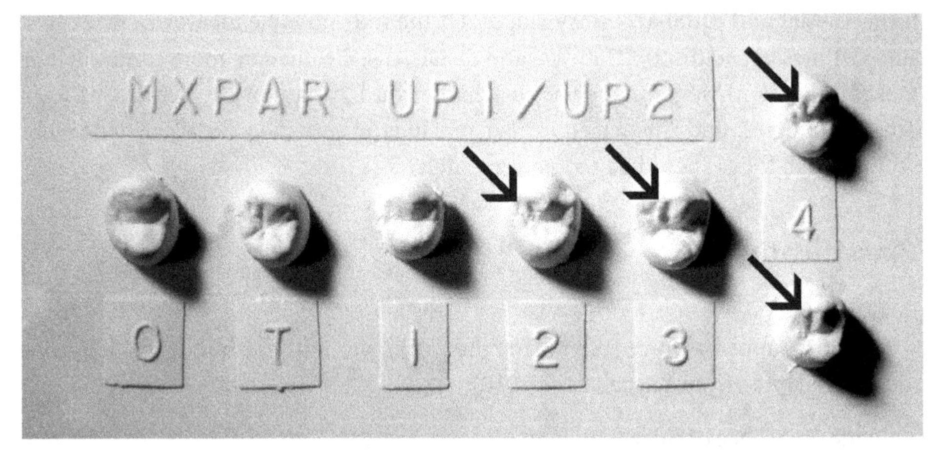


**Figure 10.1** Ranked scale for upper premolar accessory ridges (MxPAR).

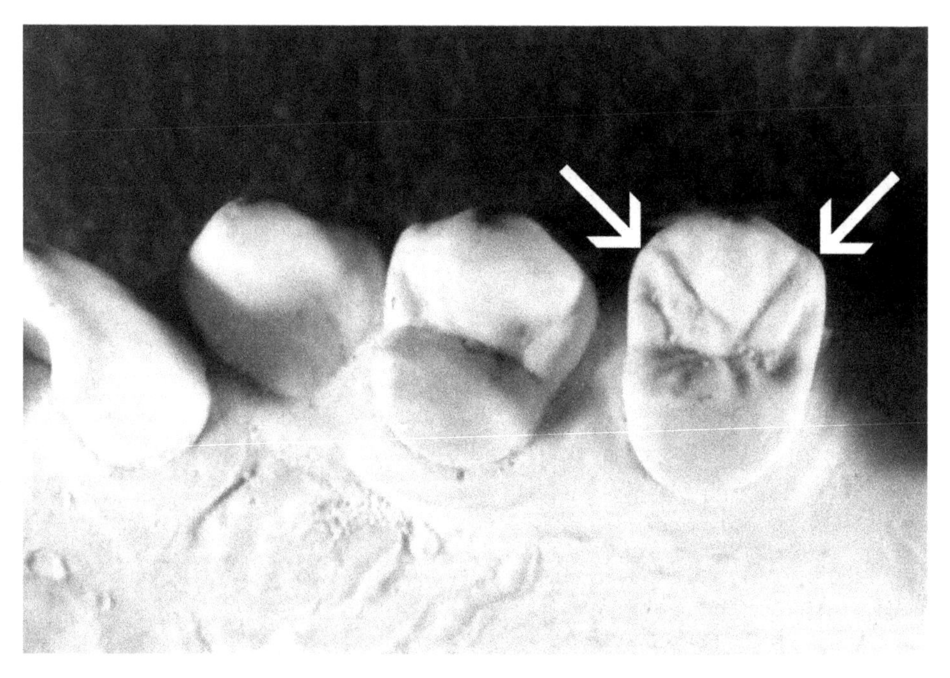

**Figure 10.2** Grade 4 mesial and distal accessory ridges on UP2; trait is either absent or faint on UP1.

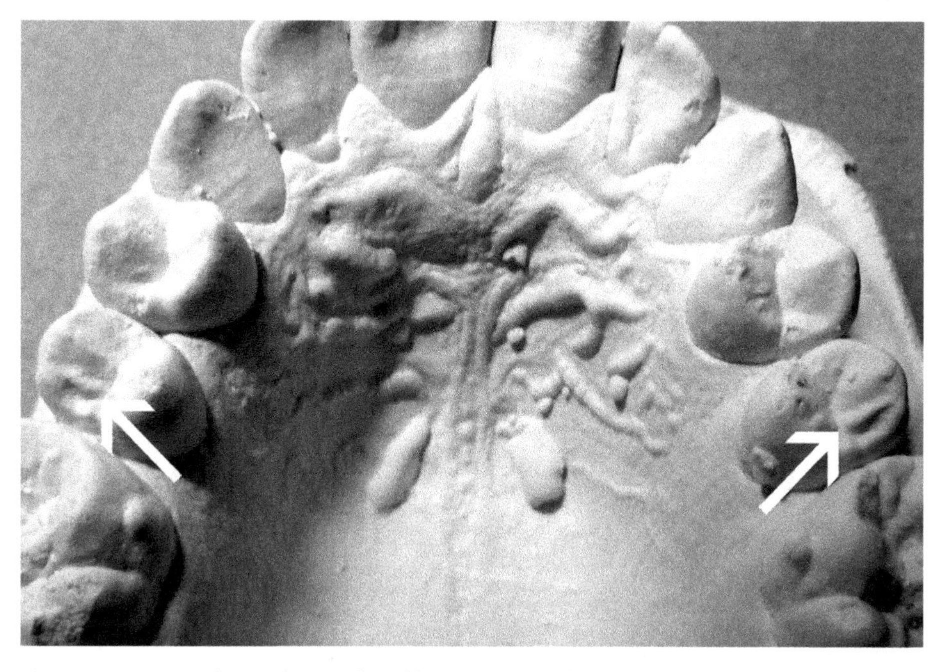

**Figure 10.3** A grade 3 ridge on distal lobe section of UP2, with ridge absence on UP1 and the mesial lobe section of UP2.

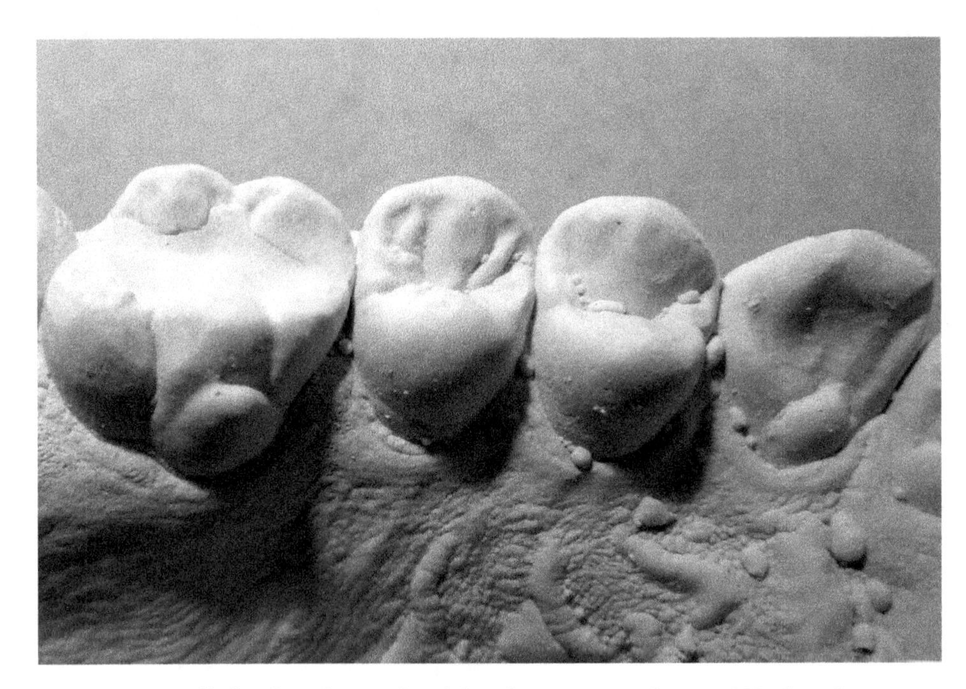

**Figure 10.4** Well-developed mesial and distal accessory ridges on UP2 but absent on UP1, a relatively common occurrence that supports the position that UP2 is the key tooth for MxPAR.

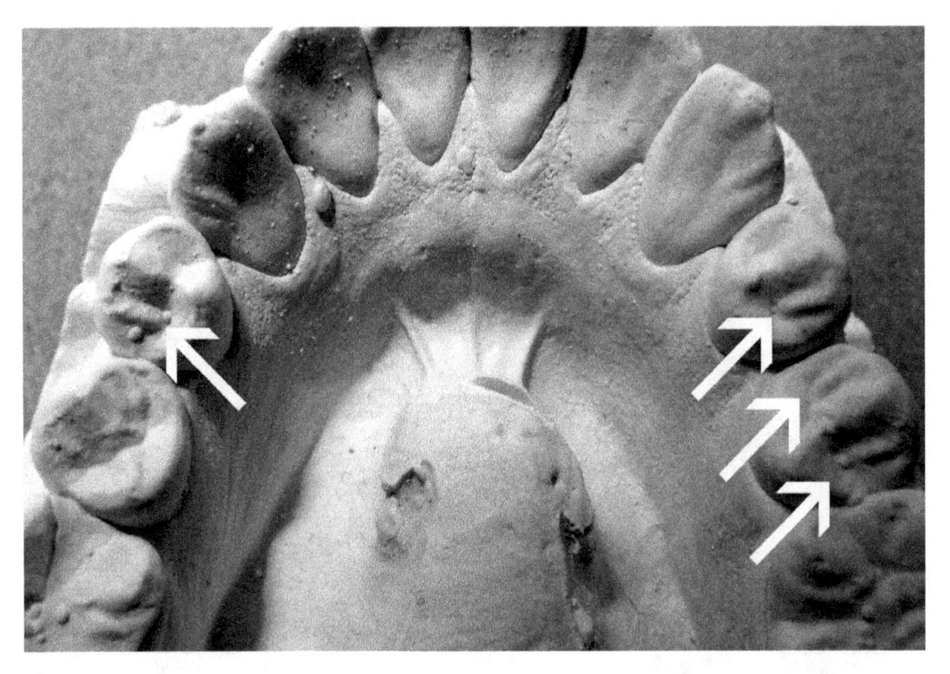

**Figure 10.5** Although the classification for premolar accessory ridges was established for the upper premolars, they are also expressed on the lower premolars and in the same pattern (distal ridges more common than mesial, ridges more common on LP2 than on LP1).

## Select Bibliography



Burnett, S.E. (1998). *Maxillary Premolar Accessory Ridges (MxPAR): Worldwide Occurrence and Utility in Population Differentiation.* MA thesis, Department of Anthropology, Arizona State University, Tempe.

Burnett, S.E., Hawkey, D.E., and Turner, C.G., II (2010). Population variation in human maxillary premolar accessory ridges (MxPAR). *American Journal of Physical Anthropology* 141, 319–324.

Gilmore RW. (1968). Epidemiology and heredity of accessory occlusal ridges on the buccal cusps of human premolar teeth. *Archives of Oral Biology* 13, 1035–1046.

Mihailidis, S., Scriven, G., Khamis, M., and Townsend, G.C. (2013). Prevalence and patterning of maxillary premolar accessory ridges (MxPARs) in several human populations. *American Journal of Physical Anthropology* 152, 19–30.

Morris DH. (1965). *The Anthropological Utility of Dental Morphology.* PhD dissertation, Department of Anthropology, University of Arizona, Tucson.

Scott, G.R. (1973). *Dental Morphology: A Genetic Study of American White Families and Variation in Living Southwest Indians.* PhD dissertation, Department of Anthropology, Arizona State University, Tempe.

# NOTES

# 11

# Upper Premolar Mesial and Distal Accessory Cusps

## Observed on

UP1, UP2

## Key Tooth

UP1

## Synonyms

Marginal tubercles (Carlsen 1987)

## Description

The buccal and lingual cusps of the upper premolars are separated by a sagittal sulcus. At either the mesial or distal margin of this sulcus, a small accessory cusp or tubercle may be present. To score this small cusp as present, it has to be clearly delineated from both major cusps. That is, it should have distinct vertical grooves outlining the cusp. It should not be a projection that is only an extension of either major cusp, a relatively common occurrence. In the original ASUDAS, any mesial or distal accessory cusp was scored as present. Although not a ranked scale, we recommend distinguishing mesial and distal accessory cusps to determine if there is patterned variation for this trait.

## Classification

Turner (1967)

Grade 0: accessory cusp absent
Grade 1: well-defined mesial accessory cusp with palpable cusp tip
Grade 2: well-defined distal accessory cusp with palpable cusp tip
Grade 3: well-defined mesial and distal accessory cusps with palpable cusp tips

## Breakpoint

Trait presence

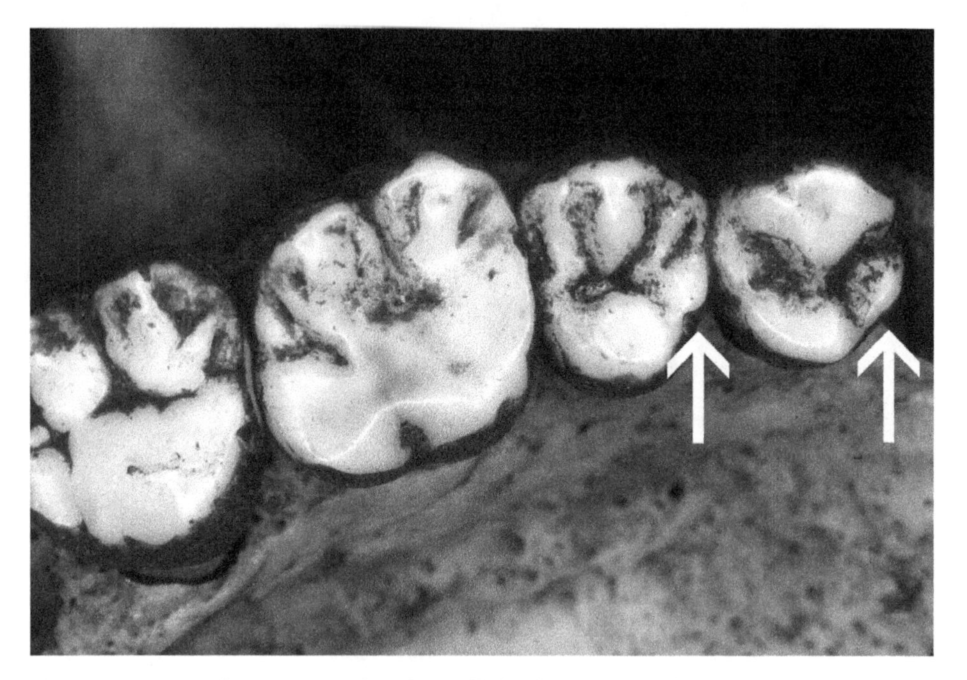

**Figure 11.1** Mesial accessory tubercles well developed on both UP1 and UP2.

## Potential Problems in Scoring

As there is no dentine component, these small cusps often wear away at an early age. As with accessory ridges, they are often positioned in such a way as to serve as the first point of contact between occluding teeth.

## Geographic Variation

Unknown.

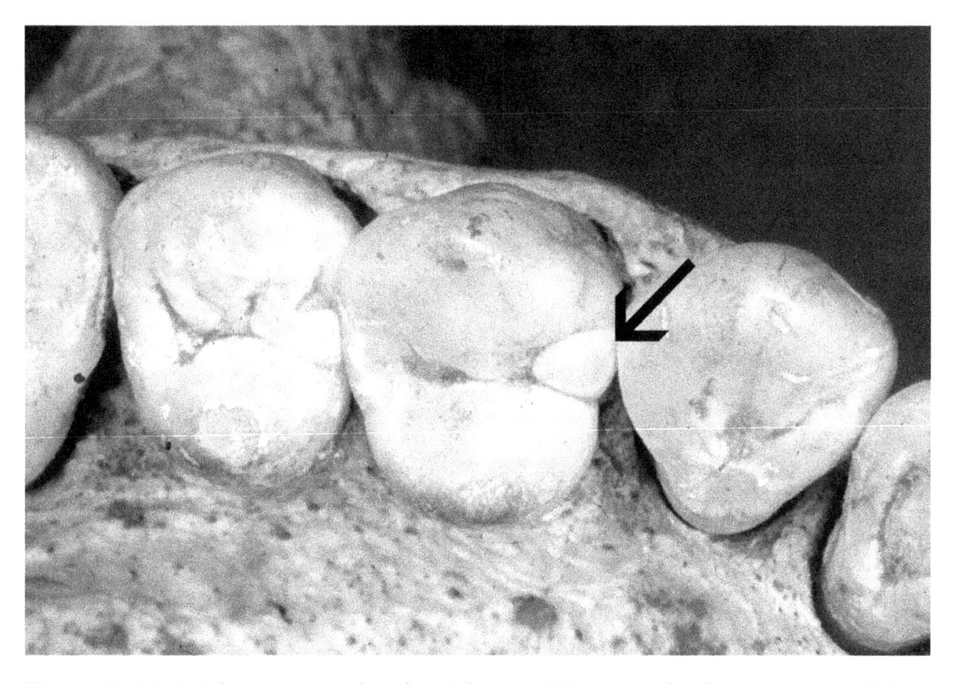

**Figure 11.2** Mesial accessory tubercle evident on UP1; may also be present on UP2.

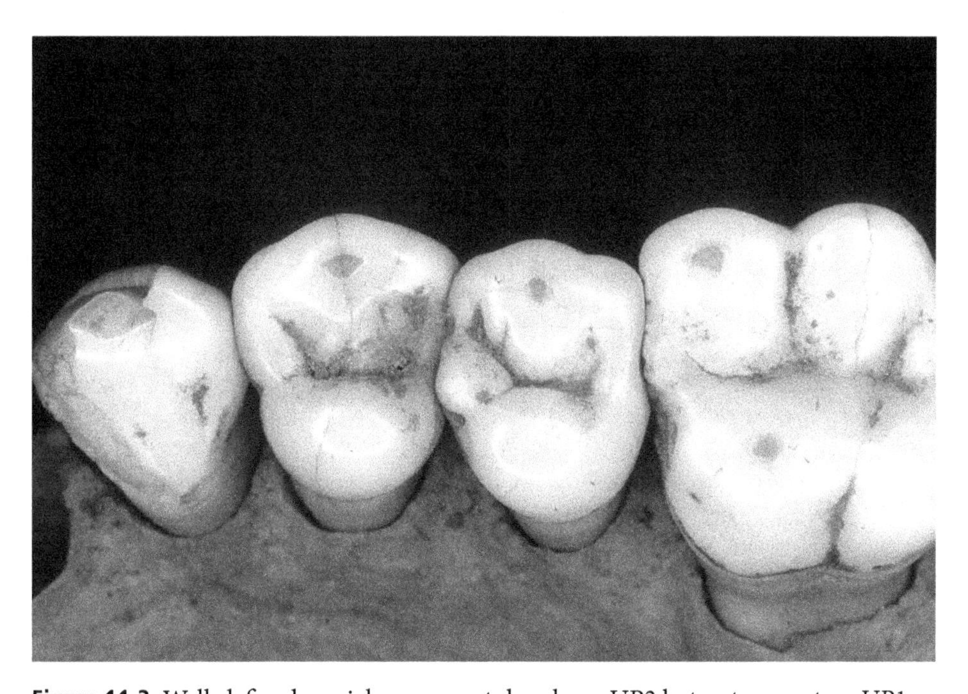

**Figure 11.3** Well-defined mesial accessory tubercle on UP2 but not present on UP1.

## Select Bibliography

Carlsen, O. (1987). *Dental Morphology*. Copenhagen: Munksgaard.

Turner, C.G., II (1967). *The Dentition of Arctic Peoples*. PhD dissertation, Department of Anthropology, University of Wisconsin, Madison.

# 12

# Uto-Aztecan Premolar

## Observed on

UP1

## Key Tooth

UP1

## Synonyms

Distosagittal ridge (Turner *et al.* 1991)

## Description

This rare variant, described first by Morris *et al.* (1978), presents a distinctive phenotype. One key to this trait's development is the general form of the upper first premolar. That is, the first premolar has a buccal and a lingual cusp whose main axes do not run in parallel. Rather, the distal margin of the buccal cusp rotates away from the sagittal sulcus. If straight lines are placed along the major axis of the buccal cusp and on the midline between the two cusps, the angle of divergence varies from 6° to 11° (Morris 1981). The Uto-Aztecan premolar is evident when this divergence is two to three times greater than normal (35–45°). The rotation is almost invariably accompanied by a pit between the distal marginal ridge of the buccal cusp and a crest from the essential ridge of the buccal cusp to the distal border.

## Classification

Grade 0: absent
Grade 1: present

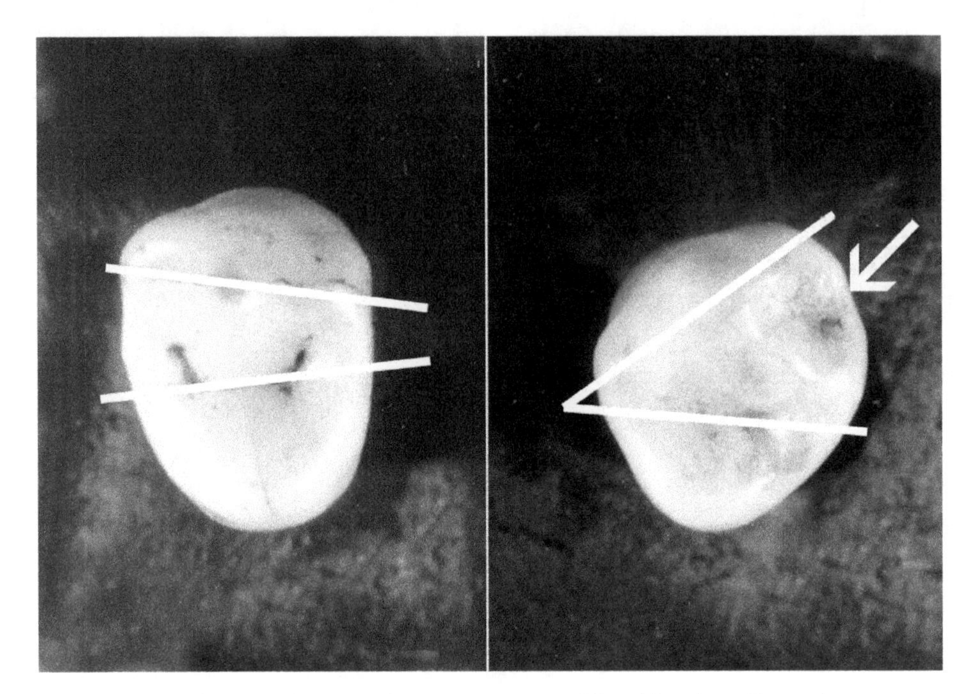

**Figure 12.1** Left: normal UP1 with some rotation of distal border of buccal cusp. Right: Uto-Aztecan premolar shows strong rotation, with ridge connecting buccal cusp to distal margin and sagittal sulcus between ridge and rotated buccal cusp.

## Breakpoint

Trait presence

## Potential Problems in Scoring

Given the pronounced angle and depth of the sulcus on the distal marginal ridge, the trait can be observed even when wear is moderately pronounced. However, some questions remain. For example, even with pronounced rotation of the buccal cusp, is the presence of a pit required to be classified as a Uto-Aztecan premolar?

## Geographic Variation

As the name implies, the researchers who first identified the trait thought it was a private variant limited to Uto-Aztecan speakers in the American Southwest (Morris *et al.* 1978). Following systematic observations on dentitions all over the world, the trait has been observed in a few Europeans and possibly an Australian, but it is by far the most common in Native North and South Americans. In some Native American groups, the frequency of this trait can be as high as 2–4%.

**Figure 12.2** An extraordinarily pronounced Uto-Aztecan premolar; UP2 is rarely affected, as the trait is found almost exclusively on UP1.

**Figure 12.3** Two rare traits on the same UP1: a Uto-Aztecan premolar with an odontome.

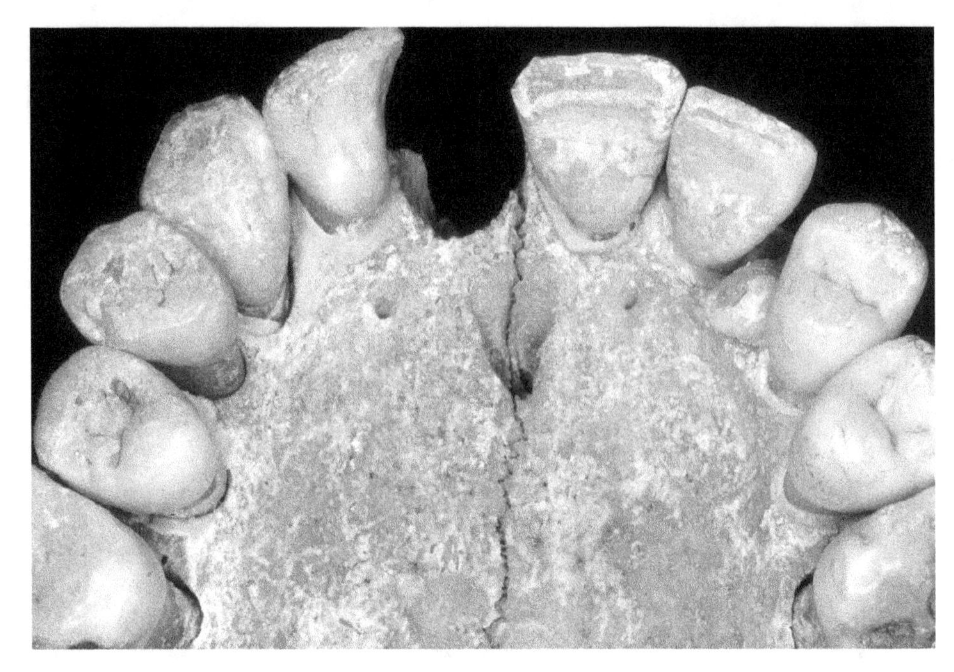

**Figure 12.4** A possible Uto-Aztecan premolar in an Australian aborigine; the only instance observed in over 1000 skeletons, and it is not as clear-cut as those exhibited by Native Americans.

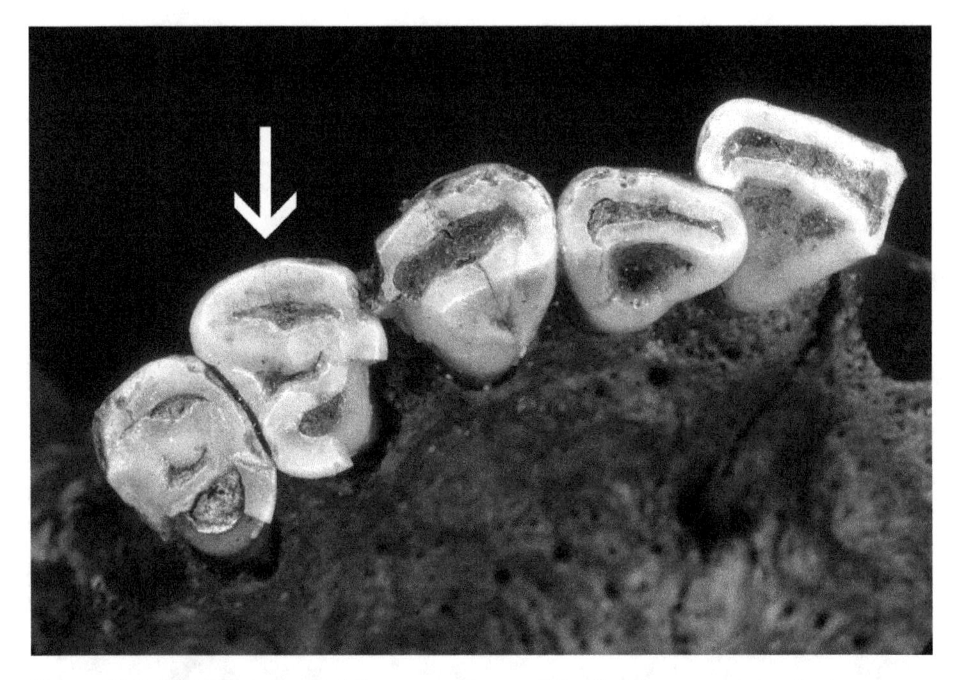

**Figure 12.5** The only recorded instance of a possible Uto-Aztecan premolar in an Eskimo from Kodiak Island (possible Indian?); the rotation of the buccal cusp is consistent with the phenotype but there is no distosaggital sulcus. Whether or not this is an actual Uto-Aztecan premolar is arguable.

## Select Bibliography


Delgado-Burbano, M.E., Scott, G.R., and Turner, C.G., II (2010). The Uto-Aztecan premolar among North and South Amerindians: geographic variation and genetics. *American Journal of Physical Anthropology* 143, 570–578.

Johnson, K.M., Stojanowski, C.M., Miyar, K.O'D., Doran, G.H., and Ricklis, R.A. (2011). New evidence on the spatiotemporal distribution and evolution of the Uto-Aztecan premolar. *American Journal of Physical Anthropology* 146, 474–480.

Johnston, C.A., and Sciulli, P.W. (1996). Technical note: Uto-Aztecan premolars in Ohio Valley populations. *American Journal of Physical Anthropology* 100, 293–294.

Morris, D.H. (1981). Maxillary first premolar angular differences between North American Indians and non-North American Indians. *American Journal of Physical Anthropology* 54, 431–433.

(1986). Maxillary molar polygons in five human samples. *American Journal of Physical Anthropology* 70, 333–338.

Morris, D.H., Glasstone Hughes, S., and Dahlberg, A.A. (1978). Uto-Aztecan premolar: the anthropology of a dental trait. In *Development, Function and Evolution of Teeth*, ed. P.M. Butler and K.A. Joysey. New York: Academic Press, pp. 69–79.

Rodríguez Flóres, C.D. (2012). Occurrence of the Uto-Aztecan premolar trait in a contemporary Colombian population. *Homo – Journal of Comparative Human Biology* 63, 396–403.

Taylor, M.S. (2012). The Uto-Aztecan premolar in early hunter-gatherers from south-central North America. *American Journal of Physical Anthropology* 149, 318–322.

Turner, C.G., II, Nichol, C.R., and Scott, G.R. (1991). Scoring procedures for key morphological traits of the permanent dentition: the Arizona State University dental anthropology system. In *Advances in Dental Anthropology*, ed. M.A. Kelley and C.S. Larsen. New York: Wiley-Liss, pp. 13–31.

# NOTES

# 13

# Metacone

## Observed on

UM1, UM2, UM3

## Key Tooth

UM3

## Synonyms

Distobuccal cusp, cusp 3, posteroexternal cusp (Butler 1941)

## Description

Although the hypocone is the most variable cusp of the upper molars, the next most variable is its adjacent neighbor on the distal border, the metacone. The metacone is rarely missing entirely (except on UM3) but can be expressed to varying degrees.

## Classification

Turner *et al.* (1991)

Grade 0: metacone absent
Grade 1: there is a ridge at the metacone site but no free apex
Grade 2: metacone expressed as faint cuspule with a free apex
Grade 3: weak cusp
Grade 3.5: intermediate-sized cusp that falls between grades 3 and 4 (interpolation necessary)
Grade 4: metacone is large
Grade 5: metacone is pronounced, equal in size to a large UM1 hypocone

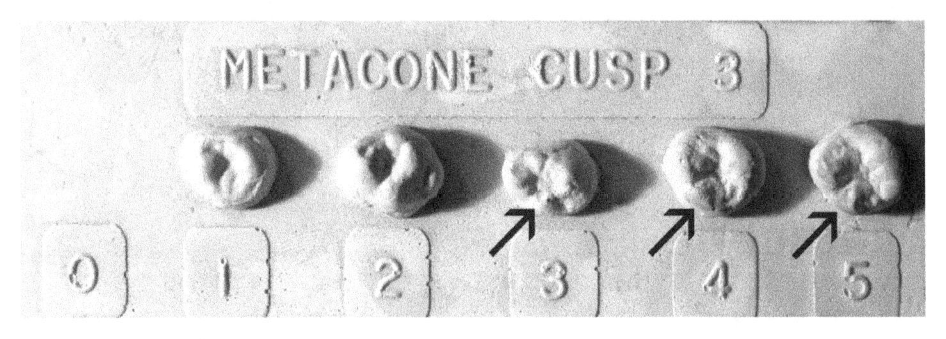


Figure 13.1 Standard plaque for metacone expression.

## Breakpoint

As the metacone is rarely missing completely, a measure of significant metacone reduction involves combining grades 0–3. This would parallel the characterization of hypocone reduction, which includes the frequency of grades 0+1 (i.e., three-cusped tooth).

## Potential Problems in Scoring

As the focus is on UM3, there is some potential for environmentally induced variation. The reduction in the metacone also anticipates the "reduced" form of pegged-reduced-missing UM3, so there is some correlation between these variables.

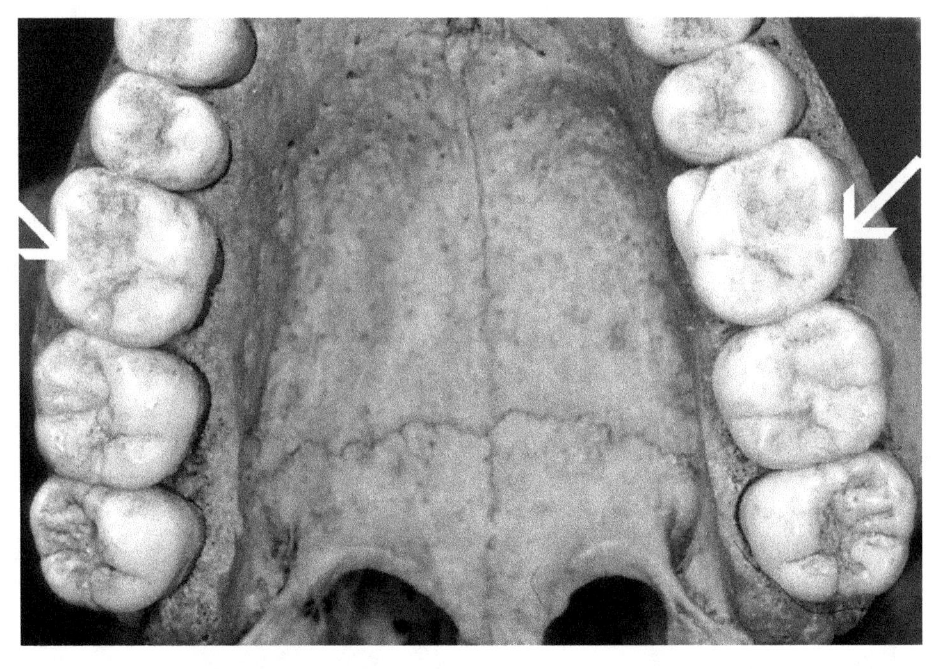

Figure 13.2 Large metacone on all three upper molars (all grade 5).

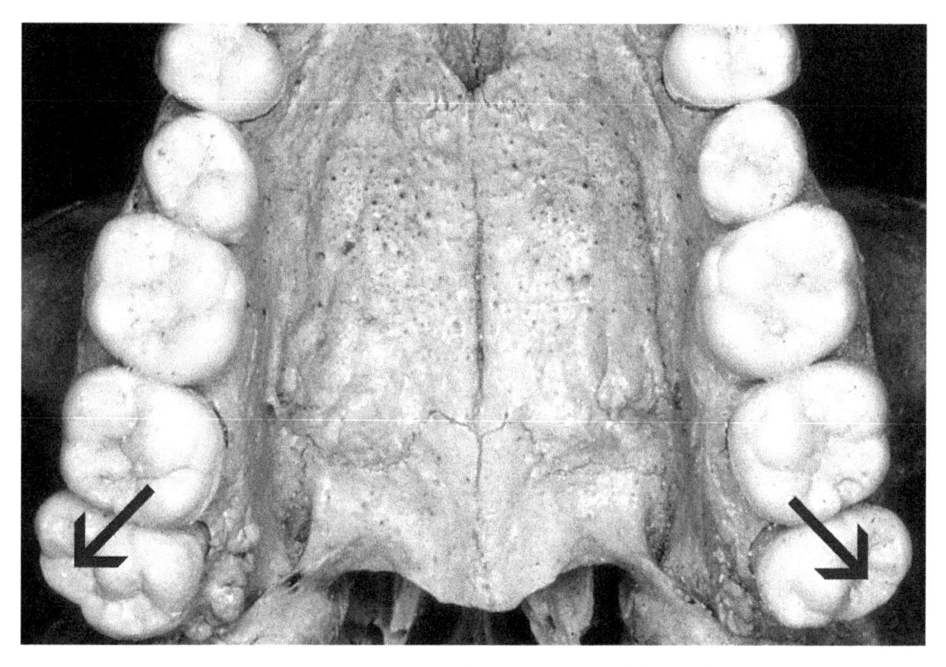

**Figure 13.3** Large metacones on UM1 and UM2 (grade 5) but reduced metacone on UM3 (grade 3.5).

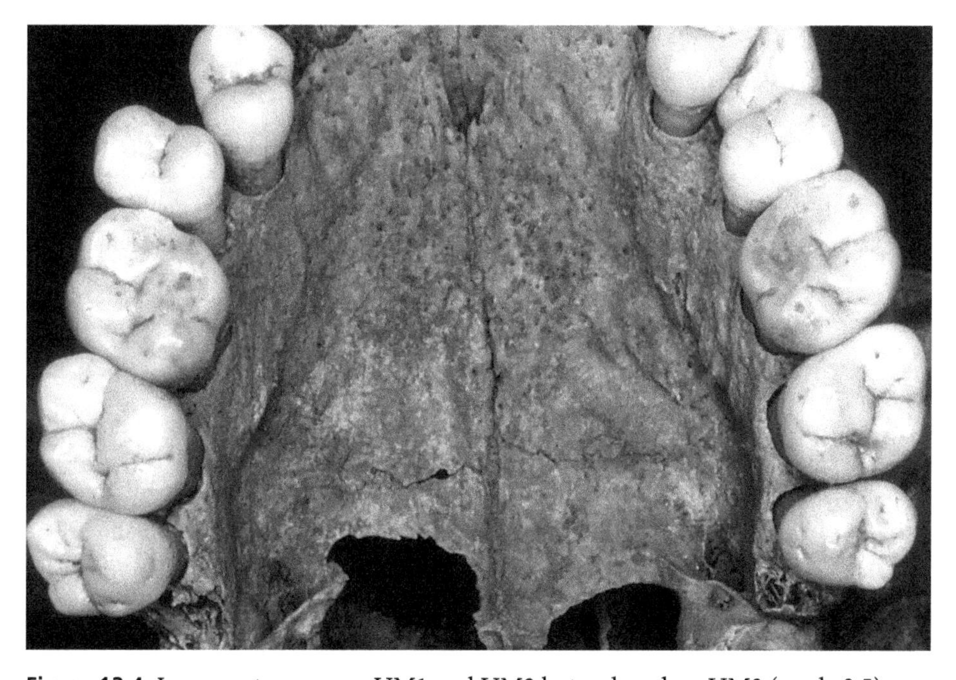

**Figure 13.4** Large metacones on UM1 and UM2 but reduced on UM3 (grade 3.5).

**Figure 13.5** Large metacones on UM1 and UM2 but significantly reduced on UM3 (despite wear, about a grade 3); with metacone reduction, the crown is strongly compressed mesiodistally.

## Geographic Variation

Although part of ASUDAS, the metacone was not one of Turner's 29 key traits, so few frequency data are available to characterize metacone variation. Rosenzweig and Zilberman (1967, 1969) found metacone reduction in Middle Eastern populations but no examples of complete absence on UM2. Although yet to be tested, it seems likely that metacone reduction and hypocone reduction are correlated.

## Select Bibliography

Butler, P.M. (1941). The theory of the evolution of mammalian molar teeth. *American Journal of Science* 239, 421–450.

(1972). Some function aspects of molar evolution. *Evolution* 26, 474–483.

Macho, G.A., and Moggi-Cecchi, J. (1992). Reduction of maxillary molars in *Homo sapiens sapiens*: a different perspective. *American Journal of Physical Anthropology* 87, 151–159.

Rosenzweig, K.A., and Zilberman, Y. (1967). Dental morphology of Jews from Yemen and Cochin. *American Journal of Physical Anthropology* 26, 15–22.

(1969). Dentition of Bedouin in Israel. II. Morphology. *American Journal of Physical Anthropology* 31, 199–204.

Sofaer, J.A., Smith, P., and Kaye, E. (1986). Affinities between contemporary and skeletal Jewish and non-Jewish groups based on tooth morphology. *American Journal of Physical Anthropology* 70, 265–275.
Turner, C.G., II, Nichol, C.R., and Scott, G.R. (1991). Scoring procedures for key morphological traits of the permanent dentition: the Arizona State University dental anthropology system. In *Advances in Dental Anthropology*, ed. M.A. Kelley and C.S. Larsen. New York: Wiley-Liss, pp. 13–31.

# NOTES

# 14

# Hypocone

## Observed on

UM1, UM2, UM3

## Key Tooth

UM2

## Synonyms

Upper molar cusp number (Gregory 1922), cusp 4, distolingual cusp

## Description

The hypocone was the last major cusp added to upper molars during the course of primate evolution. Derived from the cingulum, it is attached to the distolingual surface of the trigon. In hominoids, the trigon plus the hypocone results in a four-cusped upper molar. In apes, all three upper molars typically express a hypocone. In hominin evolution, a reduction in tooth size is associated with a reduction in cusp number. The cusp added last during primate evolution is the cusp lost first. The absence of a hypocone results in an upper molar that reverts to its ancestral form, the trigon. In accord with field effect as set forth by Butler (1939) and Dahlberg (1945), the most distal member of a tooth district is the most variable member of a field. For upper molars, the third molar is by far the most variable in terms of size, number, and morphology. In terms of morphology, the variation takes the form of hypocone reduction or loss. The most stable tooth of the upper molar tooth district is the first molar. The hypocone is almost invariably present on the first molar, retaining its ancestral four-cusped form in most individuals. The second molar varies much more than the first molar but much less than the third molar. For this reason, the key tooth for recording hypocone expression is UM2.

## Classification

Larson (1978); modified by Turner *et al.* (1991)

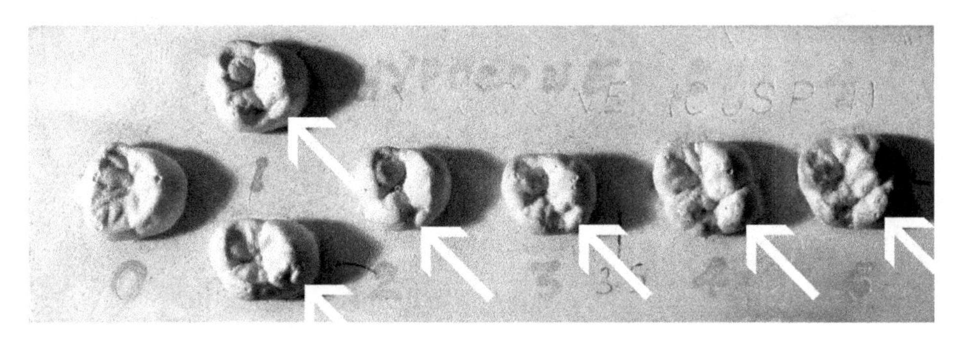

**Figure 14.1** Standard plaque for hypocone expression.

Larsen established a hypocone scale with absence and five degrees of trait presence. As she did this prior to the establishment of one major classificatory principle, the hypocone standard is in need of modification. That is, to establish a standard plaque, one should use the same tooth from the same side of the jaw. For the hypocone plaque, upper second molars were used to establish grades 0, 1, 2, and 3 but upper first molars were used for grades 4 and 5. The principle that the relative degree of expression from one grade to the next should be approximately equal was violated in this instance because there is a major gap between grade 3 (on a UM2) and grade 4 (on a UM1). This gap was so substantial that Turner inserted a grade 3.5 between the two grades. As decimals are cumbersome for a ranked scale, the classification set forth here restores the expanded scale but with all integers.

Grade 0: no hypocone expression of any form; a true three-cusped tooth

Grade 1: for this grade, there is a low-level expression of the hypocone, often expressed as no more than an outline on the distolingual aspect of the trigon. In Dahlberg's original classification, this would be scored as a three-cusped upper molar along with grade 0

Grade 2: in the Dahlberg classification, 3+ was equivalent to a small conical hypocone on the distolingual border of the trigon; grade 2 reflects this phenotype, where there is basically a conical cusp, or tubercle, with a free apex

Grade 3: the hypocone is reduced in size but assumes a normal ovate shape along with a distinct free apex

Grade 4: this grade would be equivalent to 3.5 on the modified hypocone plaque; the hypocone is reduced in size but is moderate rather than slight in expression

Grade 5: hypocone is well developed, a step beyond grade 4

Grade 6: pronounced expression of the hypocone; often equals or exceeds the size of the major cusps of the trigon

## Breakpoint

To make frequencies consistent with workers who adopted the Dahlberg standard, Scott and Turner (1997) used the frequency of grade 0 + grade 1, or three-cusped upper second molars, to characterize population variation of the hypocone. Alternatively, researchers could use grades 2–6/0–6 to estimate the frequency of four-cusped upper molars.



## Potential Problems in Scoring

The groove that separates the hypocone from the trigon can be obscured by wear at a relatively early age. This can lead observers to score a tooth as three-cusped even though a hypocone may have been present prior to obliteration of the groove. Of course, crown wear often introduces issues into scoring trait expression, but this is particularly true for this trait.

## Geographic Variation

For hypocone expression on UM2, two populations stand out as showing the highest level of three-cusped teeth: Europeans and Eskimo-Aleuts. At the other end of the spectrum, the hypocone is most often retained on the UM2s of Sub-Saharan African and Australian populations. Southeast Asians, Pacific Islanders, and Native Americans fall between these extremes.

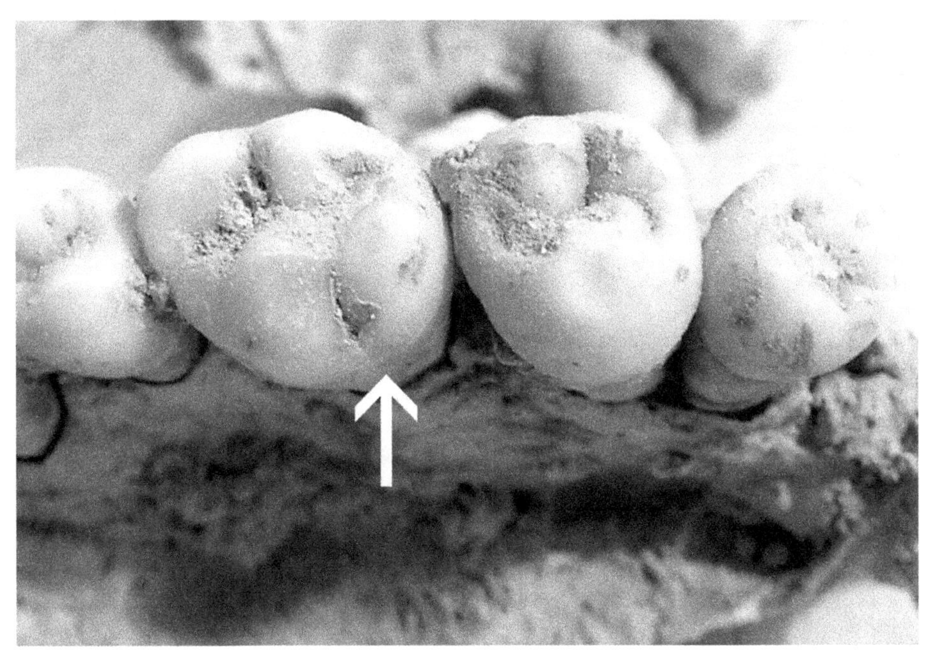

**Figure 14.2** A typical circumstance in a European dentition, where the hypocone is well developed on UM1 but is absent on UM2 and UM3.

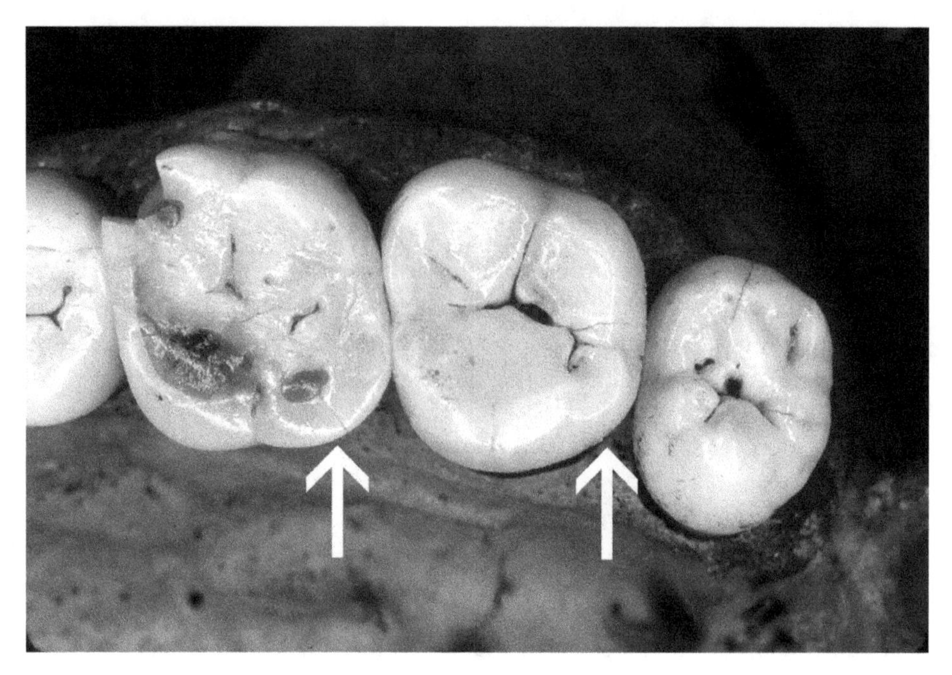

**Figure 14.3** Despite wear, a large hypocone is still evident on UM1; the hypocone is reduced (grade 3) but present on UM2, and absent on a simplified UM3.

**Figure 14.4** The hypocone is grade 5 on both UM1 and UM2 but absent on a simplified UM3. Although the hypocone is larger on UM1, its relative size is no greater than that of the hypocone on UM2.

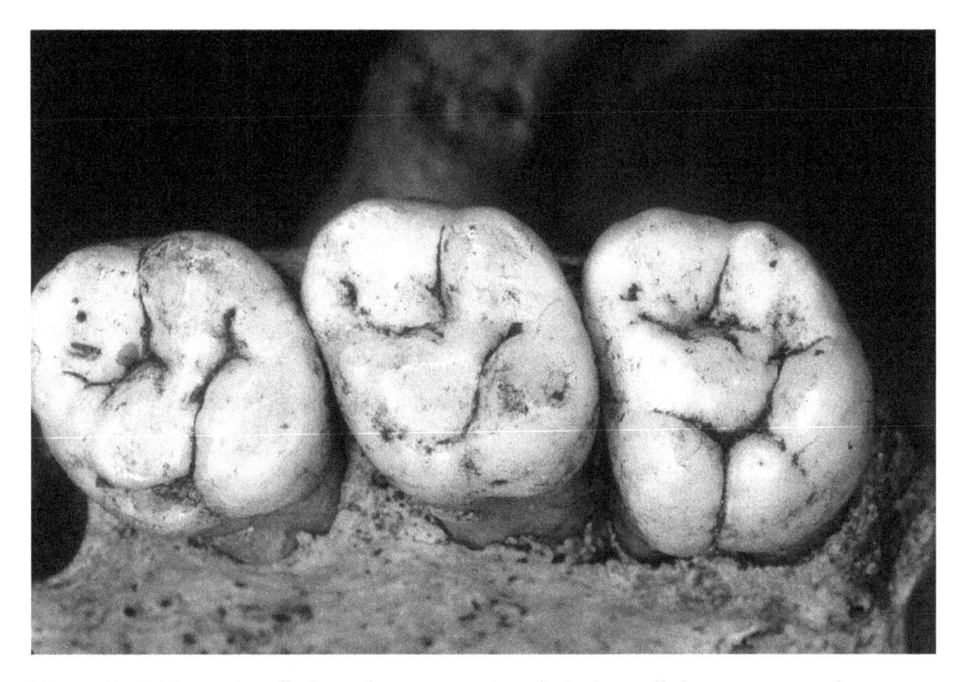

**Figure 14.5** Exceptionally large hypocones (grade 5+) on all three upper molars; hypocone of UM3 is bifurcated.

## Select Bibliography

Bermúdez de Castro, J.M., and Martinez, I. (1986). Hypocone and metaconule: identification and variability on human molars. *International Journal of Anthropology* 1, 165–168.

Butler, P.M. (1939). Studies of the mammalian dentition. Differentiation of the post-canine dentition. *Proceedings of the Zoological Society of London* B 109, 1–36.

Dahlberg, A.A. (1945). The changing dentition of man. *Journal of the American Dental Association* 32, 676–690.

Drennan, M.R. (1929). The dentition of a Bushman tribe. *Annals of the South African Museum* 24, 61–87.

Gregory, W.K. (1922). *The Origin and Evolution of the Human Dentition.* Baltimore: Williams and Wilkins.

Hunter, J.P., and Jernvall, J. (1995). The hypocone as a key innovation in mammalian evolution. *Proceedings of the National Academy of Sciences of the USA* 92, 10718–10722.

Keene, H.J. (1968). The relationship between Carabelli's trait and the size, number and morphology of the maxillary molars. *Archives of Oral Biology* 13, 1023–1025.

Khraisat, A., Alsoleihat, F., Subramani, K., *et al.* (2011). Hypocone reduction and Carabelli's traits in contemporary Jordanians and the association between Carabelli's trait and the dimensions of the maxillary first permanent molar. *Collegium Antropologicum* 35, 73–78.

Korenhof, C.A.W. (1960). *Morphogenetical Aspects of the Human Upper Molar*. Utrecht: Uitgeversmaatschappij Neerlandia.

Larson, M.A. (1978). *Dental Morphology of the Gran Quivira Indians*. MA thesis, Arizona State University, Tempe.

Macho, G.A., and Cecchi, J.M. (1992). Relationship between size of distal accessory tubercles and hypocones in permanent maxillary molar crowns of southern Africans. *Archives of Oral Biology* 37, 575–578.

Scott, G.R. (1979). Association between the hypocone and Carabelli's trait of the maxillary molars. *Journal of Dental Research* 58, 1403–1404.

Scott, G.R., and Turner, C.G., II (1997). *The Anthropology of Modern Human Teeth: Dental Morphology and its Variation in Recent Human Populations*. Cambridge: Cambridge University Press.

Takahashi, M., Kondo, S., Townsend, G.C., and Kanazawa, E. (2007). Variability in cusp size of human maxillary molars, with particular reference to the hypocone. *Archives of Oral Biology* 52, 1146–1154.

Turner, C.G., II, Nichol, C.R., and Scott, G.R. (1991). Scoring procedures for key morphological traits of the permanent dentition: the Arizona State University dental anthropology system. In *Advances in Dental Anthropology*, ed. M.A. Kelley and C.S. Larsen. New York: Wiley-Liss, pp. 13–31.

# 15

# Bifurcated Hypocone

## Observed on

UM1, UM2, UM3

## Key Tooth

UM2

## Synonyms

N/A

## Description

When the first author (GRS) scored St. Lawrence Island (Alaska) Eskimos for crown and root traits, he noted that it was not uncommon for the hypocone on UM2 to have an occlusal groove that divided the cusp into two roughly equal sections. For lack of a better term, these were referred to as bifurcated hypocones. They were not incorporated into the original ASUDAS, but systematic observations on this variant might allay difficulties in determining whether or not the centrally placed half of the hypocone represents cusp 5.

## Classification

Present study

Grade 0: no bifurcation of hypocone
Grade 1: hypocone divided roughly into two equal sections by distinct groove

## Breakpoint

Grade 1

## Potential Problems in Scoring

Multiple cusps can sometimes be found along the distal margin of the upper molars. In some cases, one of the cusps is a cusp 5; in other cases, the distal lobe segment of the metacone is divided from the essential lobe segment by a developmental groove. Bifurcated hypocones are characterized by a cusp about equally divided into two parts, not one big cusp and one small cusp (see Figures 15.1–15.5 for examples). The groove dividing the hypocone into two roughly equal sections is preserved even with moderate wear.

## Geographic Variation

Unknown at this time, although it was relatively common in Alaskan Eskimos.

### Select Bibliography

No references to date.

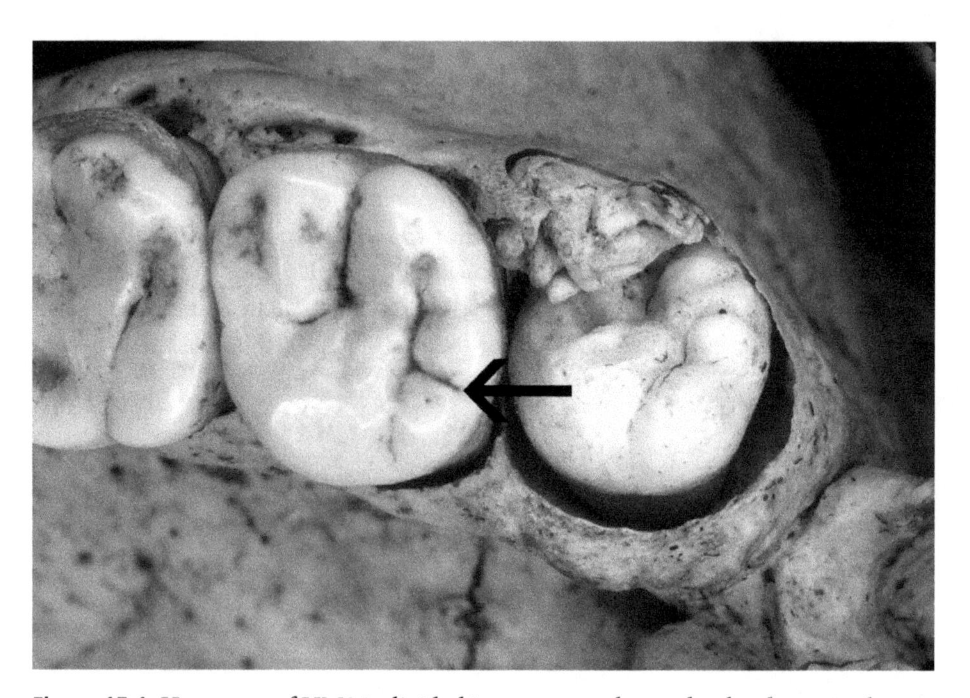

**Figure 15.1** Hypocone of UM2 is divided into two equal parts by developmental groove (not present on UM1 or UM3).

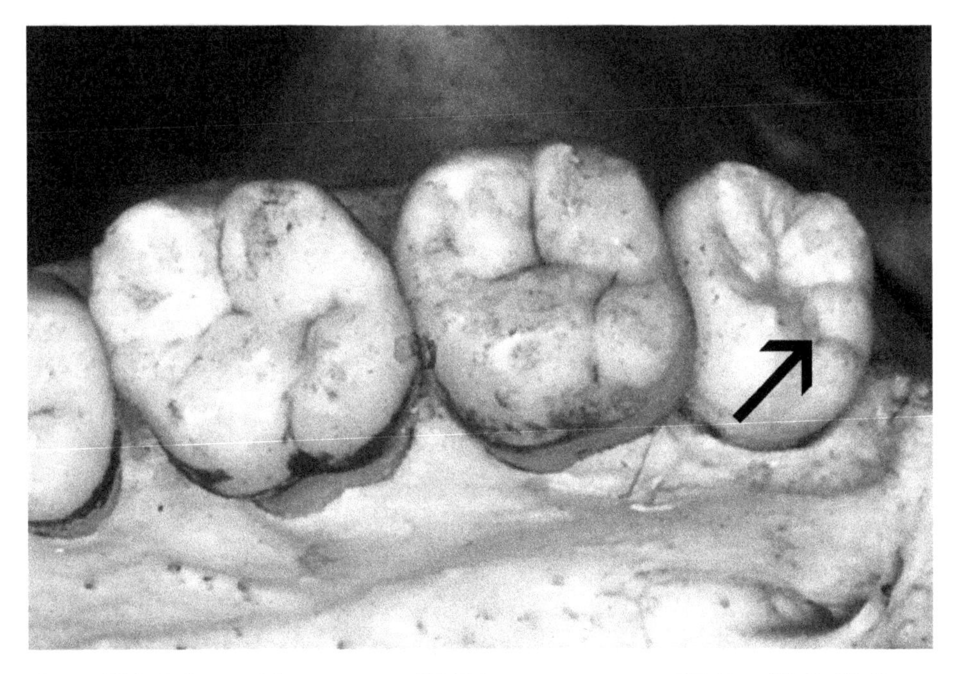

**Figure 15.2** Bifurcated hypocone on UM3 but not present on UM1 or UM2. UM3 highly variable for crown and root morphology, so caution urged for scoring this tooth.

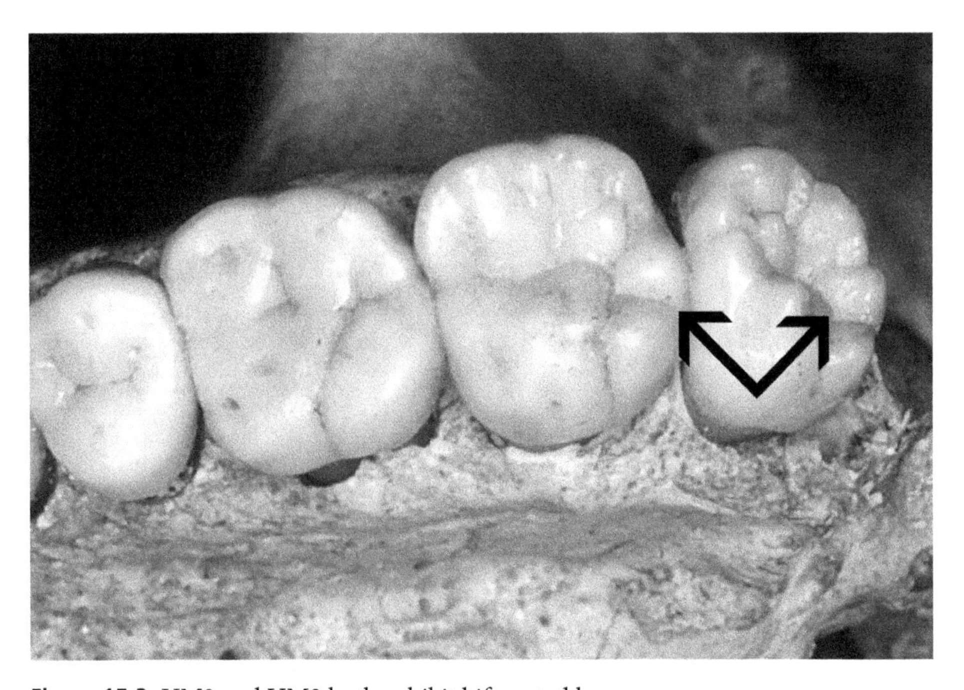

**Figure 15.3** UM2 and UM3 both exhibit bifurcated hypocones.

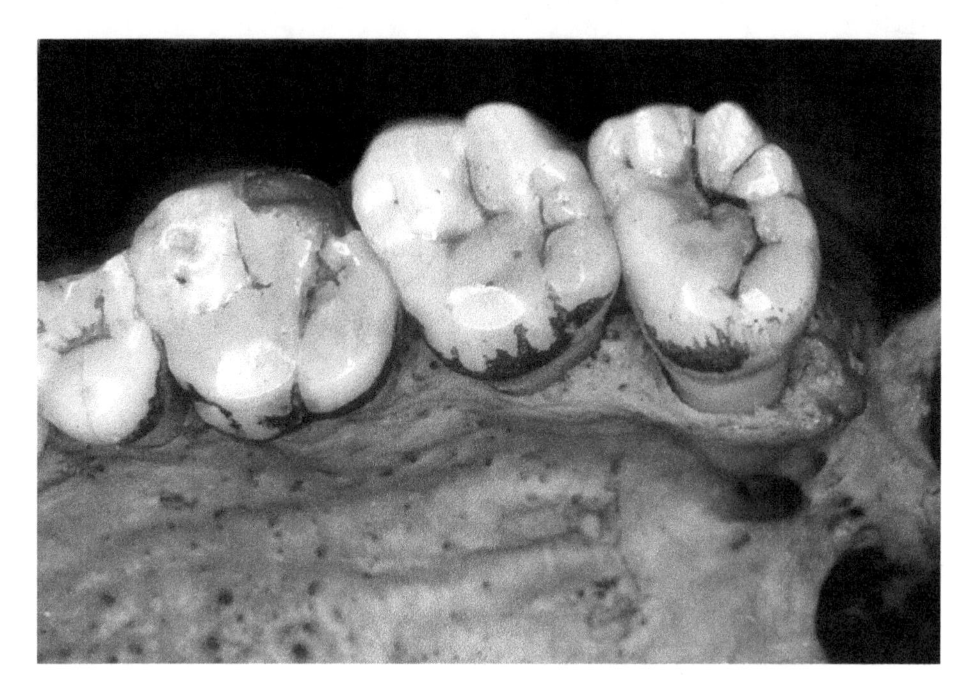

**Figure 15.4** Hypocones on UM2 and UM3 are bifurcated but not into two equal parts; in such instances, this is more likely a reduced hypocone associated with a large cusp 5. Also note that the distal lobe segment of the metacone is set off from the essential lobe segment by developmental grooves.

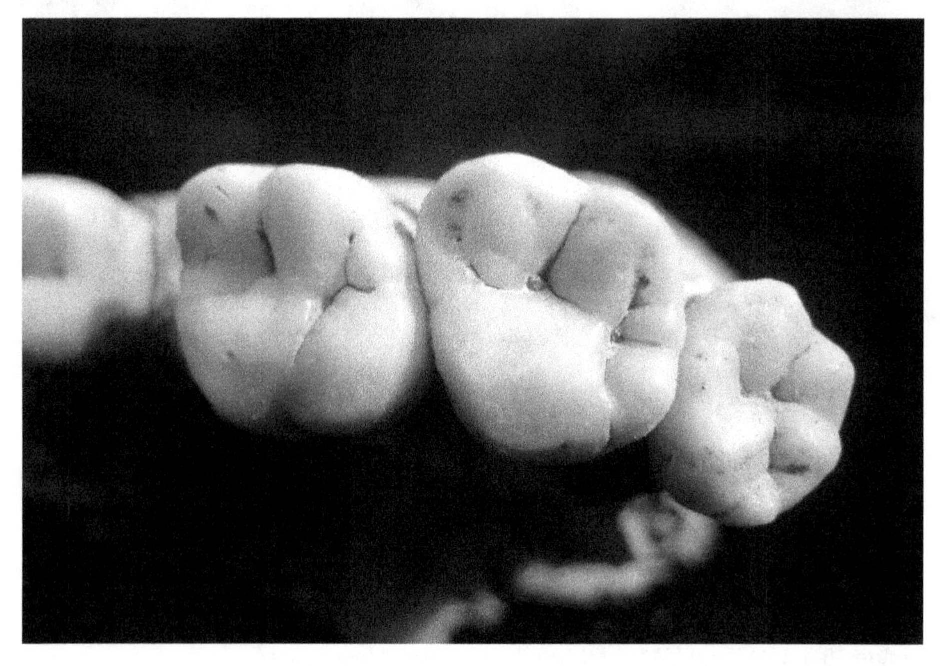

**Figure 15.5** UM2 shows bifurcation, but this is more likely to be reduced hypocone and large cusp 5 rather than bifurcated hypocone.

# 16

# Cusp 5

## Observed on

UM1, UM2, UM3

## Key Tooth

UM1

## Synonyms

Metaconule (Harris 1977), distal accessory tubercle (Kanazawa *et al.* 1990)

## Description

Cusp 5 takes the form of a conule that is expressed between the hypocone and metacone of the upper molars. To be scored as present, the cusp or conule should show two vertical grooves that run in parallel on the distal marginal ridge complex. This variant does not attain the size of other accessory cusps (e.g., cusp 6, cusp 7) so the range in expression goes from a small conule to a moderate or medium-sized cusp.

## Classification

Harris and Bailit (1980)

Grade 0: trait is absent, only one vertical groove on distal surface of upper molar between hypocone and metacone
Grade 1: slight conule
Grade 2: trace conule
Grade 3: small cuspule
Grade 4: small cusp
Grade 5: medium cusp

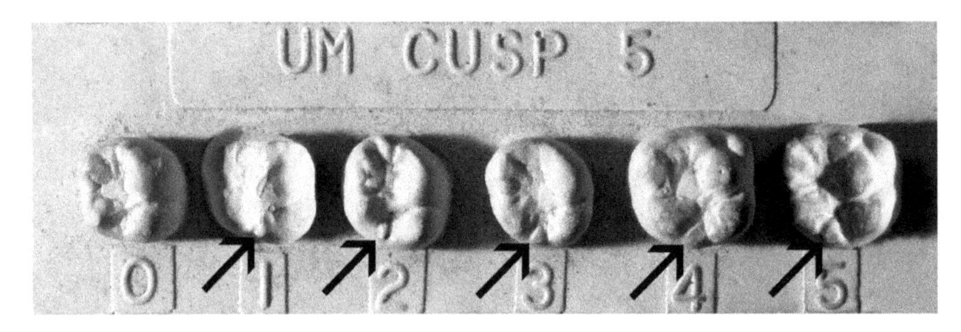


**Figure 16.1** Standard plaque for upper molar cusp 5 expression.

## Breakpoint

Grade 1+

## Potential Problems in Scoring

The locus where cusp 5 is expressed is often one of the earliest surfaces to wear on the upper molars. In some instances, the presence of two parallel vertical grooves indicates the presence of cusp 5 even with some wear. Caution should be used to not score teeth that are too worn. Another complicating factor can be unusual forms of hypocone expression; in some instances, hypocones can be bifurcated, which might lead some

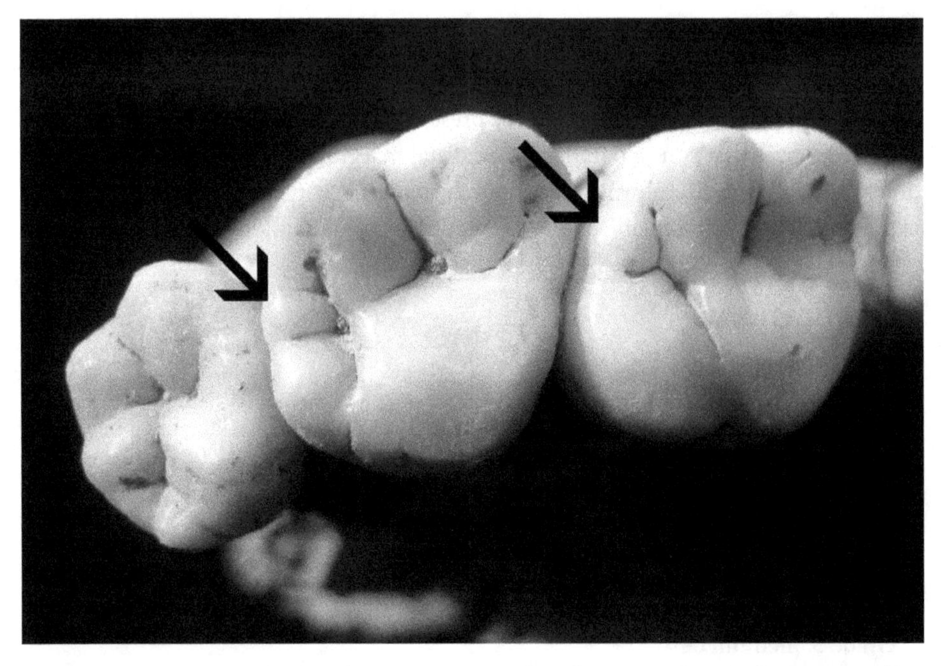

**Figure 16.2** An unusual case where UM1 is smaller than UM2 but both show a well-developed cusp 5 between the hypocone and metacone.

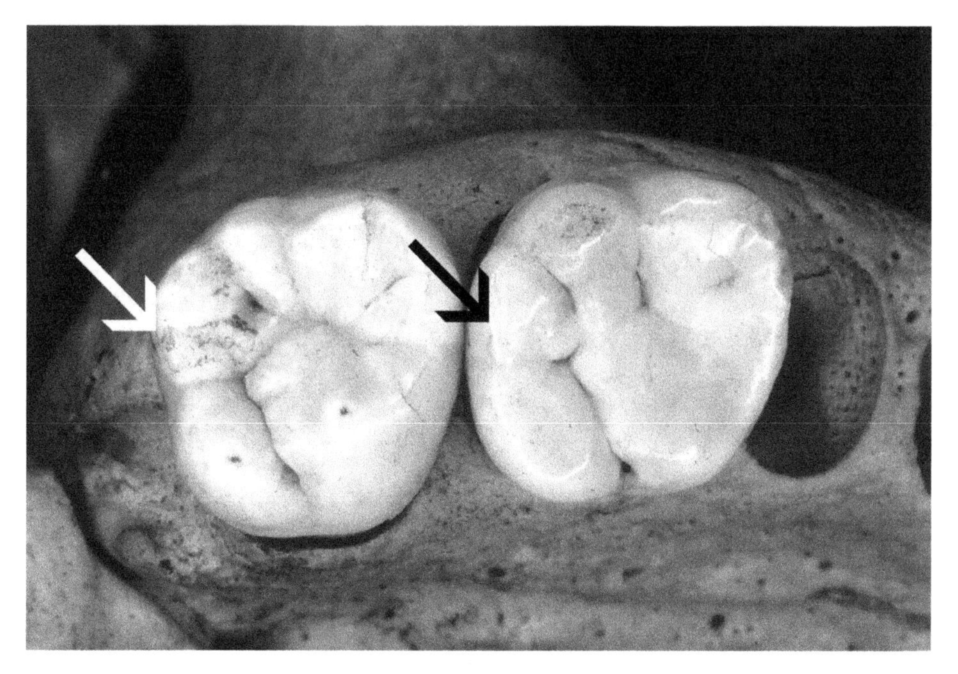

**Figure 16.3** A cusp 5 partially obscured by wear on UM1 but large and pronounced on UM2 (along with complex metacone).

**Figure 16.4** A moderate but well-defined cusp 5 between the metacone and hypocone of UM1 but not present on UM2.

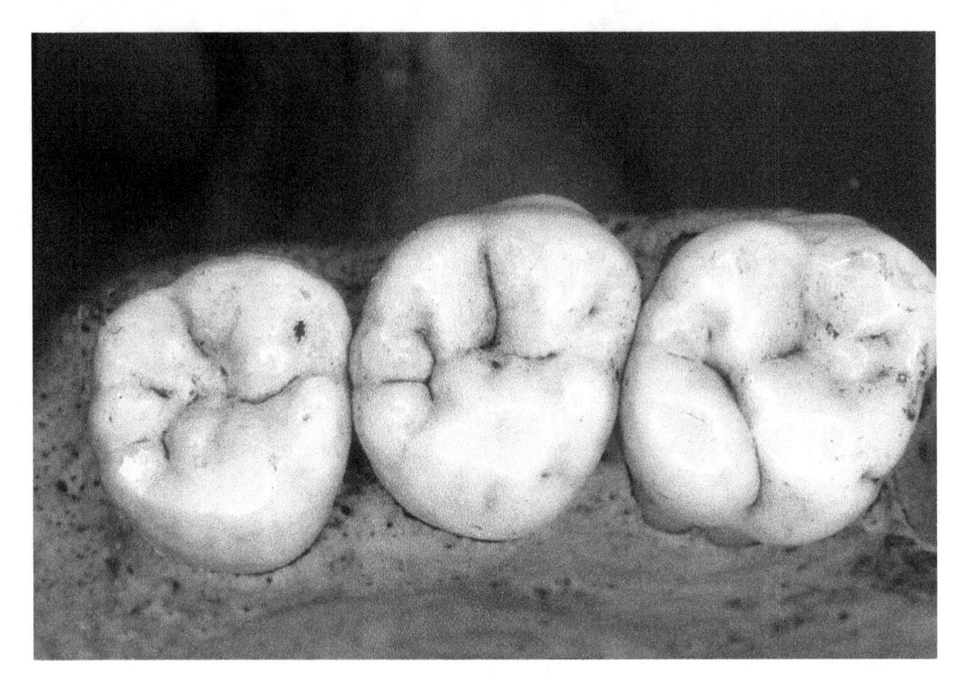

**Figure 16.5** Cusp 5 on all three upper molars; for this individual, cusp size increases from UM1 to UM2 to UM3.

workers to score such manifestations as cusp 5. One element to keep in mind is that such bifurcations often produce a cusp form that exceeds the size of the largest cusp 5.

## Geographic Variation

The highest frequencies of cusp 5 occur in Australia, New Guinea, and Africa (50–60%). Western Eurasians show the lowest frequencies, generally below 20%. Asian and Asian-derived groups in the Pacific and Americas are intermediate, with frequencies between 20% and 40%.

### Select Bibliography

Bermúdez de Castro, J.M., and Martínez, I. (1986). Hypocone and metaconule: identification and variability on human molars. *International Journal of Anthropology* 1, 165–168.

Harris, E.F. (1977). *Anthropologic and Genetic Aspects of the Dental Morphology of Solomon Islanders, Melanesia.* PhD dissertation, Department of Anthropology, Arizona State University, Tempe.

Harris, E.F., and Bailit, H.L. (1980). The metaconule: a morphologic and familial analysis of a molar cusp in humans. *American Journal of Physical Anthropology* 53, 349–358.

Kanazawa, E., Natori, M., and Ozaki, T. (1992). Anomalous tubercles on the occlusal table of upper first molars in nine populations including Pacific populations. In *Craniofacial Variation in Pacific Populations*, ed. T. Brown and S. Molnar. Adelaide: Anthropology and Genetics Lab, Department of Dentistry, University of Adelaide, pp. 53–59.

Kanazawa, E., Sekikawa, M., and Ozaki, T. (1990). A quantitative investigation of irregular cuspules in human maxillary permanent molars. *American Journal of Physical Anthropology* 83, 173–180.

Macho, G.A., and Cecchi, J.M. (1992). Relationship between size of distal accessory tubercles and hypocones in permanent maxillary molar crowns of southern Africans. *Archives of Oral Biology* 37, 575–578.

Townsend, G., Yamada, H., and Smith, P. (1986). The metaconule in Australian Aboriginals: an accessory tubercle on maxillary molar teeth. *Human Biology* 58, 851–862.

Turner, C.G., II (1967). *The Dentition of Arctic Peoples*. PhD dissertation, Department of Anthropology, University of Wisconsin, Madison.

Turner, C.G., II, and Hanihara, K. (1977). Additional features of Ainu dentition. V. Peopling of the Pacific. *American Journal of Physical Anthropology* 46, 13–24.

## NOTES

# 17

# Marginal Ridge Tubercles (Protoconule, Mesial Accessory Tubercle, Mesial Paracone Tubercle)

## Observed on

UM1

## Key Tooth

UM1

## Synonyms

Upper molar accessory cusps, irregular cuspules, anomalous tubercles (Kanazawa *et al.* 1990)

## Description

Three independent cuspules can be expressed along the mesial marginal ridge complex of the upper first molars. The two major cuspules are the protoconule (derived from the mesial lobe of the protocone) and the mesial paracone tubercle (derived from the mesial lobe of the paracone). Another cuspule, the mesial accessory tubercle, is expressed between the protoconule and mesial paracone tubercle. There are no graded scales for these tubercles.

## Classification

Kanazawa *et al.* (1990)

Grade 0: protoconule absent
Grade 1: protoconule present

Grade 0: mesial accessory tubercle absent
Grade 1: mesial accessory tubercle present
Grade 0: mesial paracone tubercle absent
Grade 1: mesial paracone tubercle present

## Breakpoint

Trait presence

## Potential Problems in Scoring

In line with other accessory tubercles (cf. UM1 cusp 5), these tubercles start showing wear soon after eruption. They are easily observed on unerupted or unworn teeth, but wear can quickly obscure the grooves separating the three tubercles. Additionally, these three tubercles are not always found in concert. There may be one or two tubercles present. In such cases, it is difficult to identify a specific tubercle.

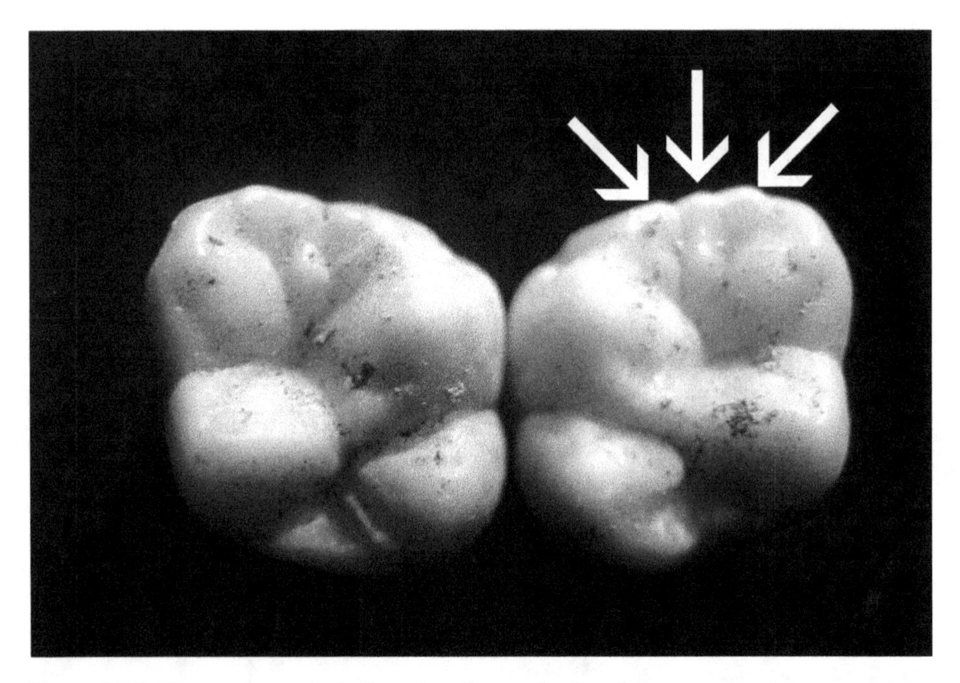

**Figure 17.1** The protoconule is the tubercle attached to the protocone, or mesiolingual cusp. The mesial paracone tubercle is attached to the paracone. The mesial accessory tubercle is situated between the protoconule and mesial paracone tubercle. In unworn teeth, the mesial marginal tubercles are often distinct and easy to recognize. This locale often receives the brunt of wear even at an early age, making it difficult, if not impossible, to identify and disentangle the three accessory tubercles.

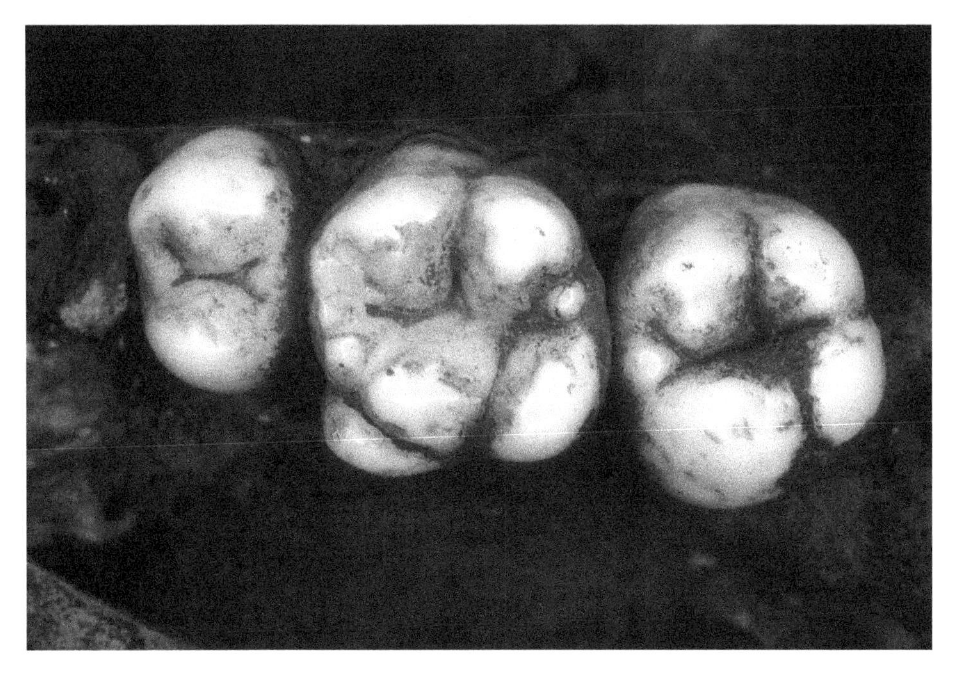

**Figure 17.2** UM1 exhibits all three marginal accessory tubercles; UM2 exhibits one tubercle, most likely a mesial accessory tubercle since it is not clearly associated with either the protocone or the paracone.

## Geographic Variation

In a small sampling of world populations, Kanazawa *et al.* (1990) report frequencies of the protoconule ranging from 5.9 to 44.4%, the mesial paracone tubercle between 23.3 and 71.1%, and the mesial accessory tubercle between 14.3 and 71.0%. Asian groups express these tubercles more commonly than Africans and Western Eurasians, but more data are required before a clear picture of geographic variation is evident.

## Select Bibliography

Kanazawa, E., Natori, M., and Ozaki, T. (1992). Anomalous tubercles on the occlusal table of upper first molars in nine populations including Pacific populations. In *Craniofacial Variation in Pacific Populations*, ed. T. Brown and S. Molnar. Adelaide: Anthropology and Genetics Lab, Department of Dentistry, University of Adelaide, pp. 53–59.

Kanazawa, E., Sekikawa, M., and Ozaki, T. (1990). A quantitative investigation of irregular cuspules in human maxillary permanent molars. *American Journal of Physical Anthropology* 83, 173–180.

Mayhall, J.T., and Kanazawa, E. (1989). A three-dimensional analysis of the maxillary molar crowns of Canadian Inuit. *American Journal of Physical Anthropology* 78, 73–78.

Turner, C.G., II (1967). *The Dentition of Arctic Peoples*. PhD dissertation, Department of Anthropology, University of Wisconsin, Madison.

# NOTES

# 18

# Carabelli's Cusp

## Observed on

UM1, UM2, UM3

## Key Tooth

UM1

## Synonyms

Carabelli's trait, the cusp of Carabelli, Carabelli tubercle, Carabelli anomaly (Kraus 1951), 5th cusp, *tuberculum* Carabelli (Carlsen 1987)

## Description

Carabelli's trait is a cingular derivative expressed on the lingual surface of the protocone. It shows a wide range of expression, from complete absence to a large free-standing tubercle that approximates the hypocone in size. Because of this broad range of expression, Dahlberg (1956) set up a classification that involved absence and seven degrees of trait presence. This system was adopted by Turner *et al.* (1991).

## Classification

Dahlberg (1956)

Grade 0: mesiolingual cusp does not exhibit any grooves or pits on the lingual surface

Grade 1: a vertical groove separates the protocone from the mesial marginal ridge complex; grade 1 expression occurs when there is a slight eminence that deflects distally from this groove

Grade 2: when expression goes beyond a slight groove or eminence and takes the form of a pit

Grade 3: expression is still slight but takes on a more distinct form than shown by grades 1 and 2

Grade 4: the most pronounced expression of Carabelli's trait that does not involve a tubercle with a free apex; grade 4 takes the classic bird-wing form

Grade 5: small tubercle with a free apex

Grade 6: moderate tubercle with a free apex

Grade 7: pronounced tubercle with a free apex

## Breakpoint

Some authors use grade 2 as a breakpoint. In part this is due to the difficulty in scoring grade 1 threshold expressions, especially on teeth that show moderate wear. Scott and Turner (1997) used the frequency of pronounced expressions (5 + 6 + 7) to characterize world variation, because observers agree on what constitutes a free-standing tubercle.

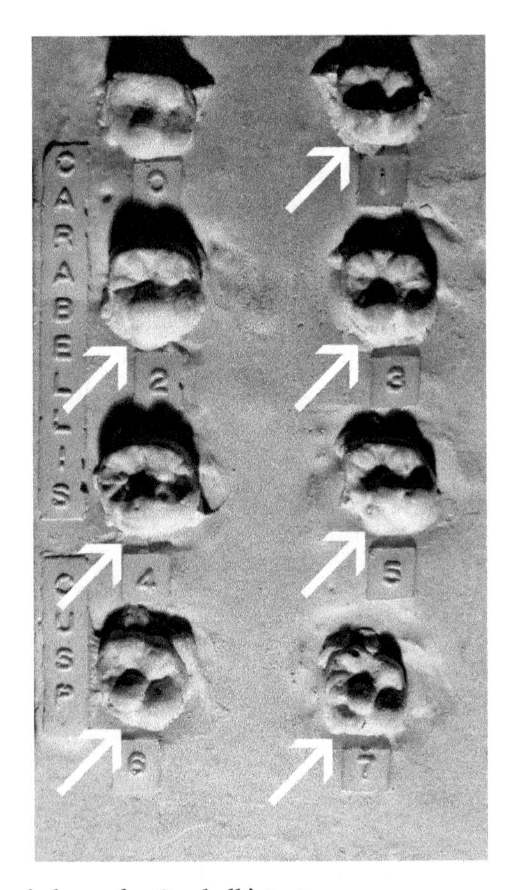


**Figure 18.1** Standard plaque for Carabelli's trait.

## Potential Problems in Scoring

Upper molars often show a helicoidal plane where wear is more pronounced on the cusp tip of the protocone than on the buccal cusps. One has to be careful to make certain the locus of Carabelli's trait expression is present and scorable. The most difficult grade to score is grade 1. For this reason, it is usually expeditious to combine grades 0 and 1 and use at least grade 2 as a breakpoint, because this expression is far more distinctive than grade 1.

## Geographic Variation

For decades, researchers assumed that Carabelli's trait was a diagnostic feature of the European dentition. Although the trait is common in Europeans, it is equally common in other regional groups. Although Europeans show a correlation between total trait frequency and mean degree of expression, that is not true for all groups. Native Americans have very high frequencies of Carabelli's trait but exhibit relatively few cusp forms. Pacific populations on the other hand show a relatively low total trait frequency but exhibit a high frequency of cusp forms (grades 5, 6, and 7), even equaling the frequency in European populations. This was noted by Scott (1980) and corroborated by Turner in many publications.

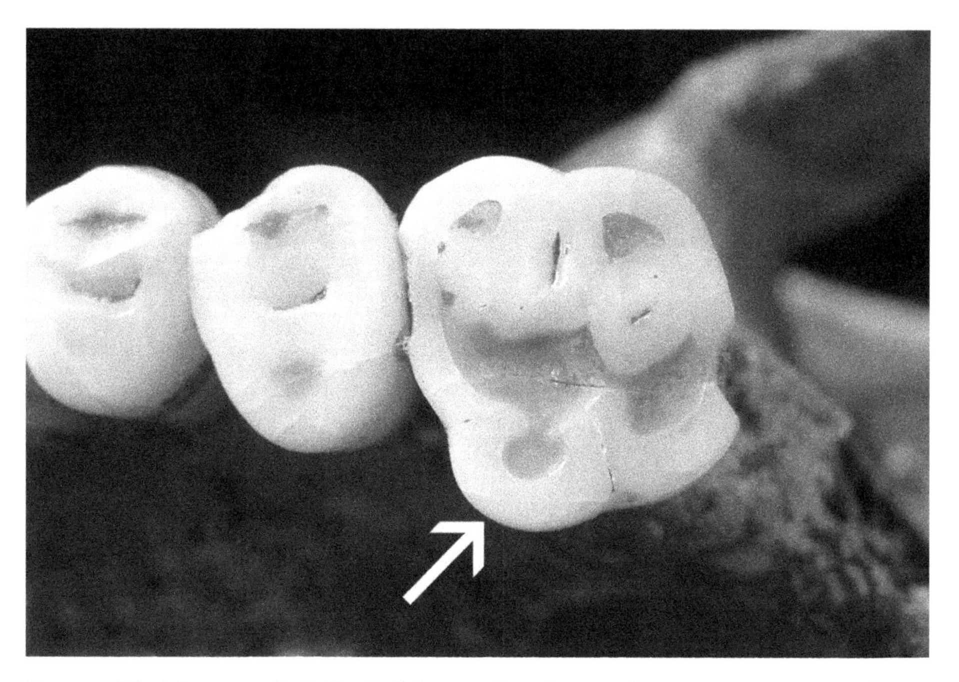

**Figure 18.2** A huge grade 7 Carabelli's cusp that shows a dentine component with pronounced wear.

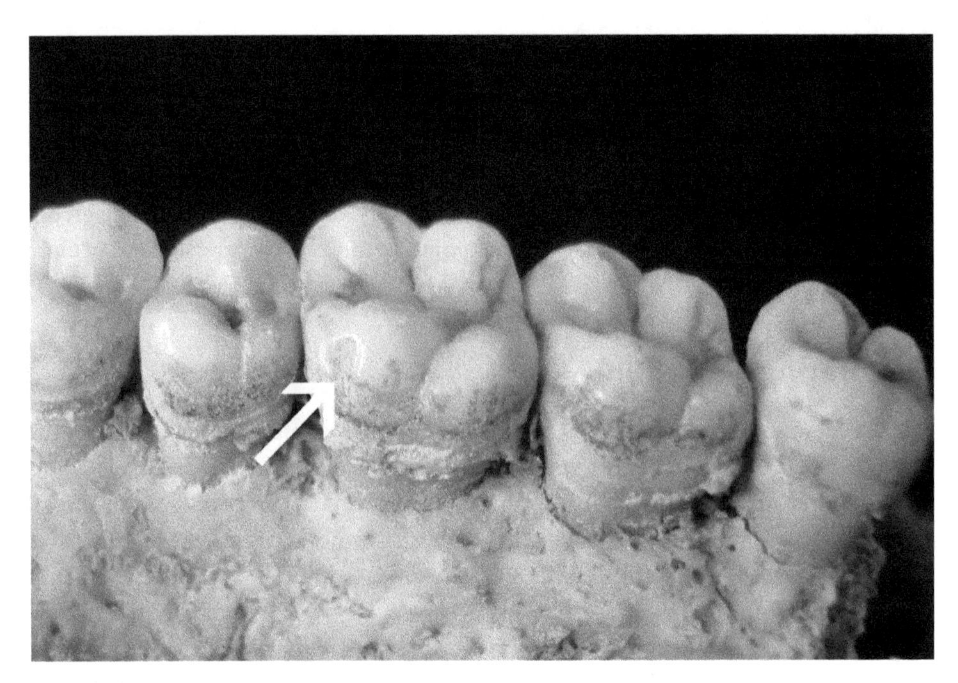

**Figure 18.3** A low-grade Carabelli's trait marked by a small pit; location is key for this trait as its presence is marked by pits, grooves, ridges, or tubercles toward the marginal ridge complex of the protocone.

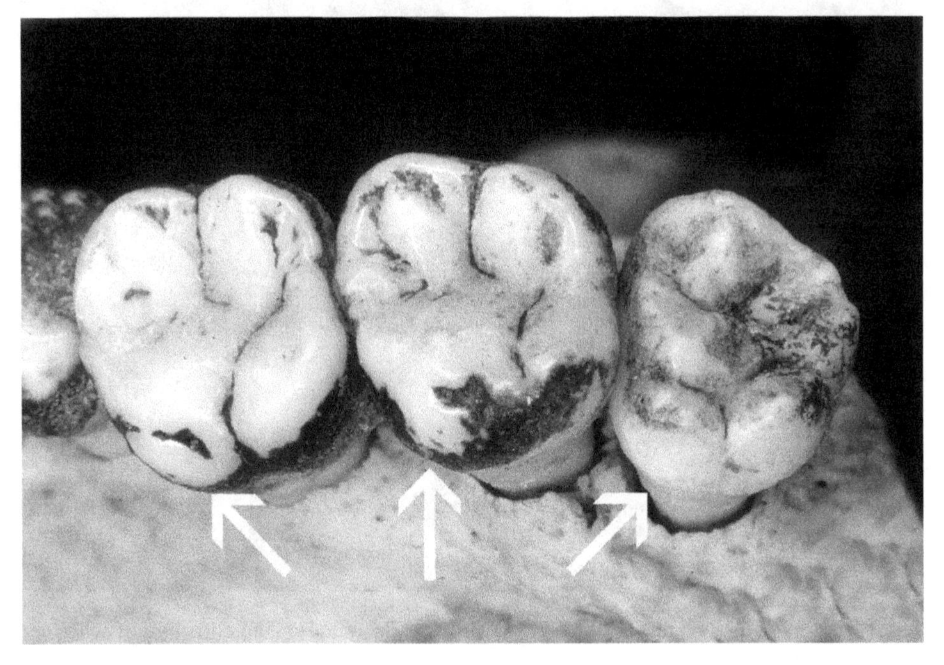

**Figure 18.4** The key tooth for Carabelli's trait is UM1, but in some instances the trait is expressed on UM2 and UM3.

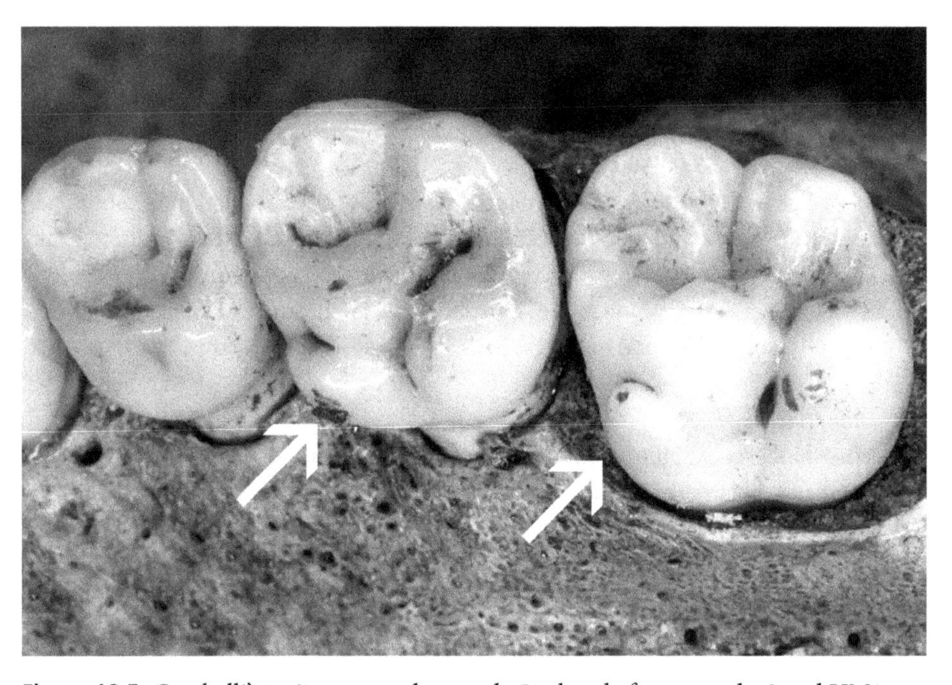

**Figure 18.5** Carabelli's trait expressed as grade 5 tubercle forms on dm2 and UM1.

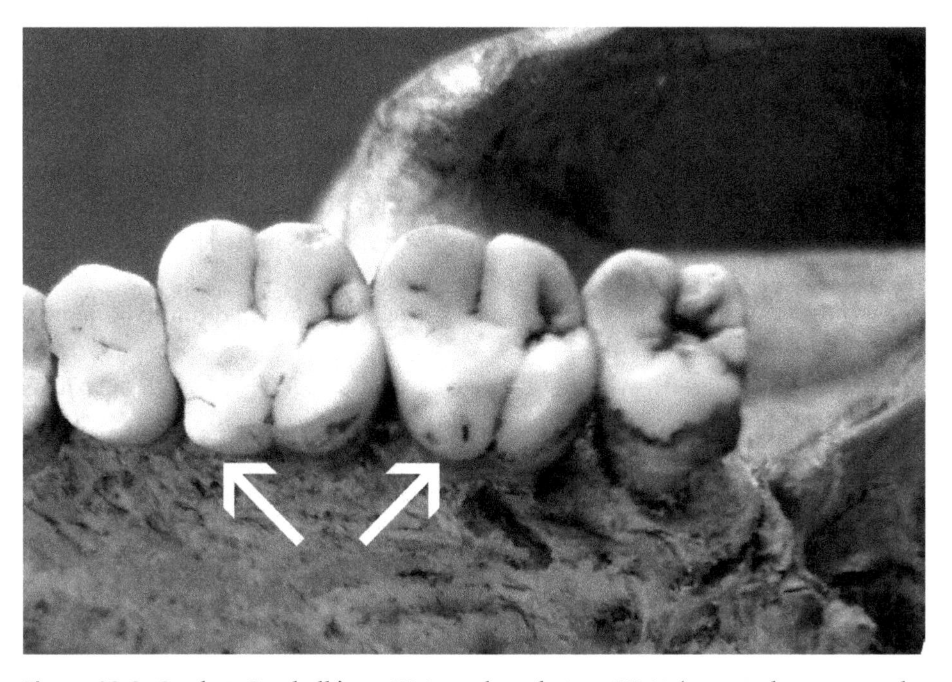

**Figure 18.6** Grade 7 Carabelli's on UM1 and grade 3 on UM2 (cusp 5 also expressed on UM1 and UM2).

## Select Bibliography

Carlsen, O. (1987). *Dental Morphology*. Copenhagen: Munksgaard.

Dahlberg, A.A. (1945). The changing dentition of man. *Journal of the American Dental Association* 32, 676–690.

(1956). Materials for the establishment of standards for classification of tooth characters, attributes, and techniques in morphological studies of the dentition. Zollar Laboratory of Dental Anthropology, University of Chicago (mimeo).

Dietz, V.H. (1944). A common dental morphotropic factor: the Carabelli cusp. *Journal of the American Dental Association* 31, 784–789.

Goose, D.H., and Lee, G.T.R. (1971). The mode of inheritance of Carabelli's trait. *Human Biology* 43, 64–69.

Guatelli-Steinberg, D., Hunter, J.P., Durner, R.M., *et al.* (2013). Teeth, morphogenesis, and levels of variation in the human Carabelli trait. In *Anthropological Perspectives on Tooth Morphology: Genetics, Evolution, Variation*, ed. G.R. Scott and J.D. Irish. Cambridge: Cambridge University Press, pp. 69–91.

Keene, H.J. (1968). The relationship between Carabelli's trait and the size, number and morphology of the maxillary molars. *Archives of Oral Biology* 13, 1023–1025.

Khraisat, A., Alsoleihat F., Subramani, K., *et al.* (2011). Hypocone reduction and Carabelli's traits in contemporary Jordanians and the association between Carabelli's trait and the dimensions of the maxillary first permanent molar. *Collegium Antropologicum* 35, 73–78.

Kondo, S., and Townsend, G.C. (2006). Association between Carabelli trait and cusp areas in permanent maxillary first molars. *American Journal of Physical Anthropology* 129, 196–208.

Kraus, B.S. (1951). Carabelli's anomaly of the maxillary molar teeth. *American Journal of Human Genetics* 3, 348–355.

Kraus, B.S. (1959). Occurrence of the Carabelli trait in Southwest ethnic groups. *American Journal of Physical Anthropology* 17, 117–123.

Meredith, H.V., and Hixon, E.H. (1953). Frequency, size, and bilateralism of Carabelli's tubercle. *Journal of Dental Research* 33, 435–440.

Scott, G.R. (1978). The relationship between Carabelli's trait and the protostylid. *Journal of Dental Research* 57, 570.

(1979). Association between the hypocone and Carabelli's trait of the maxillary molars. *Journal of Dental Research* 58, 1403–1404.

(1980). Population variation of Carabelli's trait. *Human Biology* 52, 63–78.

Scott, G.R., and Turner, C.G., II (1997). *The Anthropology of Modern Human Teeth: Dental Morphology and its Variation in Recent Human Populations*. Cambridge: Cambridge University Press.

Shapiro, M.M.J. (1949). The anatomy and morphology of the tubercle of Carabelli. *Official Journal of the Dental Association of South Africa* 4, 355–362.

Turner, C.G., II, and Hawkey, D.E. (1998). Whose teeth are these? Carabelli's trait. In *Human Dental Development, Morphology, and Pathology: A Tribute to Albert A. Dahlberg*, ed. J.R. Lukacs. Eugene: University of Oregon Anthropological Papers, Number 54, pp. 41–50.

Turner, C.G., II, Nichol, C.R., and Scott, G.R. (1991). Scoring procedures for key morphological traits of the permanent dentition: the Arizona State University dental anthropology system. In *Advances in Dental Anthropology*, ed. M.A. Kelley and C.S. Larsen. New York: Wiley-Liss, pp. 13–31.

# NOTES

# 19

# Parastyle

## Observed on

UM1, UM2, UM3

## Key Tooth

UM1

## Synonyms

Paramolar tubercle (Bolk 1916, Dahlberg 1945)

## Description

The parastyle is sometimes linked to Bolk's paramolar tubercle, but they may be two entirely different traits. Typically, the parastyle is expressed on the paracone of the upper molars. It ranges in size from a pit to a large free-standing tubercle. In some instances, it is expressed on cusp 3 (metacone). For large paramolar tubercles, the key teeth are UM2 and UM3.

## Classification

Turner *et al.* (1991)

Grade 0: buccal surfaces of cusps 2 and 3 are smooth
Grade 1: a small pit near the buccal groove between cusps 2 and 3
Grade 2: small cusp but no free apex
Grade 3: medium cusp with free apex
Grade 4: large cusp with free apex
Grade 5: very large cusp with free apex that may extend onto the surfaces of both cusps 2 and 3

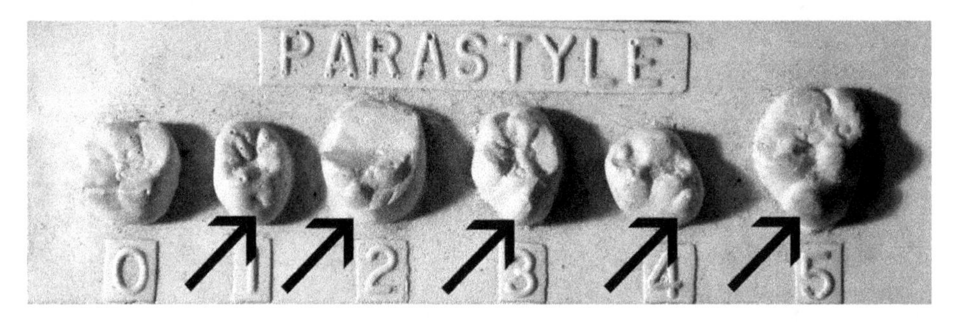


**Figure 19.1** Standard plaque for the parastyle.

Grade 6: peg-shaped crown attached to root of second or third molar. This classic form of Bolk's paramolar tubercle may represent a supernumerary tooth that is fused to the buccal surface of UM2 or UM3. Accessory cusps with all the characteristics of a paramolar tubercle have also been observed on LM2 and LM3, adding evidence to the possibility these are fused supernumerary teeth.

## Breakpoint

Grade 2+ (small to large tubercles)

## Potential Problems in Scoring

The buccal surface of the upper molars may have ridges and grooves that represent part of the parastyle complex. Grade 6, or Bolk's paramolar tubercle, is possibly a fused conical supernumerary tooth. It is always expressed on UM2 and UM3 (or LM2 and LM3), not UM1 (or LM1).

## Geographic Variation

The parastyle does not distinguish any modern human population. Total frequencies for hundreds of world samples all fall below 10%.

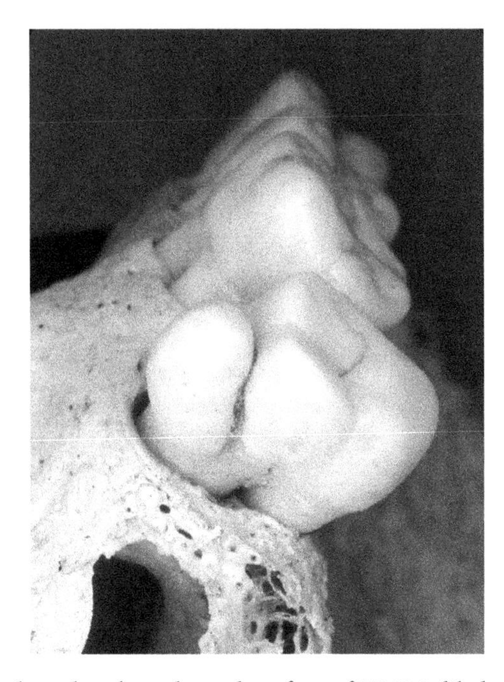

**Figure 19.2** Paramolar tubercle on buccal surface of UM2 is likely a fused supernumerary tooth.

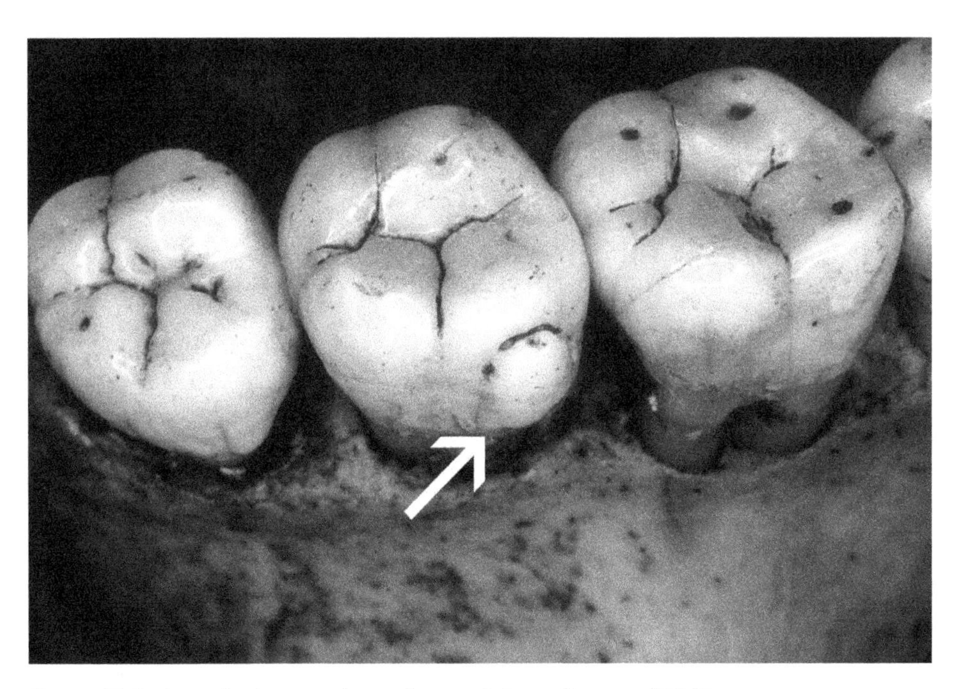

**Figure 19.3** A grade 4 parastyle on the mesiobuccal cusp of UM2.

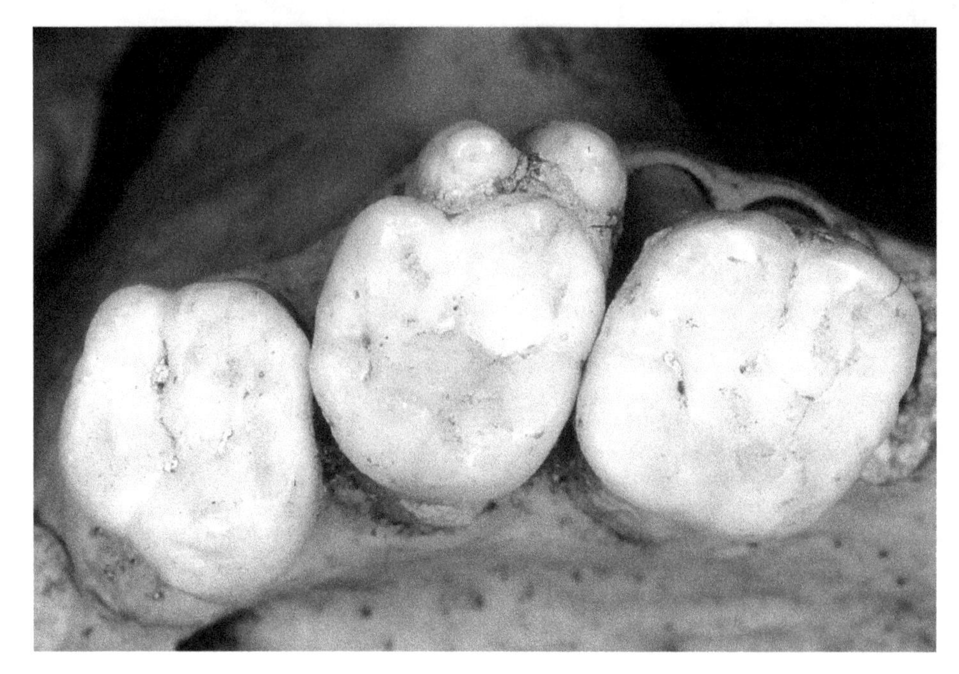

**Figure 19.4** As Bolk noted, paramolar tubercles are found on UM2 and UM3, not UM1. In this case, there is a double paramolar tubercle on UM2.

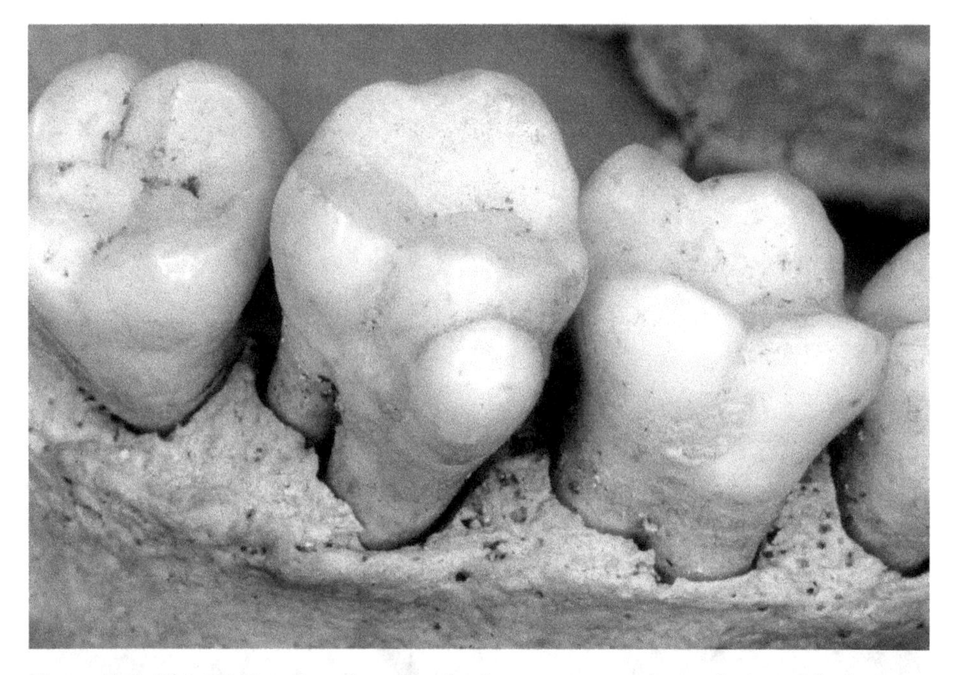

**Figure 19.5** This UM2 paramolar tubercle shows crown and root fusion with the paracone and mesiobuccal root of UM2. Cases such as this are overwhelmingly asymmetrical.

**Figure 19.6** Paramolar tubercles are most common on UM2, but in this instance it is expressed on UM3; it may be a fused supernumerary tooth.

## Select Bibliography

Bolk, L. (1916). Problems of human dentition. *American Journal of Anatomy* 19, 91–148.

Dahlberg, A.A. (1945). The paramolar tubercle (Bolk). *American Journal of Physical Anthropology* 3, 97–103.

Katich, J.F. (1975). Parastyle variation in Hawaiian maxillary molars. *American Journal of Physical Anthropology* 42, 310.

Kustaloglu, O.A. (1962). Paramolar structures of the upper dentition. *Journal of Dental Research* 41, 75–83.

Magalee, R.E., and Kramer, S. (1984). The paramolar tubercle: a morphological anomaly with clinical considerations. *New York State Dental Journal* 50, 564–566.

Nabeel, S., Danish, G., Hegde, U., and Mull, P. (2012). Parastyle: clinical significance and management of two cases. *International Journal of Oral and Maxillofacial Pathology* 3, 61–64.

Nagaveni, N.B., Umashankara, K.V., Radhika, N.B., and Garewal, R.S. (2009). "Paramolar tubercle" in the primary dentition: case reports and literature review. *International Journal of Dental Anthropology* 14, 12–18.

Nayak, G., Shetty, S., and Singh I. (2013). Paramolar tubercle: a diversity in canal configuration identified with the aid of spiral computed tomography. *European Journal of Dentistry* 7, 139–144.

Ooshima, T., Ishima, R., Mishima, K., and Sobue, S. (1996). The prevalence of developmental anomalies of teeth and their association with tooth size in the primary and deciduous dentition of 1650 Japanese children. *International Journal of Paediatric Dentistry* 6, 87–94.

Rodríguez, C., and Moreno, F. (2006). Paramolar tubercle in the left maxillary second premolar: a case report. *Dental Anthropology* 19, 65–69.

Turner, R.A., and Harris, E.F. (2004). Maxillary second premolars with paramolar tubercles. *Dental Anthropology* 17, 75–78.

Turner, C.G., II, Nichol, C.R., and Scott, G.R. (1991). Scoring procedures for key morphological traits of the permanent dentition: the Arizona State University dental anthropology system. In *Advances in Dental Anthropology*, ed. M.A. Kelley and C.S. Larsen. New York: Wiley-Liss, pp. 13–31.

# 20

# Enamel Extensions

## Observed on

Buccal surface of upper and lower molars

## Key Tooth

UM1/LM1

## Synonyms

Cervical enamel projection (de Souza *et al.* 2014)

## Description

This trait is distinct from accessory cusps, tubercles, and ridges as it involves the course of the cervical enamel line. The standard expression of this line is horizontal or straight. In some instances, however, the enamel extends toward the apex of the roots in the direction of the bifurcation of the two buccal aspects of the roots of either upper or lower molars. In some instances, the extension ends in an enamel pearl. These pearls, although interesting, are not involved in the scale for scoring enamel extensions.

## Classification

Pedersen (1949), Turner *et al.* (1991)

Grade 0: cervical enamel line is horizontal
Grade 1: enamel line extends about 1 mm toward root bifurcation
Grade 2: enamel line extends about 2 mm toward root bifurcation
Grade 3: enamel line extends 4 mm or more toward root bifurcation

## Breakpoint

Grade 2+

## Potential Problems in Scoring

The biggest issue with enamel extensions is that they cannot be scored on dental casts because they are obscured by tissue covering the alveolus. One advantage, however, is that they can be scored on molars that show advanced wear.

## Geographic Variation

Enamel extensions are rare in Western Eurasians, Sub-Saharan Africans, and Sahul-Pacific groups, where they rarely exceed 10%. Frequencies of around 50% are found in East Asians and New World populations. Southeast Asians and Polynesians show intermediate frequencies of 20–30%.

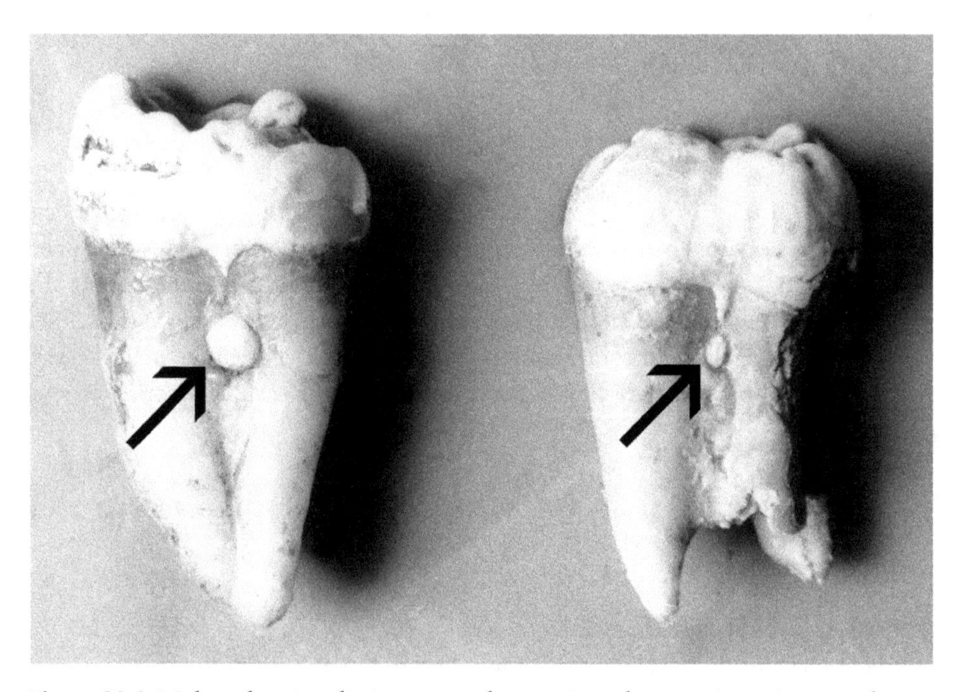

**Figure 20.1** Molars showing distinct enamel extensions that terminate in enamel pearls. Scoring extensions does not require the presence of such pearls, as they are often obscured when a tooth is firmly ensconced in its socket.

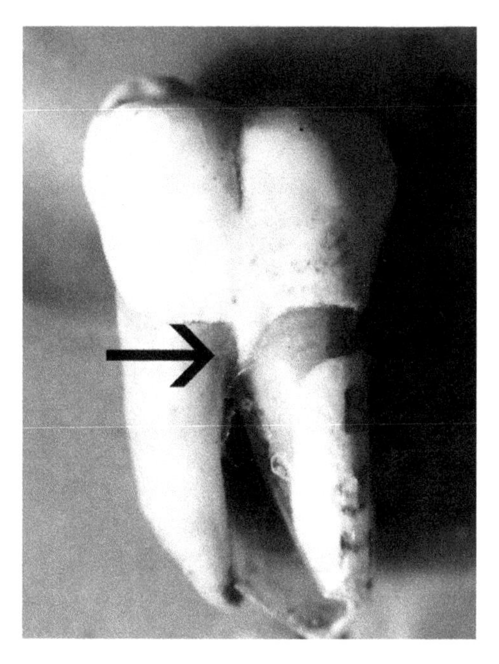

**Figure 20.2** Enamel extensions may be more common on LM1 than UM1, but Turner adopted UM1 as the key tooth for this trait.

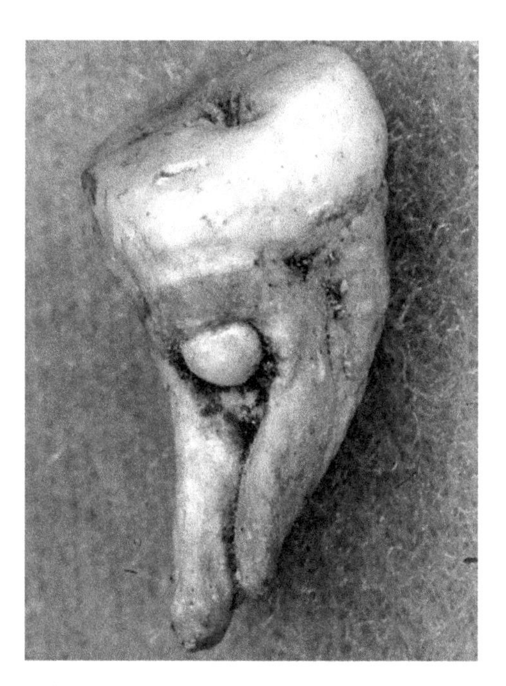

**Figure 20.3** An extraordinarily large enamel pearl tucked between the inter-radicular projections of lower molar roots.

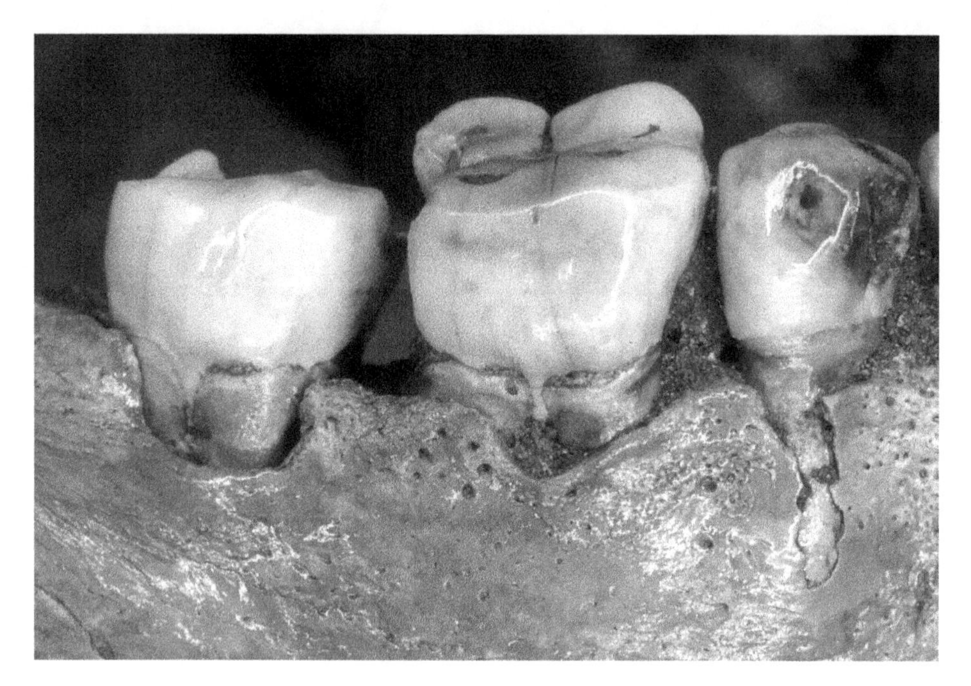

**Figure 20.4** Enamel extensions on LM1 and LM2; in this case, more pronounced on LM2.

## Select Bibliography

Birkby, W.H., Fenton, T.W., and Anderson, B.E. (2008). Identifying Southwest Hispanics using nonmetric traits and the cultural profile. *Journal of Forensic Sciences* 53, 29–33.

Brabant, H., and Ketelbant, R. (1975). Observations sur la frequence de certains caracteres Mongoloides dans la denture permanente de la population Belge. *Bulletin du Groupement International pour l Recherche Scientifique en Stomatologie et Odontologie* 18, 121–34.

de Souza, M.R.L., Marques, A.A.F., Sponchiado, E.C., de Vargas, T.A., and Garcia, L.F.R. (2014). Prevalence of cervical enamel projection in human molars. *Dental Hypotheses* 5, 21–4.

Hanihara, K. (1969). Mongoloid dental complex in the permanent dentition. *Proceedings of the VIIIth International Congress of Anthropological and Ethnological Sciences, Tokyo and Kyoto*, pp. 298–300.

Lasker, G.W. (1950). Genetic analysis of racial traits of the teeth. *Cold Spring Harbor Symposia on Quantitative Biology* XV, 191–203.

Leigh, R.W. (1928). Dental pathology of aboriginal California. *University of California Publications in American Archaeology and Ethnology* 23, 399–440.

Pedersen, P.O. (1949). The East Greenland Eskimo dentition. In *Indian Tribes of Aboriginal America*, ed. S. Tax. Selected papers of the 29th International Congress of Americanists. Chicago: University of Chicago Press.

Risnes, S. (1974). The prevalence and distribution of cervical enamel projections reaching into the bifurcation on human molars. *Scandinavian Journal of Dental Research* 82, 413–9.

Turner, C.G. II (1990). Major features of Sundadonty and Sinodonty, including suggestions about East Asian microevolution, population history, and Late Pleistocene relationships with Australians Aboriginals. *American Journal of Physical Anthropology* 82, 295–317.

Turner, C.G., II, Nichol, C.R., and Scott, G.R. (1991). Scoring procedures for key morphological traits of the permanent dentition: the Arizona State University dental anthropology system. In *Advances in Dental Anthropology*, ed. M.A. Kelley and C.S. Larsen New York: Wiley-Liss, pp. 13–31.

# NOTES

# 21

# Upper Premolar Root Number

## Observed on

UP1, UP2

## Key Tooth

UP1

## Synonyms

Two-rooted upper premolar, two-rooted UP1

## Description

Upper premolars have either two or three root cones. Typically, there are two. One cone is buccal (below buccal cusp) and one is lingual (below lingual cusp). The root cones may or may not be separated by an inter-radicular projection (i.e., root bifurcation). When there is no separation and you can only observe a vertical groove that separates the buccal and lingual root cones, the tooth is one-rooted. If there is a bifurcation that extends from ¼ to ⅓ of total root length, the tooth is two-rooted. Very rarely, there are two buccal roots. When these are bifurcated along with a bifurcation of the lingual root, the result is a three-rooted UP1. Although UP1 is the key tooth for upper premolar root number, UP2 is occasionally two-rooted, but this is much less common than a two-rooted UP1 (in fact, root number often aids in distinguishing UP1 from UP2, because upper premolars show many similarities, especially when the principal cusps are worn).

## Classification

Turner (1981)

Grade 1: one-rooted UP1 (root grooves separate cones but no inter-radicular projection)
Grade 2: two-rooted UP1 (inter-radicular projection separates buccal and lingual root cones for ¼ to ⅓ of total root length)

Grade 3: three-rooted UP1 (there is an inter-radicular projection that separates the buccal root into two distinct roots, and another projection separating the two buccal roots from a single lingual root)

## Breakpoint

Presence of two or three roots

## Potential Problems in Scoring

Although one can often score upper premolar root number even when the teeth are missing, one has to be careful and make certain there is a distinct root bifurcation indicated in the socket. There should be a thin section of bone that separates the two roots.

## Geographic Variation

UP1 root number exhibits pronounced variation around the world. Africans have the highest frequency of two-rooted UP1, at about 60%. East Asians have a lower frequency (25%), but the lowest frequencies are in New World populations, where two-rooted UP1 are rare (<10%). Southeast Asians, Austral-Melanesians, and Europeans have intermediate frequencies (ca. 40%).

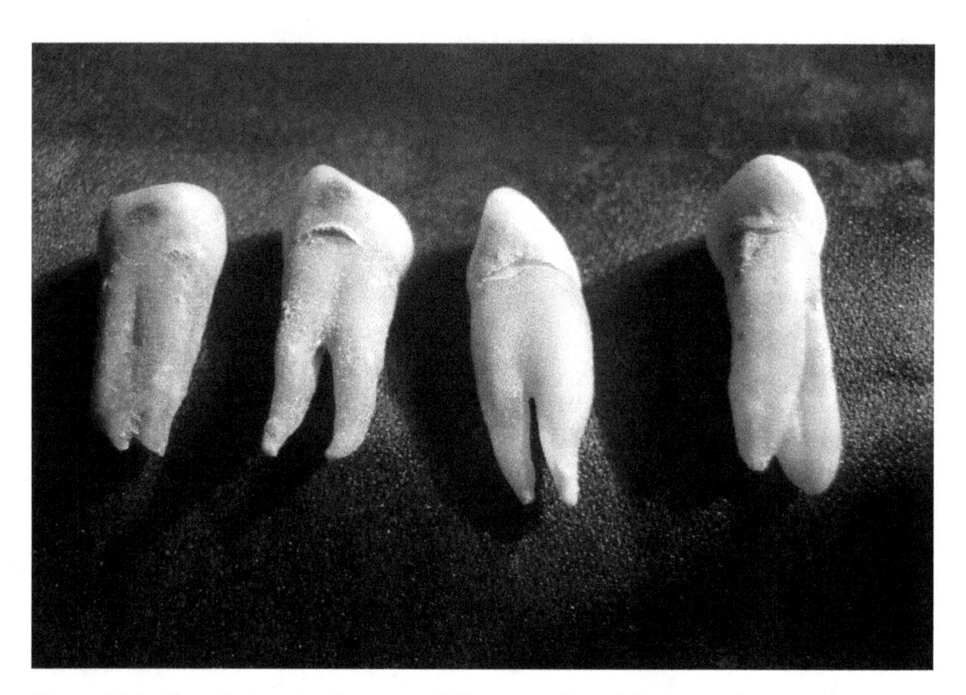

**Figure 21.1** The relationship between different teeth with bifurcated roots is unknown, but for this single individual, there is (from left to right) a single-rooted UP2 with distinct root cones, a two-rooted UP1, a two-rooted LC, and an LP1 with a pronounced Tomes' root.

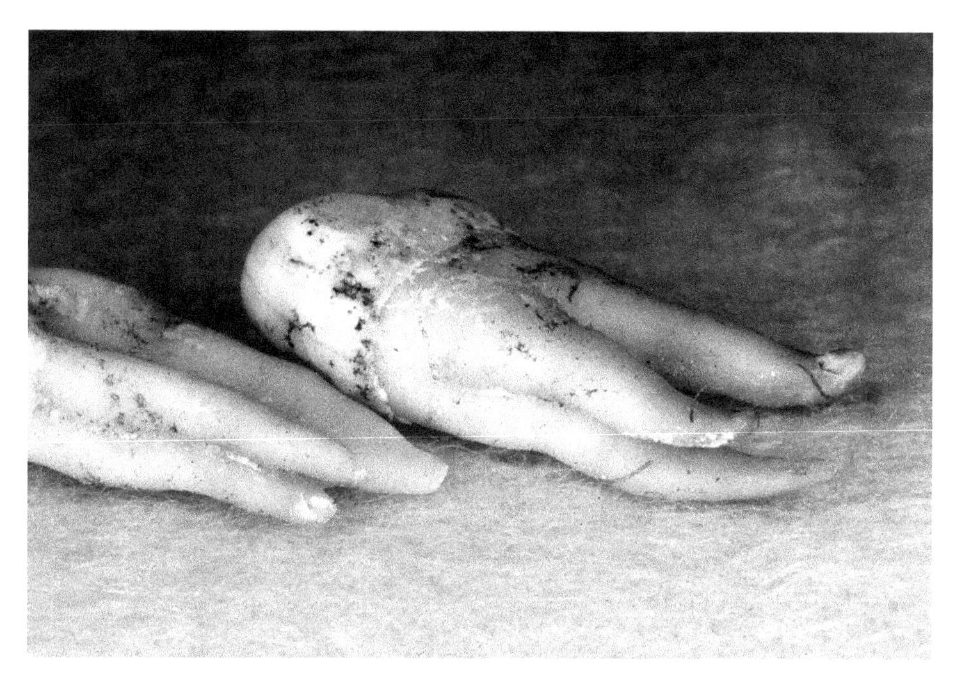

**Figure 21.2** Three-rooted UP1 (note buccal root bifurcation).

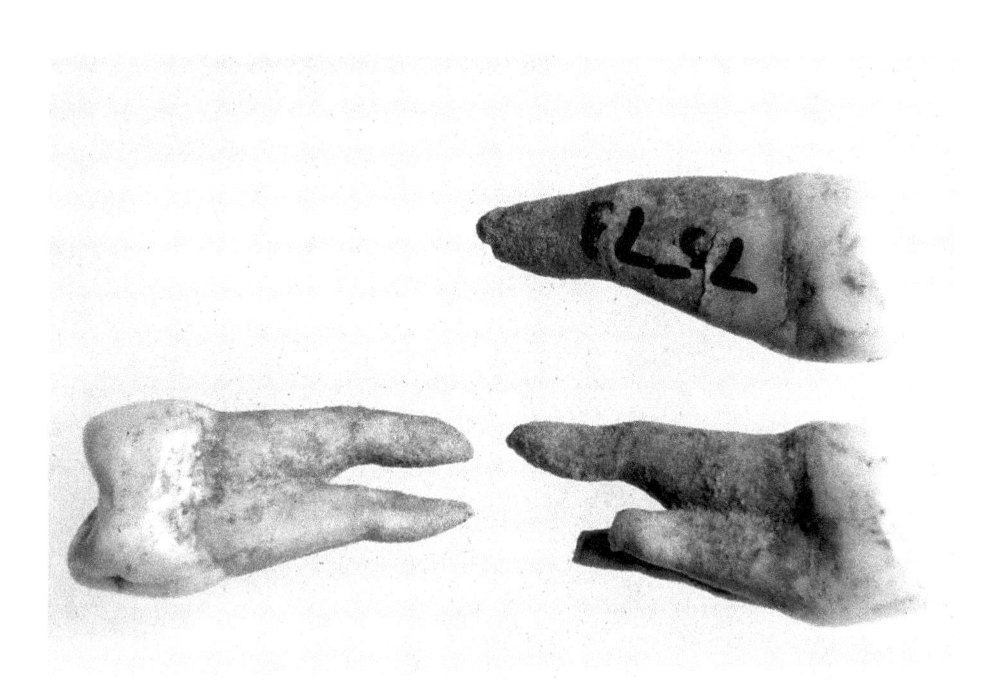

**Figure 21.3** Two-rooted and three-rooted UP1s below a one-rooted UP2.

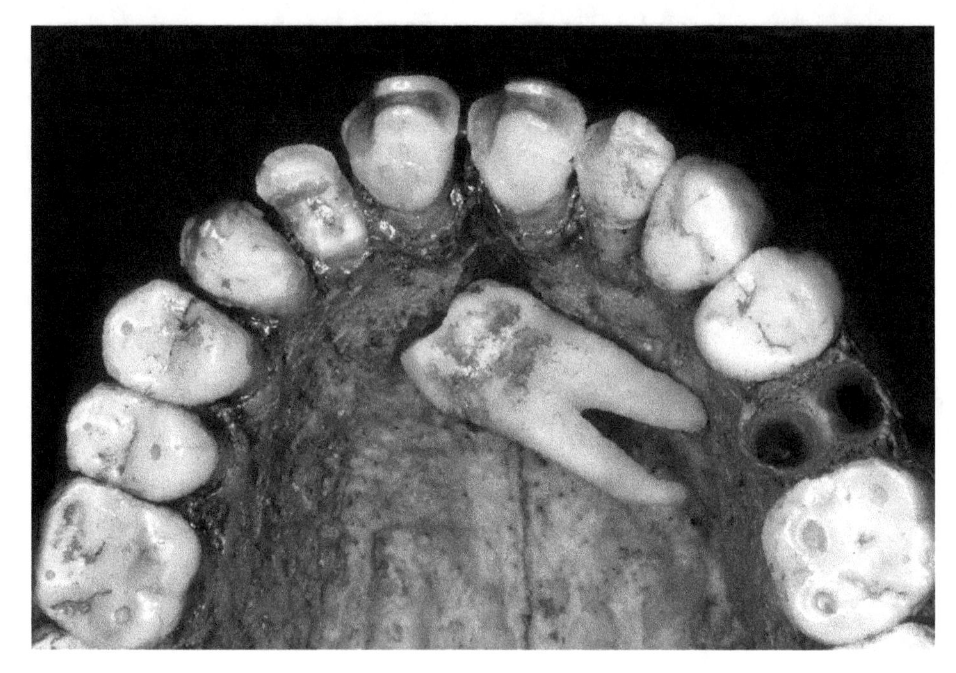

**Figure 21.4** Two-rooted UP2; the second premolar is rarely two-rooted, but this individual is African and that geographic group has the world's highest frequencies of multi-rooted upper premolars so it is less surprising than it would be in a group with a low frequency of two-rooted UP1.

## Select Bibliography

Kovacs, I. (1967). Contribution to the ontogenetic morphology of roots of human teeth. *Journal of Dental Research* 46, 865–873.

Loh, H.S. (1998). Root morphology of the maxillary first premolar in Singaporeans. *Australian Dental Journal* 43, 399–402.

Turner C.G., II (1967). *The Dentition of Arctic Peoples.* PhD dissertation, Department of Anthropology, University of Wisconsin, Madison.

Turner, C.G., II (1981). Root number determination in maxillary first premolars for modern human populations. *American Journal of Physical Anthropology* 54, 59–62.

Wheeler, R.C. (1965). *A Textbook of Dental Anatomy and Physiology*, 4th edn. Philadelphia and London: W.B. Saunders.

# 22

# Upper Molar Root Number

## Observed on

UM1, UM2, UM3

## Key Tooth

UM2

## Synonyms

Three-rooted upper molar

## Description

In hominoids and early hominins, upper molars had three roots, each of which was associated with a major cusp of the trigon. The two buccal roots are linked to the paracone and metacone, the two buccal cusps. The third and typically largest root is lingual and is associated with the protocone. As with upper and lower molar cusp number, the most stable tooth in the upper molar field is UM1. In modern human populations, most UM1s still have three primary roots. Because of space limitations, it is quite common for UM3 to exhibit three root cones but with no inter-radicular projections separating them. For that reason, UM3 often has but one root. The tooth between the highly conserved UM1 and highly variable UM3 is the second molar. This tooth can exhibit either three roots like UM1 or several types of root cone fusion, producing either a one-rooted or a two-rooted UM2. Sometimes the two buccal roots are fused while the lingual cusp is independent, producing a two-rooted tooth. Alternatively, the mesiobuccal root can fuse with the lingual cusp along with an independent distobuccal root, or the distobuccal root can fuse with the lingual root along with an independent mesiobuccal root. These three patterns result in a two-rooted UM2. Finally, all three root cones can be fused, producing a one-rooted UM2. Because of the several alternative arrangements that produced one-rooted and two-rooted UM2, using three-rooted UM2 as the key phenotype helps allay confusion on root fusion and focuses on what is basically the retained condition of UM2. Independent roots have to show a bifurcation of at least ¼ to ⅓ of total root length.

## Classification

Grade 1: one-rooted UM2 (root cones separated by grooves but there are no inter-radicular projections)
Grade 2: two-rooted UM2 (one inter-radicular projection separates one root from two fused roots)
Grade 3: three-rooted UM2 (three inter-radicular projections separate all three roots for at least ¼ to ⅓ of total root length)

## Breakpoint

Total frequency of three-rooted UM2

## Potential Problems in Scoring

This trait can be scored from the sockets even with post-mortem tooth loss. Caution should be used to be certain that thin layers of cancellous bone separate three independent roots. If this is not clearly evident, the trait should not be scored.

## Geographic Variation

Africans, Australians, and Melanesians show the highest frequencies of three-rooted UM2, falling in the range between 80% and 90%. Northeast Siberians and Native Americans show the most root fusion, with three-rooted UM2 frequencies of 40–50%. Europeans, East Asians, and Southeast Asians fall between these extremes at 60–75%.

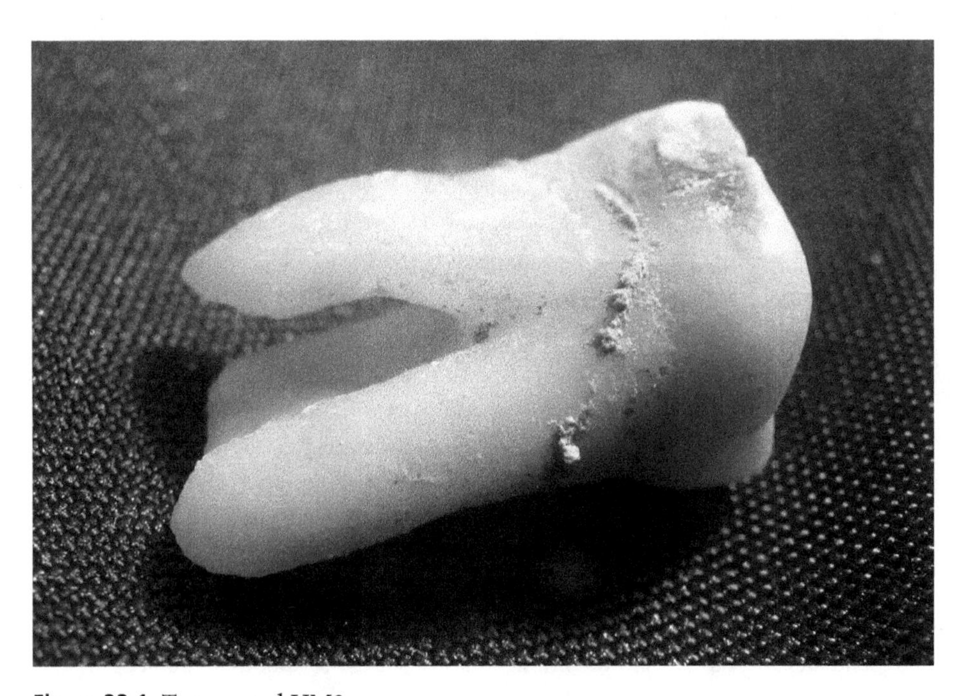

**Figure 22.1** Two-rooted UM2.

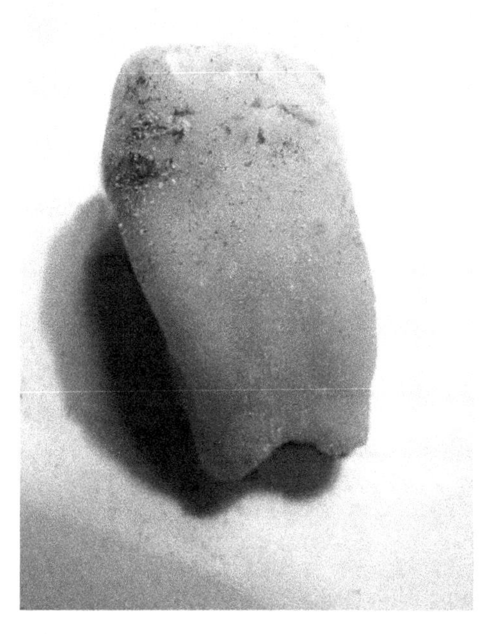

**Figure 22.2** One-rooted UM2.

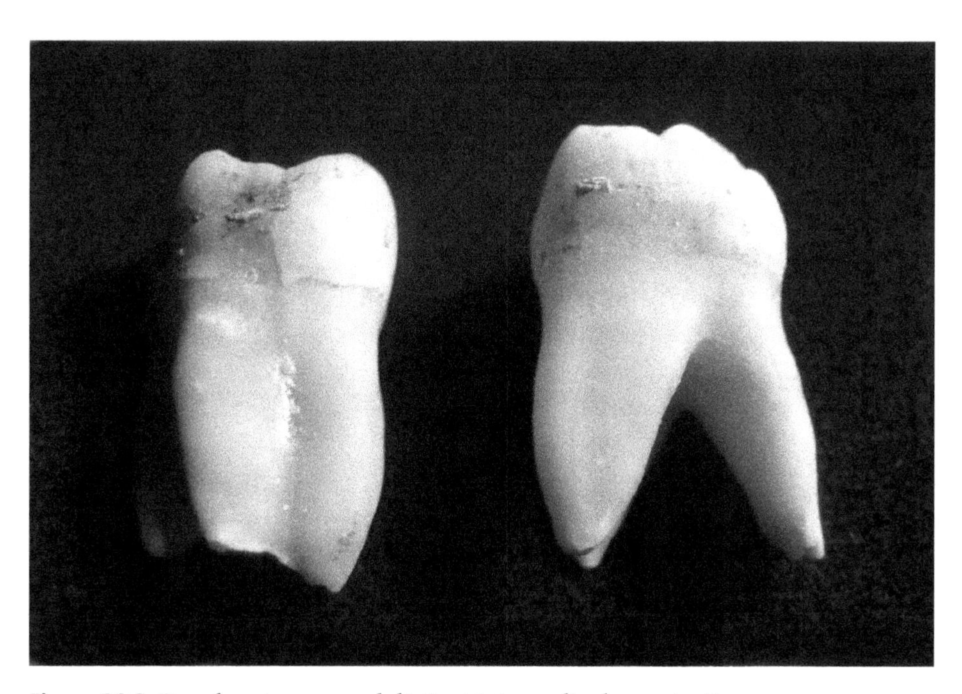

**Figure 22.3** Fused root cones and distinct inter-radicular projection.

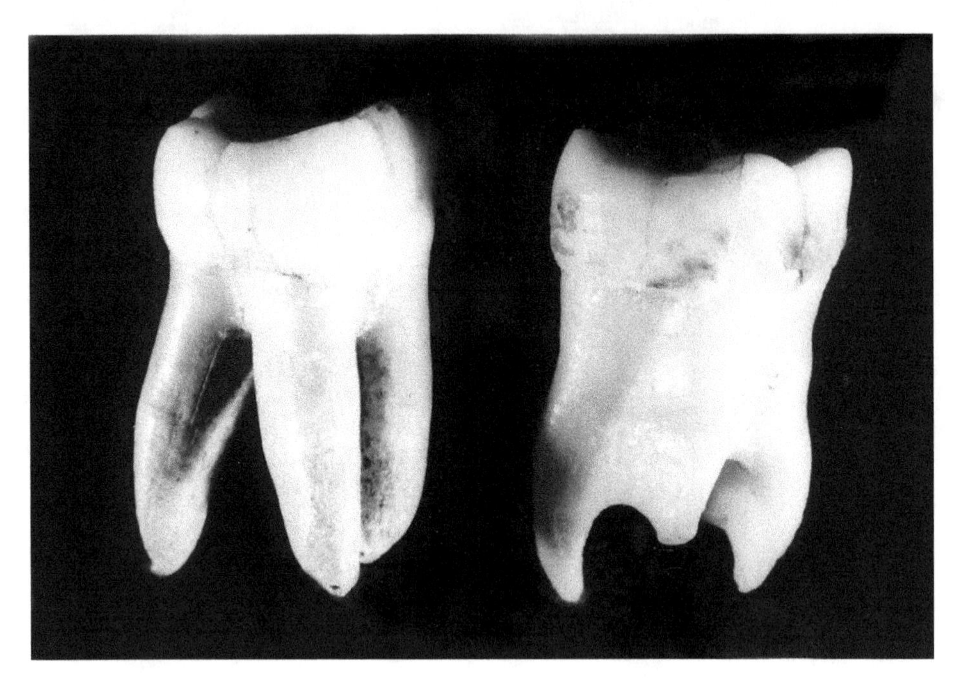

**Figure 22.4** Three-rooted UM1 and two-rooted UM2 (bifurcation less that ¼ total root length).

## Select Bibliography

Caliskan, M.K., Pehlivan, Y., Sepetcioğlu, F., Türkün, M., and Tuncer, S.S. (1995). Root canal morphology of human permanent teeth in a Turkish population. *Journal of Endodontics* 21, 200–204.

Kovacs, I. (1967). Contribution to the ontogenetic morphology of roots of human teeth. *Journal of Dental Research* 46, 865–873.

Pecora, J.D., Woelfel, J.B., Sousa Neto, M.D., and Issa, E.P. (1992). Morphologic study of the maxillary molars. Part II. Internal anatomy. *Brazilian Dental Journal* 3, 53–57.

Wheeler, R.C. (1965). *A Textbook of Dental Anatomy and Physiology*, 4th edn. Philadelphia and London: W.B. Saunders.

# 23

# Upper Lateral Incisor Variants

## Observed on

UI2

## Key Tooth

UI2

## Synonyms

Pegged UI2, diminutive UI2, missing UI2, talon cusp, triform

## Description

Upper lateral incisors can assume a variety of forms and exhibit many unusual morphological features. Given their rarity, however, it is more efficient to include the array of potential variants under one trait, rather than designating each of them as a separate trait.

## Classification

Present study

Grade 0: UI2 normal in form and size
Grade 1: UI2 normal in form but diminutive in size (less than ½ mesiodistal diameter of UI1)
Grade 2: congenital absence
Grade 3: peg-shaped UI2, conical in form, often with no morphological features
Grade 4: talon cusp (in same location but much more pronounced and distinctive than a *tuberculum dentale*)
Grade 5: triform UI2 with a large lingual structure that runs from basal cingulum to incisal edge
Grade 6: unusual UI2 forms that do not fit any of the above categories

## Breakpoint

The classification is not ranked, but it is still possible to calculate the total frequency for all UI2 variants combined, or for specific UI2 variants.

## Potential Problems in Scoring

The problems are minor, given that there are no ranked scores to contend with. Not all UI2 variants are exactly the same, but the figures provide examples for each kind of variant. Some variables are so unusual they do not fit in any category.

## Geographic Variation

Western Eurasians rarely exhibit talon cusps or have UI2s that are peg-shaped or triform. They do, however, often show diminutive lateral incisors. For Sub-Saharan Africans, the most common lateral incisor variant is the peg-shaped form. In Native Americans, both talon cusps and triforms appear, although they are not common. These variants are typically in a frequency of less than 5% so their utility may be more in intra-cemetery analysis than in biodistance.

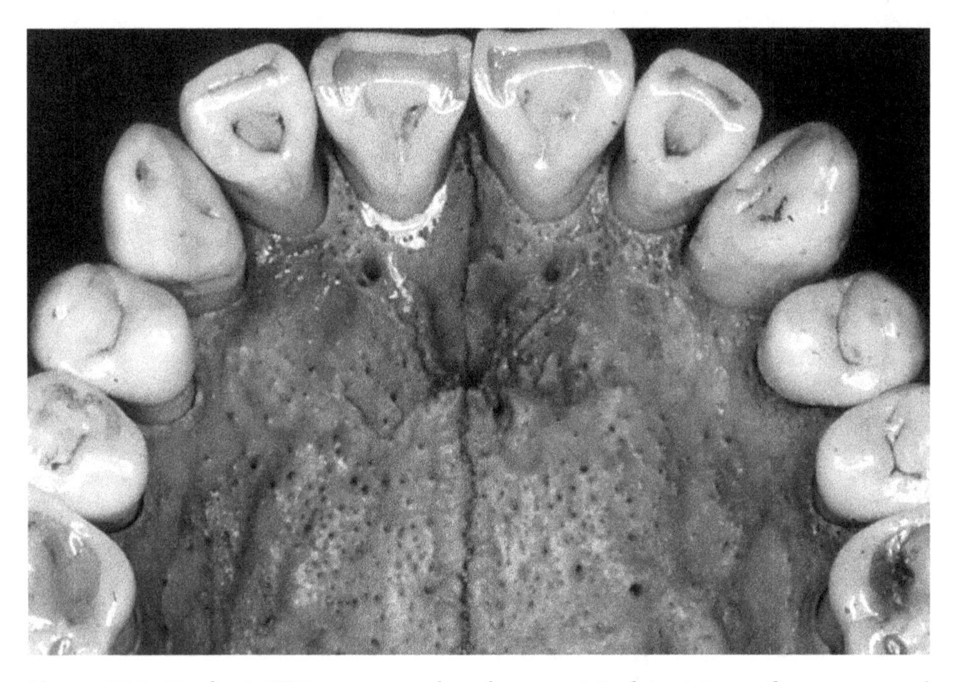

**Figure 23.1** Grade 0: UI2 are normal and symmetrical in terms of presence and morphology.

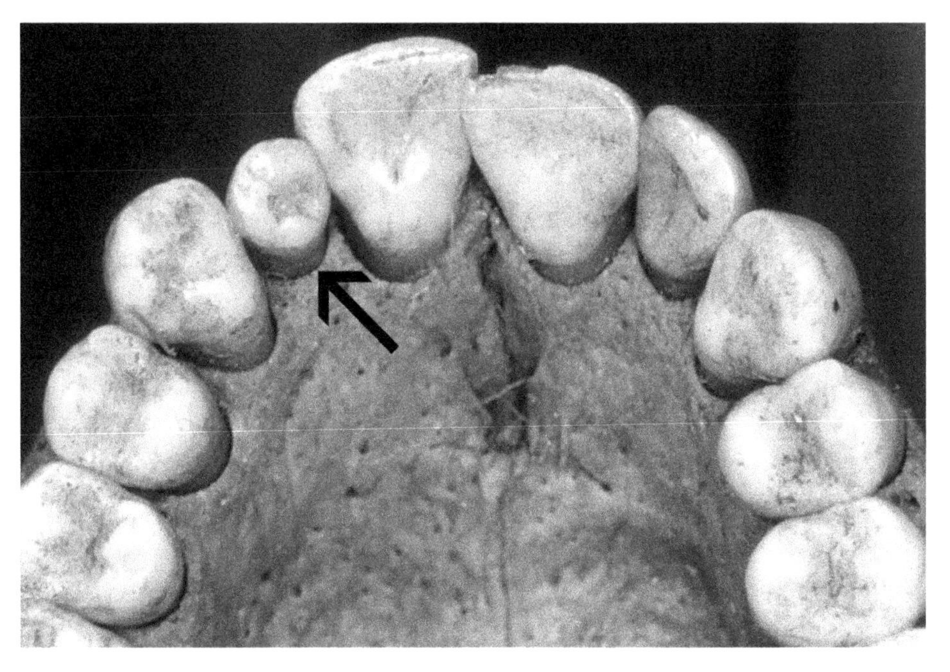

**Figure 23.2** Grade 1: right UI2 is diminutive while left UI2 is normal.

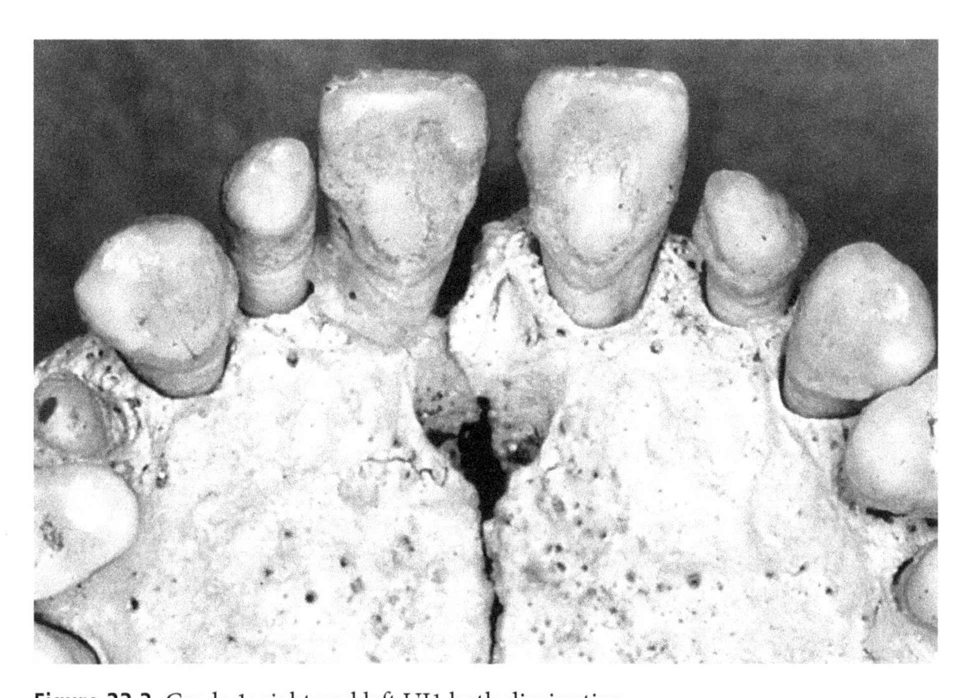

**Figure 23.3** Grade 1: right and left UI1 both diminutive.

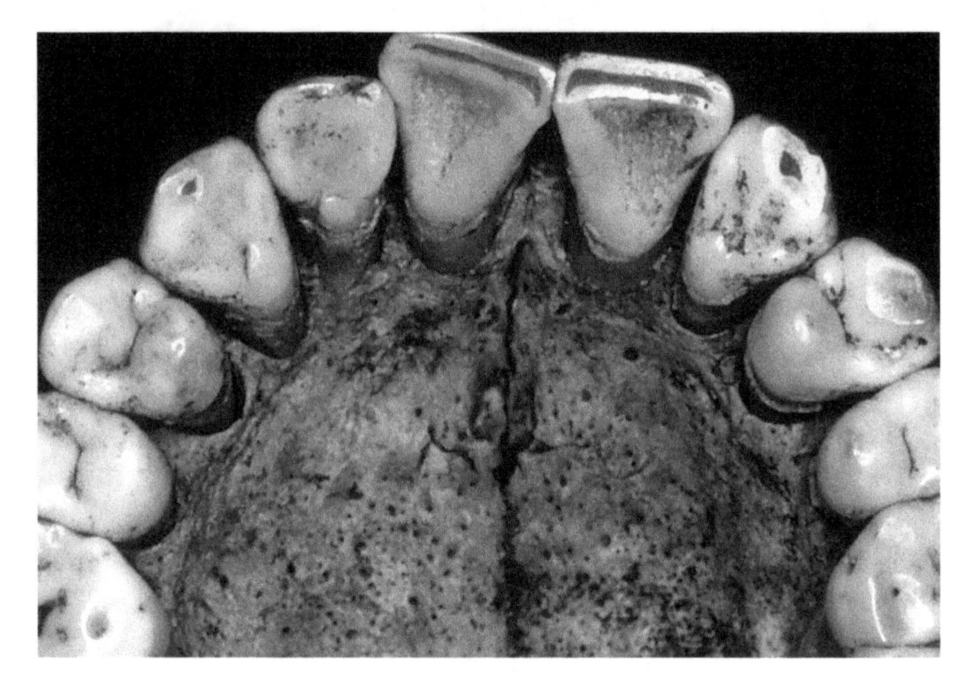

**Figure 23.4** Grade 2: young adult with minor tooth wear has right UI2, but left UI2 is missing, presumably congenitally absent.

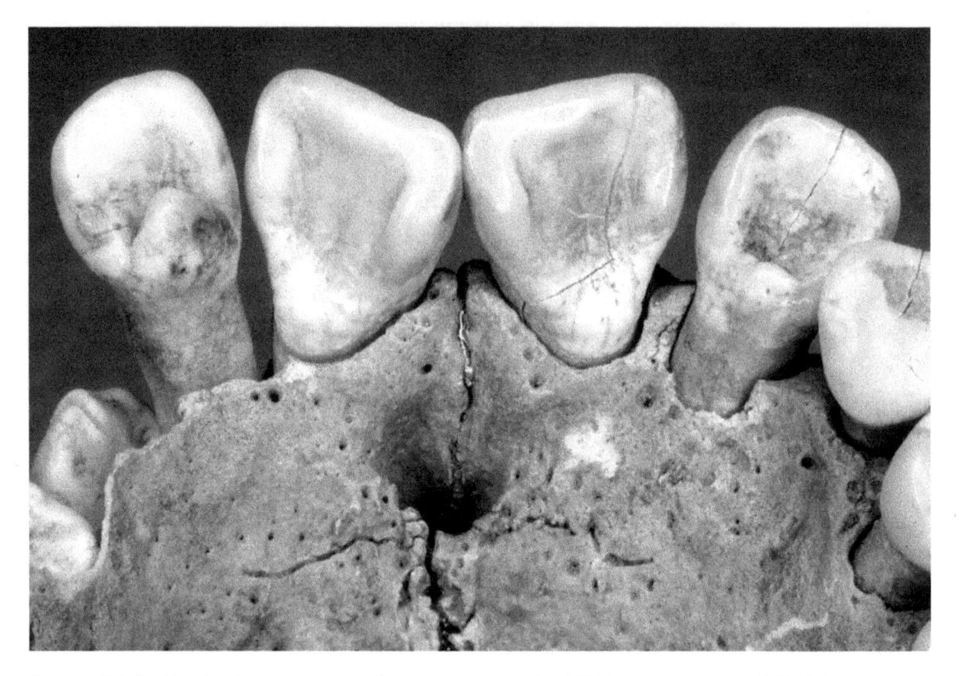

**Figure 23.5** Grade 4: moderate talon cusp on right UI2 but absent on left UI2.

**Figure 23.6** Grade 4: talon cusp on left UI2 but absent on right UI2 (asymmetry is the norm for talon cusps).

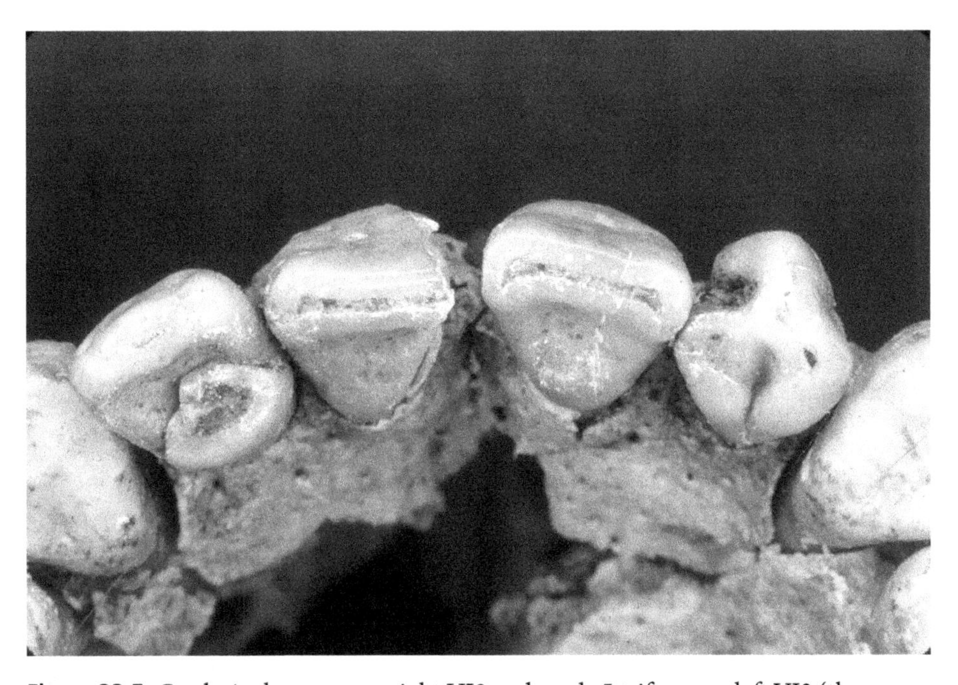

**Figure 23.7** Grade 4 talon cusp on right UI2 and grade 5 triform on left UI2 (these may be related forms).

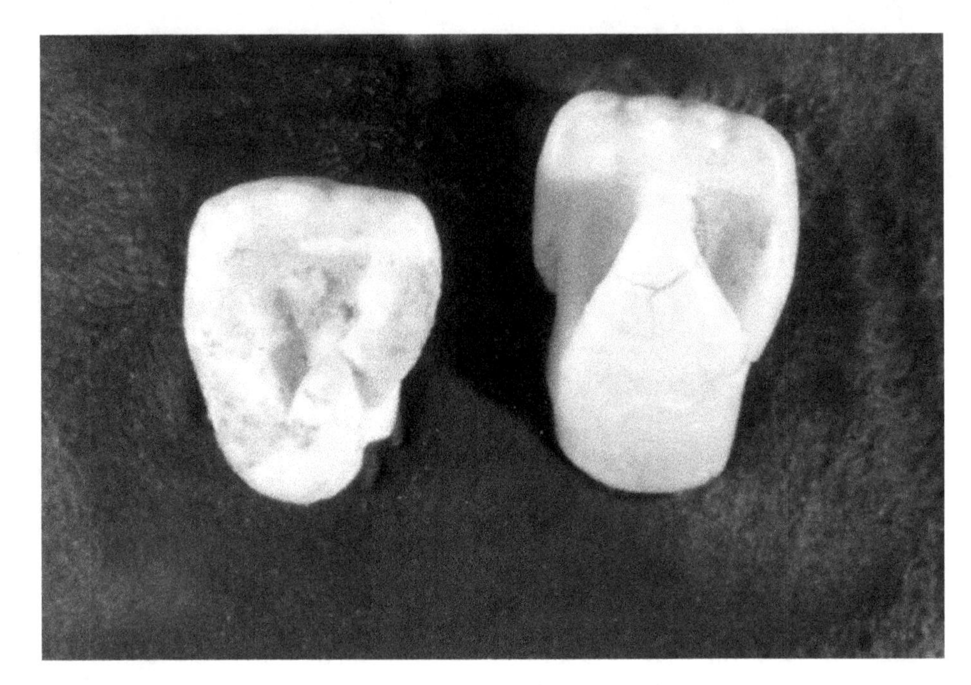

**Figure 23.8** Normal UI2 on left and grade 5 triform on right.

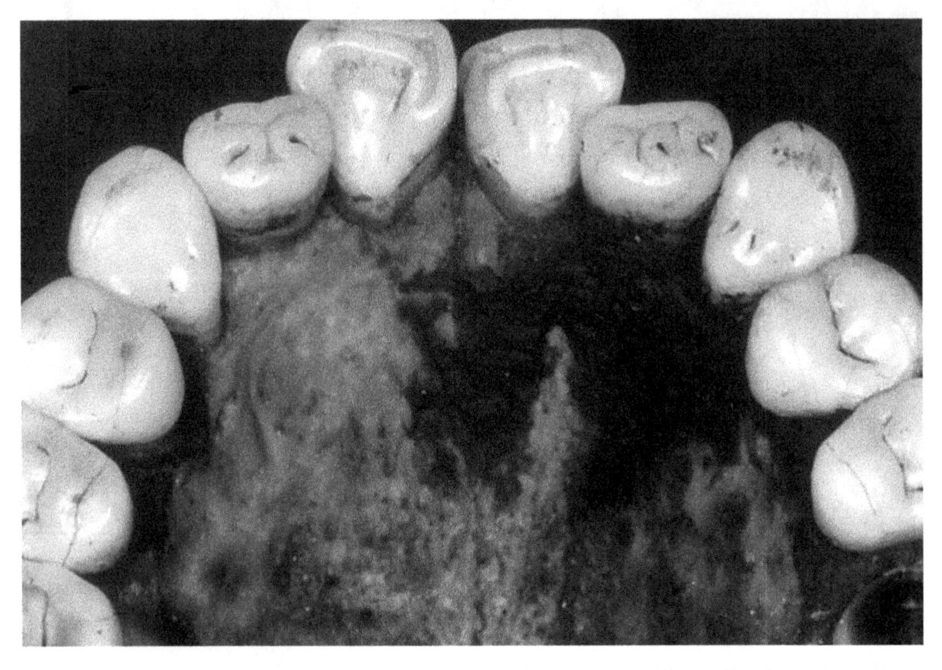

**Figure 23.9** Grade 6 forms for both right and left UI2; would be difficult to classify even without moderate wear.

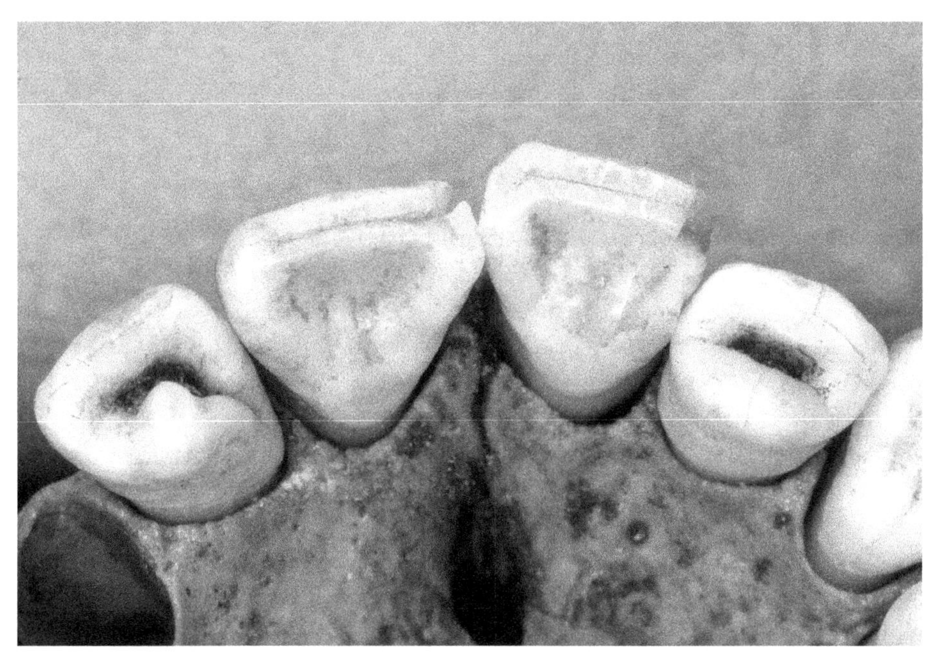

**Figure 23.10** Two grade 6 UI2 that defy classification and fall into the general category of miscellaneous unusual forms.

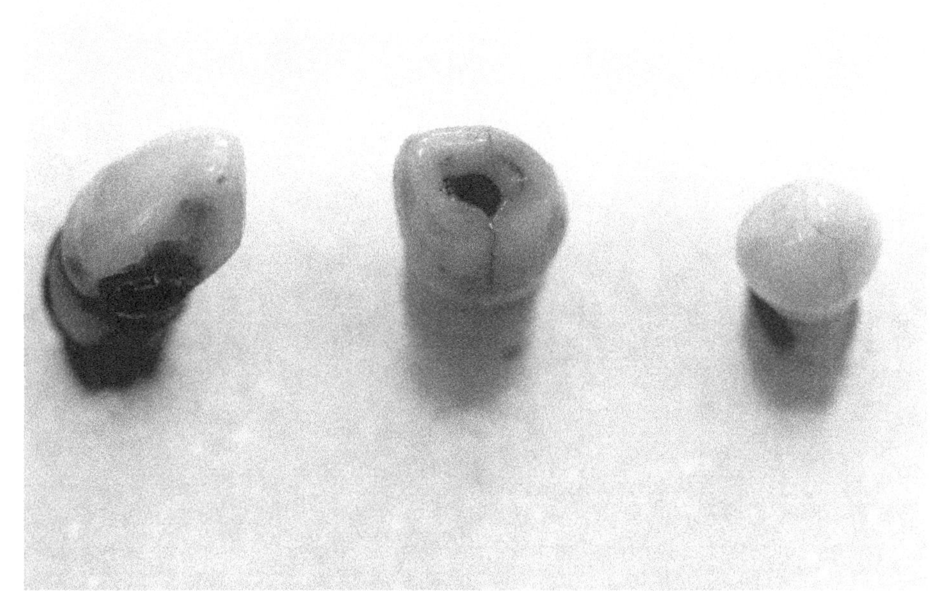

**Figure 23.11** Three UI2 variants. The middle example is a grade 7 barreled UI2 (see Trait 4) (or unclassifiable grade 6 form), while the tooth on the far right is grade 3 peg-shaped UI2 (courtesy of Christopher Stojanowski).

## Select Bibliography

Bailey-Schmidt, S.E. (1995). *Population Distribution of the Tuberculum Dentale Complex and Anomalies of the Maxillary Anterior Teeth.* MA thesis, Department of Anthropology, Arizona State University, Tempe.

Harris, E.F., and Owsley, D.W. (1991). Talon cusp: a review with three cases from native North America. *Journal of the Tennessee Dental Association* 71, 20–22.

Mayes, A.T. (2007). Labial talon cusp. *Journal of the American Dental Association* 138, 515–518.

Meon, R. (2009). Talon cusp in Malaysia. *Australian Dental Journal* 36, 11–14.

Meskin, L.H., and Gorlin, R.J. (1963). Agenesis and peg-shaped permanent maxillary incisors. *Journal of Dental Research* 42, 1476–1479.

Montagu, M.F.A. (1940). The significance of the variability of the upper lateral incisor teeth in man. *Human Biology* 12, 323–358.

Pinho, T., Tavares, P., Maciel, P., and Pollmann, C. (2005). Developmental absence of maxillary lateral incisors in the Portuguese population. *European Journal of Orthodontics* 27, 443–449.

Prabhu, R.V., Rao, P.K., Veena, K.M., *et al.* (2012). Prevalence of talon cusp in Indian population. *Journal of Clinical and Experimental Dentistry* 4, e23–e27.

Stojanowski, C.M., Johnson, K.M., Doran, G.H., and Ricklis, R.A. (2011). Talon cusp from two archaic period cemeteries in North America: implications for comparative evolutionary morphology. *American Journal of Physical Anthropology* 144, 411–420.

Symons, A.L., Stritzel, F., and Stamation, J. (1993). Anomalies associated with hypodontia of the permanent lateral incisor and second premolar. *Journal of Clinical Pediatric Dentistry* 17, 109–111.

# 24

# Pegged/Reduced/Missing Third Molars

## Observed on

UM3, LM3

## Key Tooth

UM3, LM3

## Synonyms

Third molar hypodontia, agenesis, congenital absence

## Description

Third molars, the most distal element of the molar field, are the most variable in terms of size, morphology, and number and are the least stable from the standpoint of evolution. Their position makes them the most likely molar to be reduced in size or lost. As loss and reduction are elements of the same phenomenon, they are grouped together to form a single trait that involves pegged or reduced forms plus congenital absence. Although the general approach adopted in this guidebook avoids observations on the crown or root traits of third molars, third molar reduction/hypodontia has a genetic component and shows patterned geographic variation.

## Classification

Grade 0: third molar present and normal
Grade 1: third molar significantly reduced in size (ca. ½ normal size, with two or more cusps)
Grade 2: third molar peg-shaped (only a single cusp evident)
Grade 3: third molar congenitally absent

## Breakpoint

Researchers are encouraged to note whether a third molar is reduced in size, peg-shaped, or congenitally absent. For population comparisons, the categories are often combined. The focus is on UM3, although that should not preclude observations on LM3.

## Potential Problems in Scoring

If the third molar is absent but there is a visible socket, one should be cautious in scoring pegged/reduced/missing (PRM) third molars. If the root socket is consistent with, albeit slightly smaller than, the sockets of M1 and M2, one could infer that the M3 was not pegged or reduced. Agenesis can usually be scored with a high degree of accuracy even in the absence of x-rays. Only adults in their 20s or older should be scored for agenesis, as M3s are variable for eruption times. However, only in unusual cases would you have perfectly sound first and second molars and a third molar that was lost for some reason. If there are second molars, they should be carefully inspected for interproximal wear on the distal surface of the crown. Pegged and reduced third molars are readily discerned, as they are less than half the size of a normal third molar.

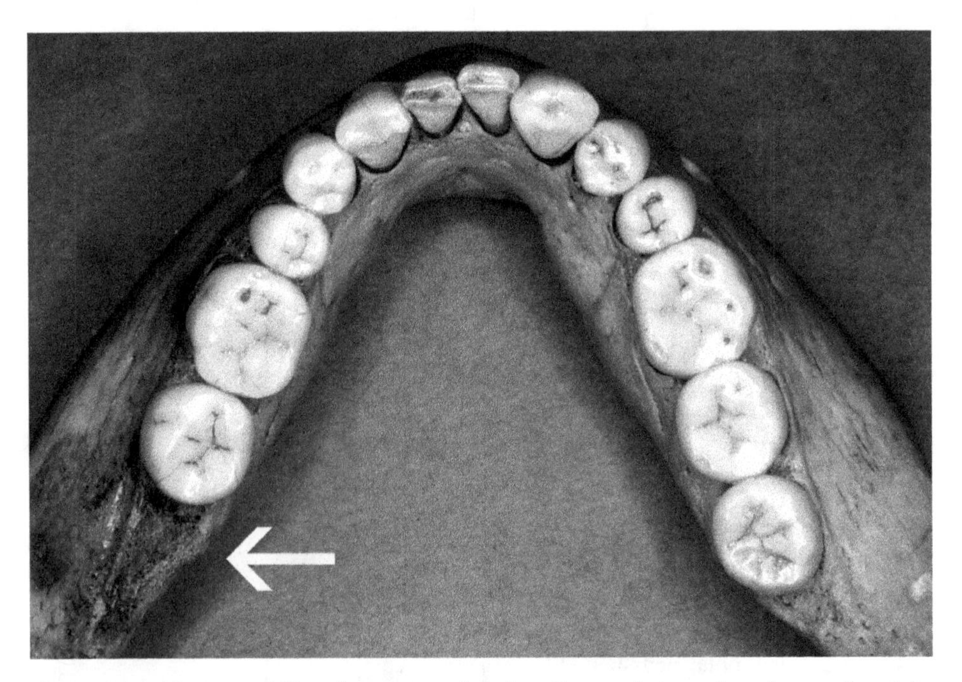

**Figure 24.1** This mandible of a young adult has first and second molars and a right LM3. The likelihood is that the left LM3 is congenitally absent.

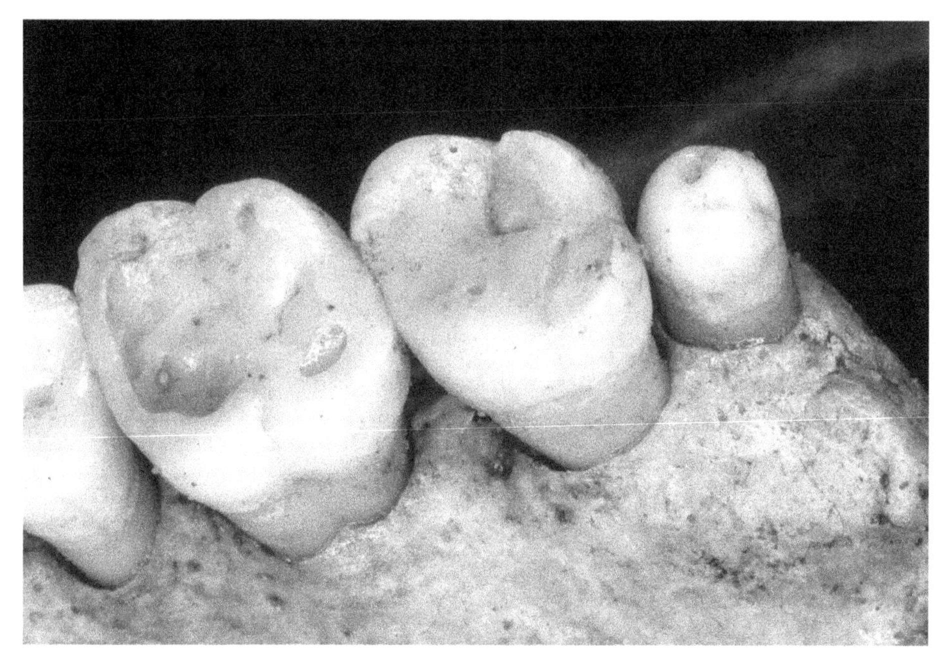

**Figure 24.2** UM3 assumes a peg-shaped form.

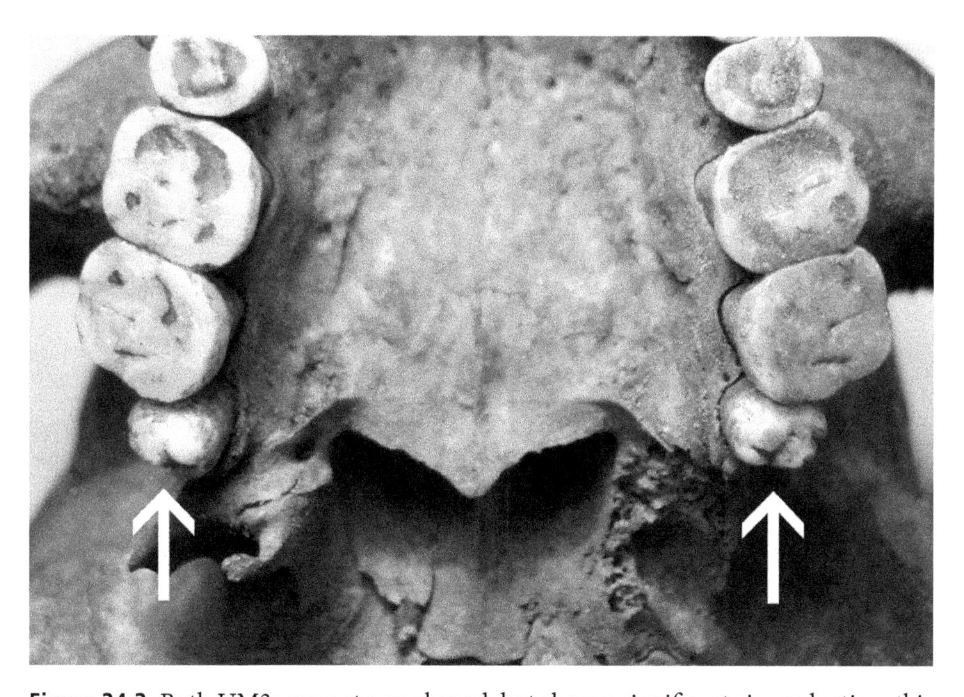

**Figure 24.3** Both UM3s are not peg-shaped, but show a significant size reduction; this expression would be scored as reduced.

## Geographic Variation

A general misconception about third molar agenesis is that it is most common in European populations. PRM UM3 is about twice as common in East Asians (25%) as in Western Eurasians (12%). African and Australian populations show the lowest frequencies of third molar reduction and/or loss at around 6%.

## Select Bibliography

Bermúdez de Castro, J.M. (1989). Third molar agenesis in human prehistoric populations of the Canary Islands. *American Journal of Physical Anthropology* 79, 207–215.

Brook, A.H. (1984). A unifying aetiological explanation for anomalies of human tooth number and size. *Archives of Oral Biology* 29, 373–378.

Chung, C.J., Han, J.-H., and Kim, K-H. (2008). The pattern and prevalence of hypodontia in Koreans. *Oral Diseases* 14, 620–625.

Garn, S.M., Lewis, A.B., and Vicinus, J.H. (1962). Third molar agenesis and reduction in the number of other teeth. *Journal of Dental Research* 41, 717.

(1963). Third molar polymorphism and its significance to dental genetics. *Journal of Dental Research* 42 (suppl. to no. 6), 1344–1363.

Irish, J.D. (1997). Characteristic high- and low-frequency dental traits in Sub-Saharan African populations. *American Journal of Physical Anthropology* 102, 455–467.

Keene, H.J. (1965). The relationship between third molar agenesis and the morphologic variability of the molar teeth. *Angle Orthodontist* 35, 289–298.

(1968). The relationship between Carabelli's trait and the size, number and morphology of the maxillary molars. *Archives of Oral Biology* 13, 1023–1025.

Mattheeuws, N., Dermaut, L., and Martens, G. (2004). Has hypodontia increased in Caucasians during the 20th century? A meta-analysis. *European Journal of Orthodontics* 26, 99–103.

Nanda, R.S. (1954). Agenesis of the third molar in man. *American Journal of Orthodontics* 40, 698–706.

Nieminen, P. (2009). Genetic basis of tooth agenesis. *Journal of Experimental Zoology (Molecular and Developmental Evolution)* 312B, 320–342.

Parkin, N., Elcock, C., Smith, R.N., Griffin, R.C., and Brook, A.H. (2009). The aetiology of hypodontia: the prevalence, severity and location of hypodontia within families. *Archives of Oral Biology* 54s, s52–s56.

Rolling, S., and Poulsen, S. (2009). Agenesis of permanent teeth in 8138 Danish schoolchildren: prevalence and intra-oral distribution according to gender. *International Journal of Paediatric Dentistry* 19, 172–175.

# 25

# Premolar Odontomes

## Observed on

UP1, UP2, LP1, LP2

## Key Tooth

UP1, UP2, LP1, LP2

## Synonyms

Tuberculated premolars, occlusal tubercles, *dens evaginatus*, interstitial cusp, Leong's premolar, odontome of the axial core type, occlusal enamel pearl, cone-shaped supernumerary cusp, evaginated odontome, composite dilated odontome (Merrill 1964), supernumerary coronal structure (Carlsen 1987)

## Description

The upper and lower premolars can both express odontomes in the central sulcus. Typically, odontomes are conical in shape. Given their location on the premolar crown, it is surprising these tubercles do not break off quickly. However, this does not seem to be the case. They do wear, and even show dentine involvement in their formation. Although the trait is too rare to be easily assessed for inter-trait correlation, in populations where the trait is fairly common (e.g., Eskimos), odontomes can be evident on both the upper and lower premolars of the same individual.

## Classification

Grade 0: absence
Grade 1: odontome present in central sulcus (score all eight premolars)

## Breakpoint

Presence/absence. This is one of Turner's key 29 traits. An individual is scored as positive for odontome presence when the trait is visible on any upper or lower premolar.

## Potential Problems in Scoring

Odontomes are readily scored in unworn or slightly worn dentitions. On rare occasions, there are small occlusal tubercles that may represent a different trait or possibly a threshold expression of the odontome.

## Geographic Variation

Odontomes are rare in Sub-Saharan Africans and Western Eurasians. Although rare, they are most common in Asian and Asian-derived populations. The highest frequencies of odontomes occur in Eskimos; in St. Lawrence Island (Alaska) Eskimos, they reached a frequency of 17% (Scott and Gillispie 2002). Most populations have frequencies under 5%.

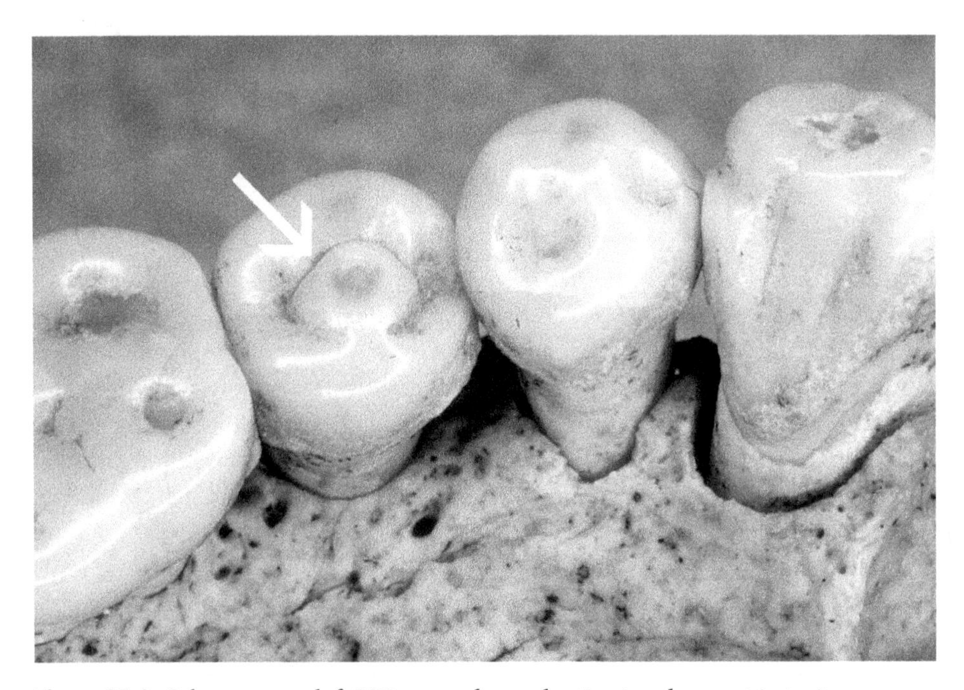

**Figure 25.1** Odontome on left LP2; wear shows dentine involvement in trait formation.

**Figure 25.2** Odontome on right LP2; worn but not broken.

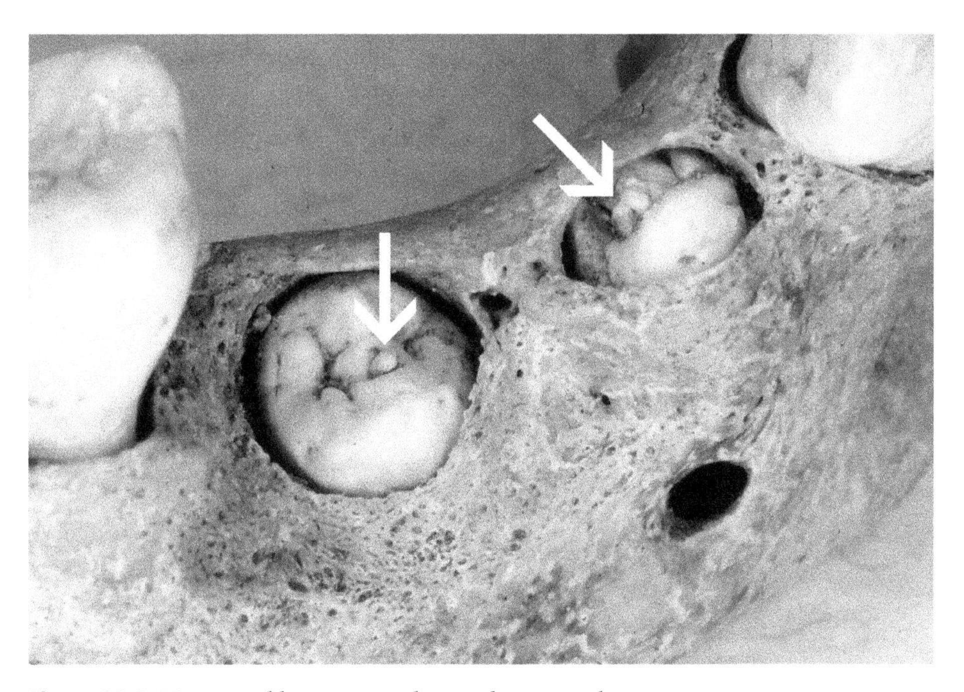

**Figure 25.3** Unerupted lower premolars with intact odontomes.

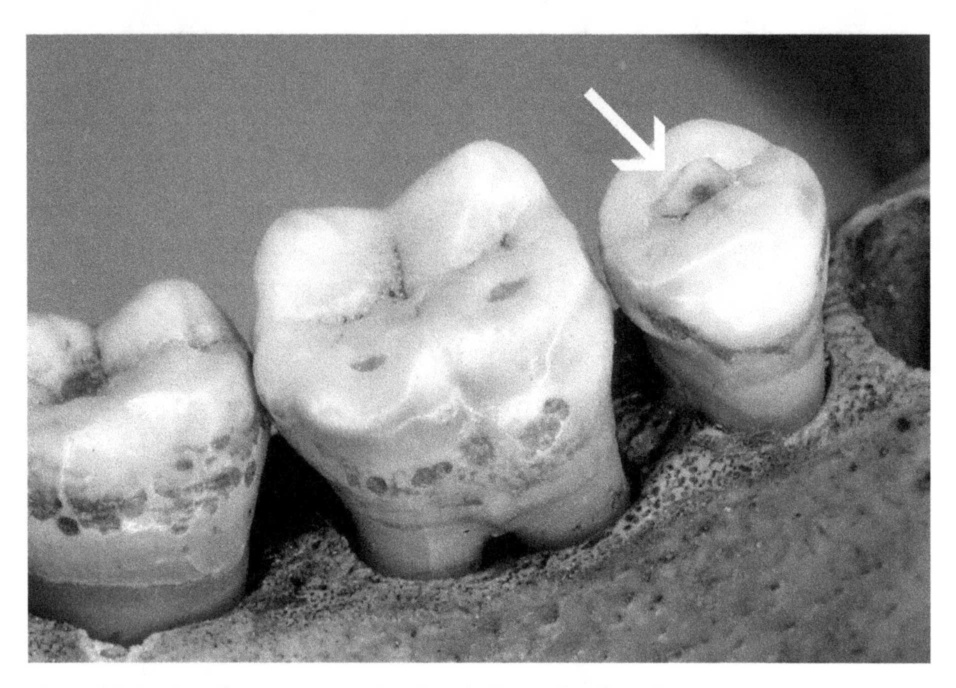

**Figure 25.4** LP2 odontome worn level with the occlusal surface.

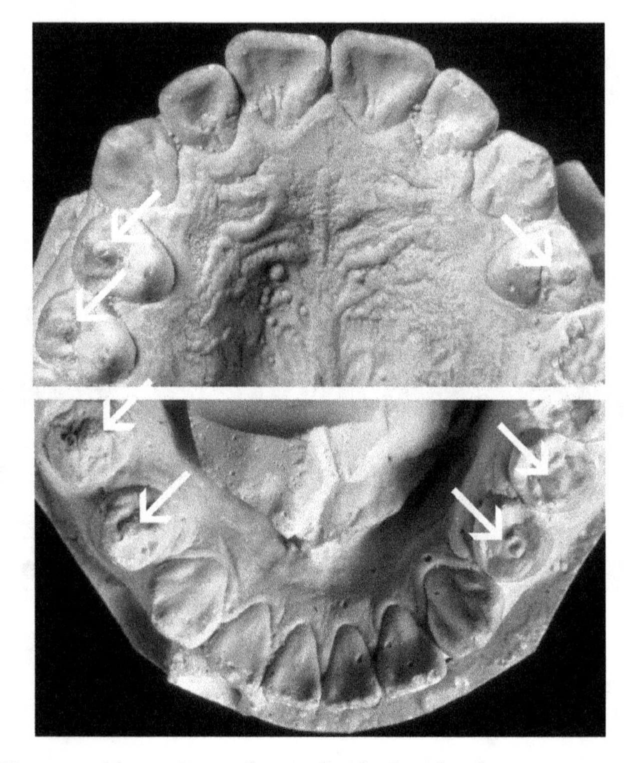

**Figure 25.5** Upper and lower jaws of an individual with odontomes on at least three upper and three lower premolars (supporting the idea that the odontome "field" extends across jaws).

## Select Bibliography



Carlsen, O. (1987). *Dental Morphology*. Copenhagen: Munksgaard.

Curzon, M.E.J., Curzon, J.A., and Poyton, H.G. (1970). Evaginated odontomes in the Keewatin Eskimo. *British Dental Journal* 129, 324–328.

Douglas, P., Sloan, P., and G.V. Gillbe. (1991). A developing complex odontome associated with delayed premolar formation. *British Journal of Oral and Maxillofacial Surgery* 29, 61–63.

Mayhall, J.T. (1979). The dental morphology of the Inuit of the Canadian central Arctic. *OSSA* 6, 199–218.

Merrill, R.G. (1964). Occlusal anomalous tubercles on premolars of Alaskan Eskimos and Indians. *Oral Surgery, Oral Medicine, Oral Pathology* 17, 484–496.

Palmer, M.E. (1973). Case reports of evaginated premolars in Caucasians. *Oral Surgery, Oral Medicine, Oral Pathology* 35, 772–779.

Ponnambalam, Y., and R. M. Love. (2006). Dens evaginatus: case reports and review of the literature. *New Zealand Dental Journal* 102, 30–34.

Rao, Y., Gu, L., and Hu, T. (2010). Multiple den evaginatus of premolars and molars in Chinese dentition: a case report and literature review. *International Journal of Oral Science* 2, 177–180.

Scott, G.R., and Gillispie, T.E. (2002). The dentition of prehistoric St. Lawrence Island Eskimos: variation, health, and behavior. *Anthropological Papers of the University of Alaska* 2, 50–72.

Yip, W.-K. (1974). The presence of dens evaginatus. *Oral Surgery, Oral Medicine, Oral Pathology* 38, 80–87.

# NOTES

# 26

# Midline Diastema

## Observed on

UI1

## Key Tooth

UI1

## Synonyms

Median diastema, "gap teeth," MLD

## Description

In addition to the standard ASUDAS traits, the occurrence of the UI1 midline diastema (MLD) is a useful trait that varies in frequency among world populations. Previous research by the second author (JDI) suggests that a presence/absence level of dichotomization is sufficient to identify and record this feature. To determine presence, a measurement with needle-point calipers should be taken of the space, if any, between the upper central incisors midway between the base (or neck) of the tooth and the (unworn) incisal edge. Given the location for measurement, the crown can be worn down approximately ¼ of its original (estimated) height and still allow recording (Figure 26.1).

## Classification

Lavelle (1970)

Grade 0: no diastema (space < 0.5 mm)
Grade 1: diastema (space > 0.5 mm)

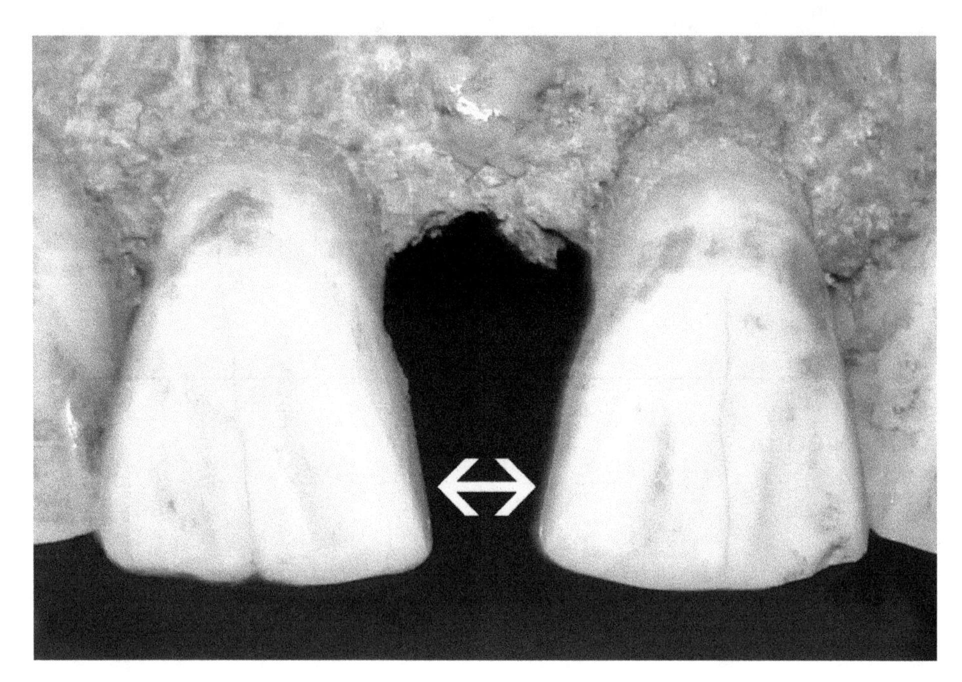

**Figure 26.1** Close up of midline diastema in a "Bantu."

## Breakpoint

Trait presence

## Potential Problems in Scoring

The MLD may be scored with slight to moderate incisal wear. However, it should not be scored when there is more severe wear. In instances of UI1 malocclusion, e.g., asymmetrical rotation, a between-UI1 gap may be present, but it should not be scored as an MLD unless it is obvious. Moreover, the trait should not be scored if the lateral incisors are reduced, peg-shaped, or congenitally absent. In such cases, the expression of the MLD could be related to "extra room" from the reduced or missing UI2s. Lastly, MLD should only be recorded in individuals with a fully erupted set of permanent teeth (or when only the third molars are unerupted). An MLD may be temporarily present as part of normal development in young individuals with a mixed dentition.

## Geographic Variation

The MLD can be found in individuals from all populations, and is often seen as an esthetic nuisance to be addressed via clinical intervention. Indeed, the first author (GRS) formerly had an exceptional space between his UI1s that, regrettably, has

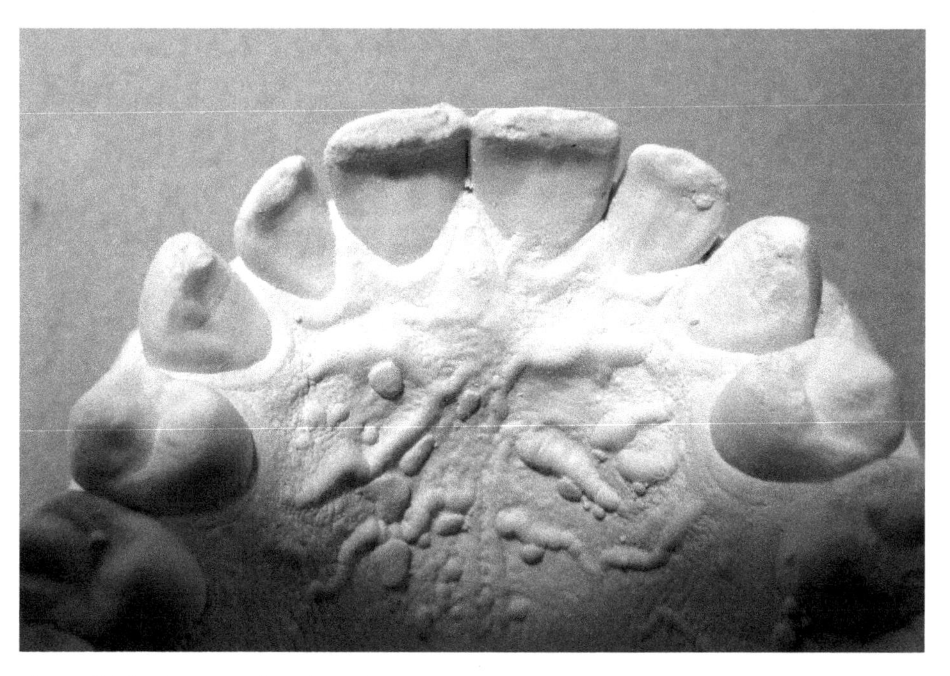

**Figure 26.2** Absence of midline diastema.

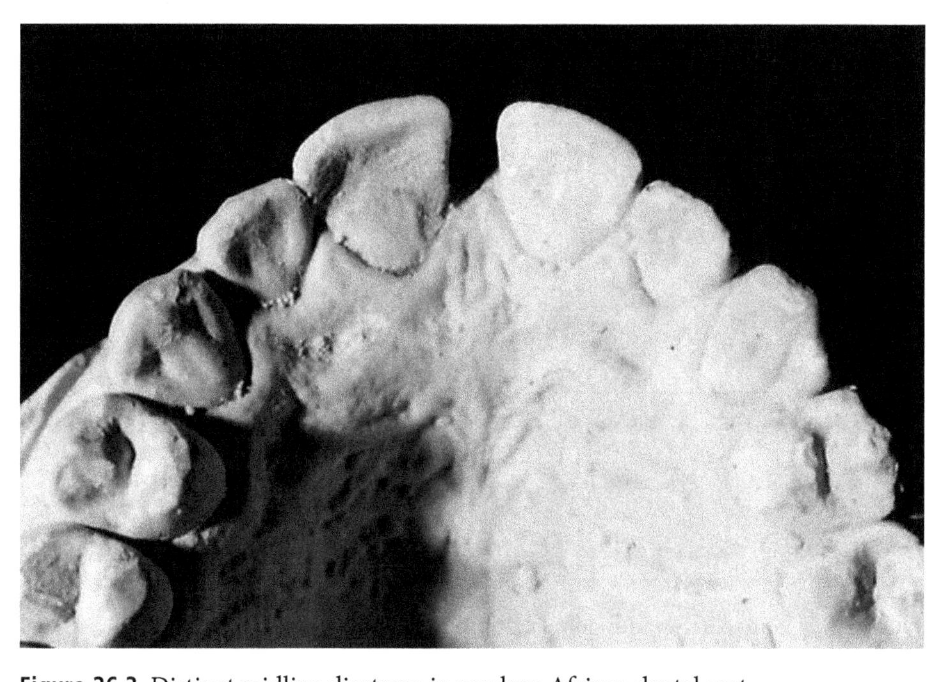

**Figure 26.3** Distinct midline diastema in modern African dental cast.

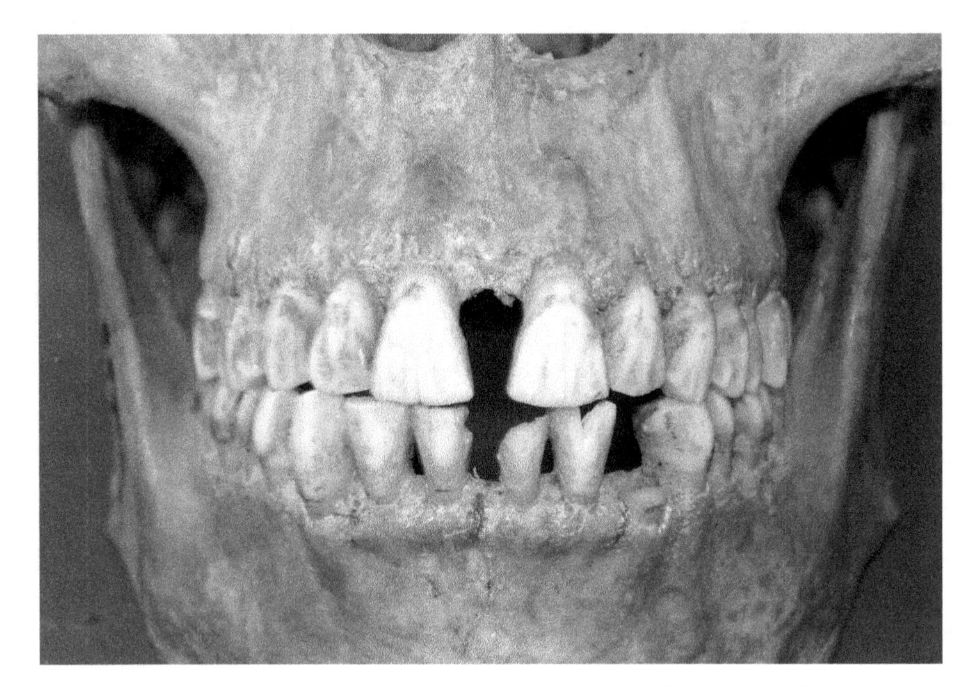

**Figure 26.4** "Typical" midline diastema in historic South African "Bantu."

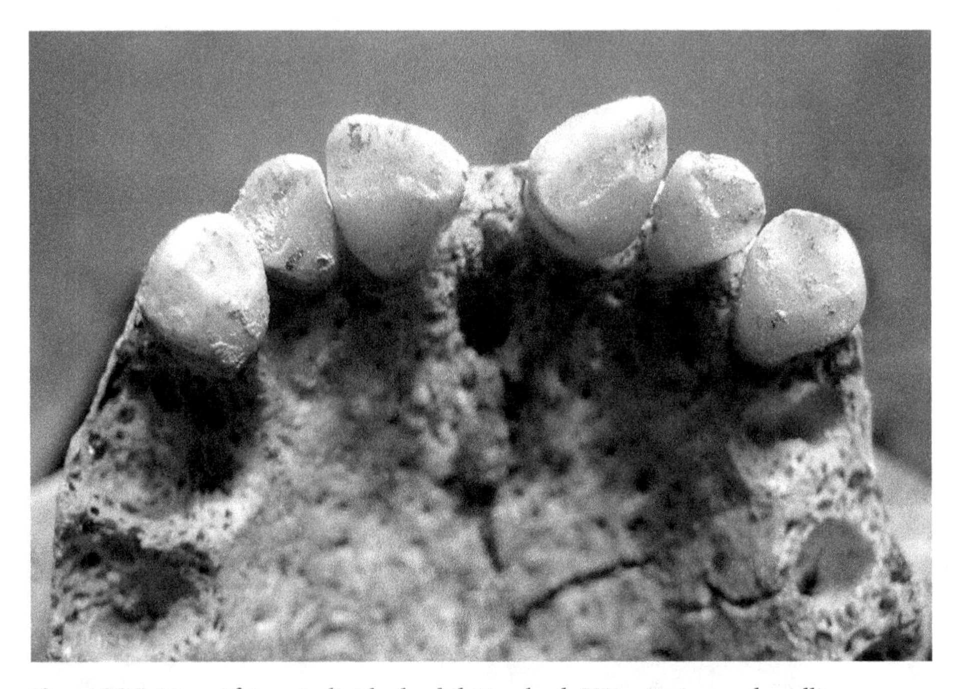

**Figure 26.5** Non-African individual exhibiting both UI1 winging and midline diastema.

since been "closed." Causes may be multiple, including reduced adjacent teeth (above), the presence of a mesiodens (within the alveolus between the UI1s), and unusual oral habits (e.g., thumb sucking), among others. The identification of an MLD as an actual epigenetic variant related to among-population variation can be discerned by its presence in a dentition that otherwise exhibits no signs of the above exceptions.

The MLD has been reported to occur in high frequencies in many Sub-Saharan African groups. For example, studies from the mid-twentieth century found MLD to be: (1) "common" in a sample of Babinga Pygmy; (2) present in 11% of southern African San; and (3) present in 35% of 206 eastern African Masai. The second author (JDI), in a more comprehensive African study, found the presence of MLD to range from 2.8% to 44% in 14 Sub-Saharan samples (mean 12.7%), but 0.0–14.3% in 12 North African samples (mean 6.1%). Elsewhere in the world, the occurrence of this trait has not been routinely recorded in anthropological studies, yet anecdotally it appears to be much less common outside Africa. Thus, the feature may prove useful as an African marker.

## Select Bibliography


Deverall, A. (1949). Kanye nutrition experiment report on dental survey. In *The Feeding and Health of African School Children*, ed. B.T. Squires. Cape Town: University of Cape Town. Anexure I.

Hassanali, J. (1982). Incidence of Carabelli's trait in Kenyan Africans and Asians. *American Journal of Physical Anthropology* 59, 317–319.

Huang, W.-J., and Creath, C.J. (1995). The midline diastema: a review of its etiology and treatment. *Pediatric Dentistry* 17, 171–179.

Irish, J.D. (1993). *Biological Affinities of Late Pleistocene through Modern African Aboriginal Populations: The Dental Evidence*. PhD dissertation, Department of Anthropology, Arizona State University, Tempe.

Jacobson, A. (1982). *The Dentition of the South African Negro*. Anniston, AL: Higgenbotham, Inc.

Keene, H.J. (1963). Distribution of diastemas in the dentition of man. *American Journal of Physical Anthropology* 21, 437–441.

Kumar, S., and Gandotra, D. (2013). An aesthetic and rapid approach to treat midline diastema. *Journal of Cranio-Maxillary Diseases* 2, 175–178.

Lavelle, C.L.B. (1970). The distribution of diastemas in different human population samples *European Journal of Oral Sciences* 78, 530–534.

Pales, L. (1938). Contribution a l'étude anthropologique des Babinga de l'Afrique équatoriale française. *L'Anthropologie* 48, 503–520.

Shaw, J.C.M. (1931). *The Teeth, the Bony Palate and the Mandible in Bantu Races of South Africa*. London: John Bale, Sons & Danielson.

Sperber, G.H. (1958). The palate and dental arcade of the Transvaal Bushman, the Auni-Khomani Bushman and the Bantu speaking Negroes of the Zulu Tribe. *South African Journal of Medical Science* 23, 147–154.

Tanaka, O.M., Morino, A.Y.K., Machuca, O.F., and Schneider, N.A. (2015). When the midline diastema is not characteristic of the "ugly duckling" stage. *Case Reports in Dentistry*, 2015, 924743.

Van Reenen, J.F. (1964). Dentition, jaws and palate of the Kalahari Bushmen. *Journal of the Dental Association of South Africa* 19, 1–15.

# 27

# Lower Premolar Cusp Number

## Observed on

LP1, LP2

## Key Tooth

LP2

## Synonyms

Number of lingual cusps (Kraus and Furr 1953, Ludwig 1957), multiple lingual cusps (Scott 1973)

## Description

Multiple lingual cusps take different forms on LP1 and LP2 and have separate classifications. Rarely, a lower premolar exhibits no lingual cusp with a free apex, but this phenotype is not part of the ASUDAS classifications. These expanded classifications include absence and nine degrees of trait presence. In contrast to the upper premolars, the lower premolars have lingual cusps that are always smaller than the buccal cusps. Typically, the lingual cusp of the lower premolars has a mesial placement relative to the buccal cusp. When there are accessory cusps, they are usually smaller and distal to the larger mesial lingual cusp.

## Classification

Scott (1973)

The original classifications were established during an early phase of ASUDAS (Scott 1973), and the expanded scales proved to be too complex. Turner simplified the scales to 0, 1, 2, and 3 defined on the basis of actual lingual cusp number. As world data are reported in those terms, the ASUDAS classifications are here reduced to this system.

## LP1

Grade 0: lingual cusp has no free apex
Grade 1: single lingual cusp (on plaque, grades 0–1)
Grade 2: two lingual cusps (on plaque, grades 2–7)
Grade 3: three lingual cusps (on plaque, grades 8–9)

## LP2

Grade 0: lingual cusp has no free apex
Grade 1: single lingual cusp (on plaque, grades 0–1)
Grade 2: two lingual cusps (on plaque, grades 2–7)
Grade 3: three lingual cusps (on plaque, grades 8–9)

## Breakpoint

Grades 2–9 on plaques. This would include all premolars with two or three lingual cusps.

## Potential Problems in Scoring

With the simplified scale, scoring multiple lingual cusps is relatively straightforward. The classic expression of multiple lingual cusps falls between grades 2 and 5. When there are more than two lingual cusps, placement can be variable.

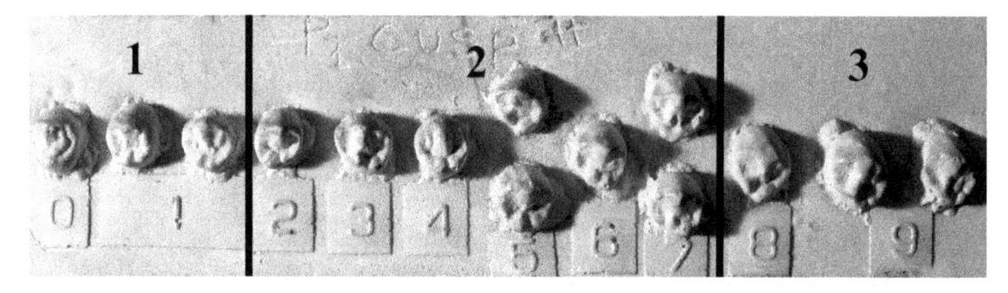


**Figure 27.1** Standard plaque for LP1 lingual cusp number.

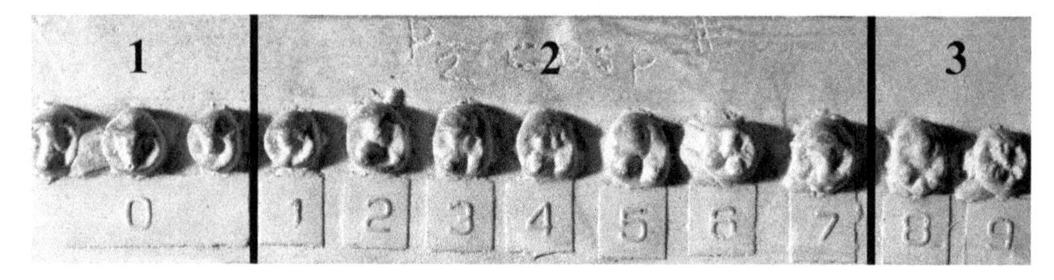


**Figure 27.2** Standard plaque for LP2 lingual cusp number.

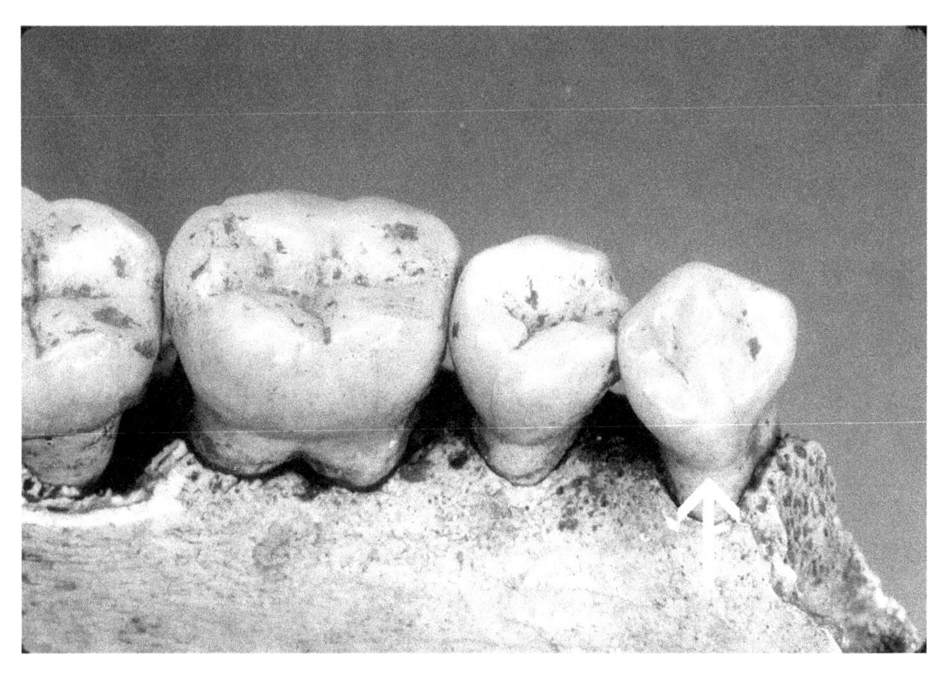

**Figure 27.3** The lingual cusp of LP1 blends directly into the essential ridge of the buccal cusp with no free apex; in this instance, lingual cusp number would be 0.

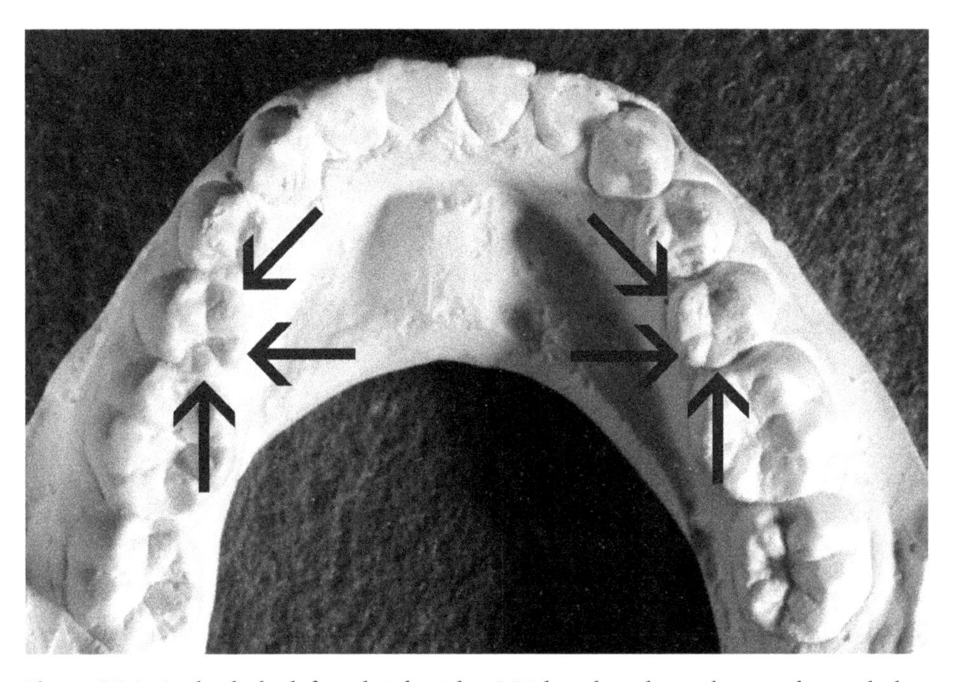

**Figure 27.4** On both the left and right sides, LP2 has three lingual cusps that include one large mesial cusp and two smaller distal cusps.

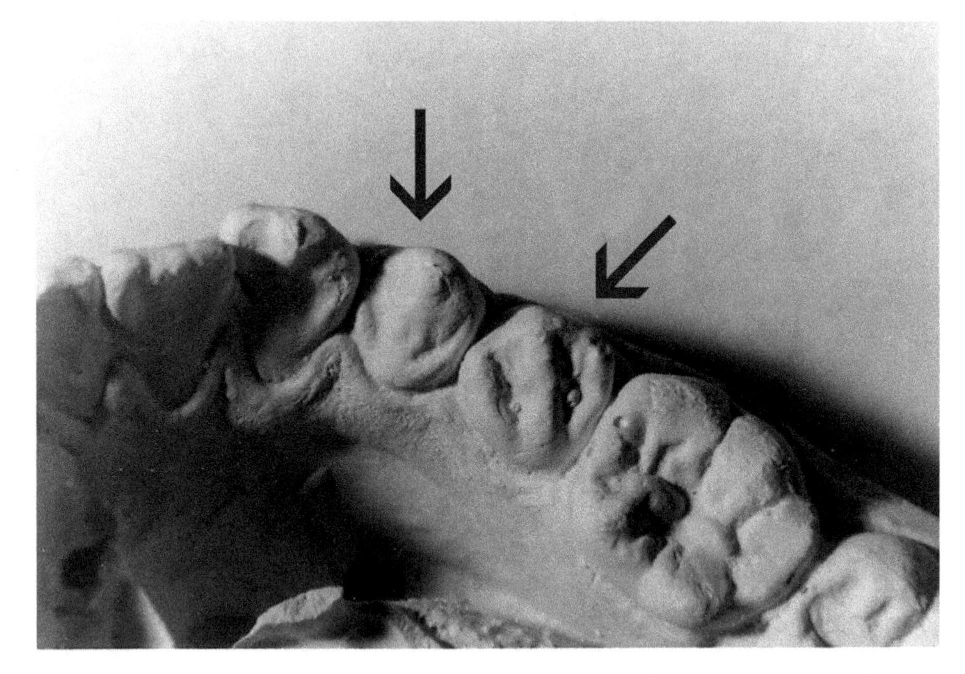

**Figure 27.5** This is a relatively rare phenotype where LP1 shows almost no evidence of a lingual cusp (hence grade 0) and the same applies to LP2, where the single lingual cusp has no free apex and blends into the essential ridge of the buccal cusp.

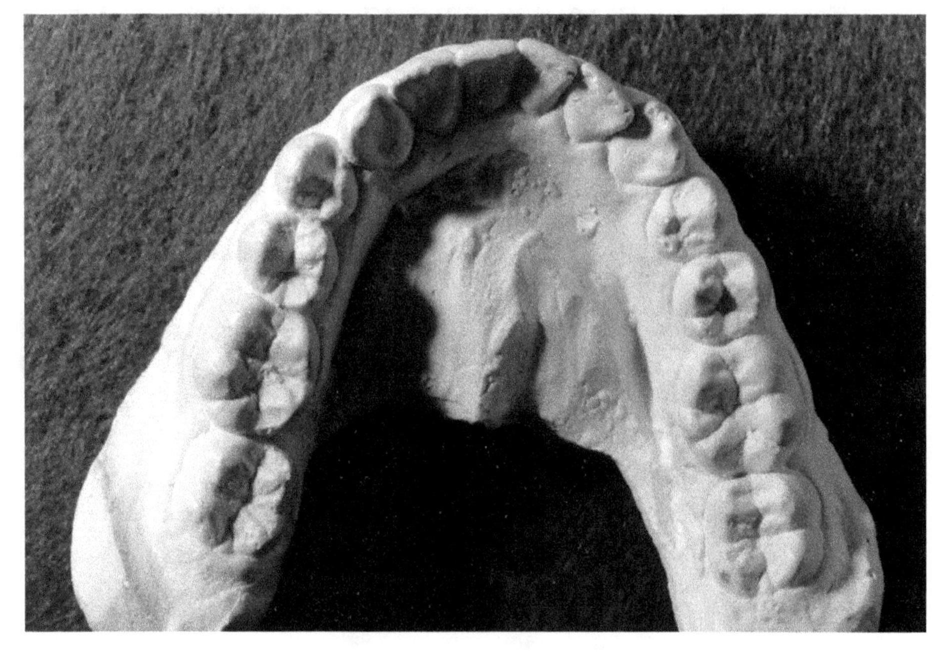

**Figure 27.6** For multiple lingual cusps, the mesial cusp is typically larger than the distal cusp; this is evident on both LP1 and LP2.

## Geographic Variation

Multiple lingual cusps are least common in Asian and Asian-derived populations (30–40%). They are most common in Australians and Africans (70–90%). Western Eurasians have intermediate frequencies (50–60%).

## Select Bibliography


Carlsen, O., and Alexandersen, V. (1994). Mandibular premolar differentiation. *Scandinavian Journal of Dental Research* 102, 81–87.

Hrdlička, A. (1921). Further studies of tooth morphology. *American Journal of Physical Anthropology* 4, 141–176.

Kraus, B.S., and Furr, M.L. (1953). Lower first premolars. Part I. A definition and classification of discrete morphologic traits. *Journal of Dental Research* 32, 554–564.

Ludwig, F.J. (1957). The mandibular second premolars: morphologic variation and inheritance. *Journal of Dental Research* 36, 263–273.

Lunt, D.A. (1976). Molarization of the mandibular second premolars. *Journal of Dentistry* 4, 83–86.

Martinón-Torres, M., Bastir, M., Bermúdez de Castro, J.M., et al. (2006). Hominin lower second premolar morphology: evolutionary inferences through geometric morphometric analysis. *Journal of Human Evolution* 50, 523–533.

Nagai, A., and Kanazawa, E. (1998). Morphological variations of the lower premolars in Asian and Pacific populations. In *Proceedings of the 11th International Symposium on Dental Morphology*, ed. J.T. Mayhall and T. Heikkinen. Oulu University Press, pp. 157–166.

Reenen, F.V., Reid, C., and Butler, P.M. (1998). Morphological variations of the lower premolars in Asian and Pacific populations. In *Proceedings of the 11th International Symposium on Dental Morphology*, ed. J.T. Mayhall and T. Heikkinen. Oulu: Oulu University Press, pp. 192–205.

Reid, C., and Reenan, F.V. (1998). Morphological variations of the lower premolars in Asian and Pacific populations. In *Proceedings of the 11th International Symposium on Dental Morphology*, ed. J.T. Mayhall and T. Heikkinen. Oulu: Oulu University Press, pp. 85–91.

Schroer, K., and Wood, B. (2015). The role of character displacement in the molarization of hominin mandibular premolars. *Evolution* 69, 1630–1642.

Scott, G.R. (1973). *Dental Morphology: A Genetic Study of American White Families and Variation in Living Southwest Indians.* PhD dissertation, Department of Anthropology, Arizona State University, Tempe.

Suwa, G. (1988). Evolution of the "robust" australopithecines in the Omo succession: evidence from mandibular premolar morphology. In *Evolutionary History of the "Robust" Australopithecines*, ed. F.E. Grine. New York: Aldine de Gruyter, pp. 199–222.

Wood, B.F., and Green, L.J. (1969). Second premolar morphologic trait similarities in twins. *Journal of Dental Research* 48, 74–87.

Wood, B.A., and Uytterschaut, H. (1987). Analysis of the dental morphology of Plio-Pleistocene hominids. III Mandibular premolar crowns. *Journal of Anatomy*, 154, 121–156.

# NOTES

# 28

# Anterior Fovea

## Observed on

LM1, LM2, LM3

## Key Tooth

LM1 or LM2

## Synonyms

Precuspidal fossa (Hrdlička 1924)

## Description

The anterior fovea is a polymorphic trait expressed on the mesial aspect of the trigonid of the lower molars in hominoids and hominins. It involves three primary elements: distinct essential ridges on the protoconid and metaconid that meet close to the center of the trigonid, and a mesial marginal ridge that is expressed to varying degrees. The conjoining of these three features produces a fovea, or depression, on the mesial section of the trigonid.

## Classification

Turner *et al.* (1991)

The fovea is graded relative to the size and configuration of ridges on the protoconid and metaconid in concert with a mesial marginal ridge. Hrdlička (1924) included a ranked classification that involved: absence, plain trace, small but well formed, and large fossa, well formed. These categories correspond closely to the ranks of the ASUDAS classification.

Grade 0: absence
Grade 1: trace, with slight development of mesial marginal ridge
Grade 2: essential ridges on trigonid better developed, as is marginal ridge

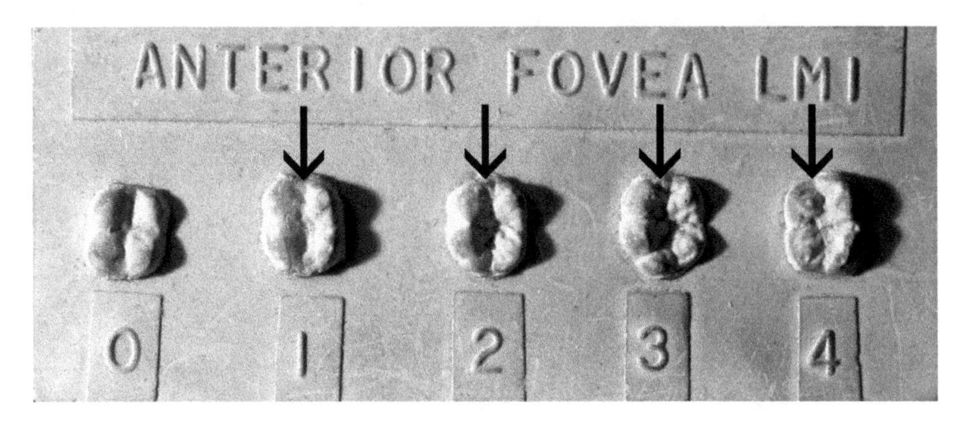


**Figure 28.1** Standard plaque for lower molar anterior fovea.

Grade 3: essential ridges pronounced and marginal ridge well developed, producing a distinctive fovea on the anterior portion of the trigonid

Grade 4: pronounced essential ridges and marginal ridge produce a well-defined fovea

## Breakpoint

Because wear can make the anterior fovea difficult to score, grades 3–4 should constitute the breakpoint, as they are associated with a large, well-formed fossa.

## Potential Problems in Scoring

Scoring the anterior fovea on LM1 is a problem in modern adults because of wear and caries. The trait can be scored on all three lower molars, but identifying the key tooth for the anterior fovea is still problematic. This trait appears to be more common in primates and earlier hominins (Hrdlička 1924); its utility in the study of modern human variation is still in question.

## Geographic Variation

The anterior fovea was not one of Turner's key 29 traits, so data for this variable have not been tabulated on a world scale. From Hrdlička's (1924) early observations, the key tooth is highly variable in five modern human groups. In some cases, it is most common on LM1, in other cases LM2 or LM3. There is apparently some variation in this trait, as the frequencies for the three lower molars are higher in Europeans and African-Americans than in Eskimos or Native Americans. Melanesians exhibit the lowest frequency.

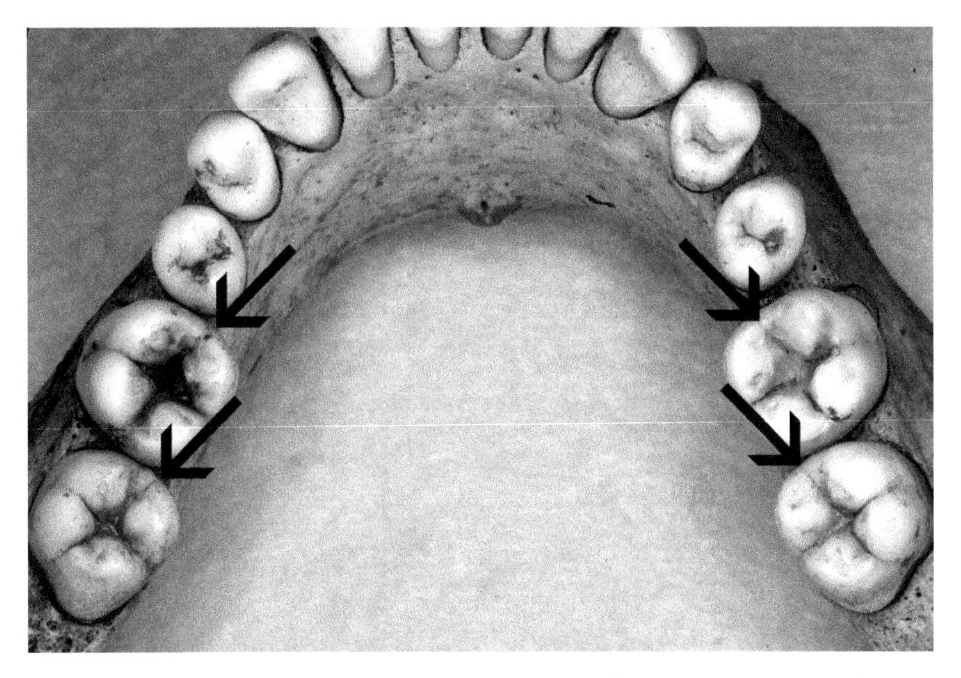

**Figure 28.2** Simplified LM1 and LM2 with no anterior fovea; there is no mesial marginal ridge and the groove separating the protoconid and metaconid extends to the mesial margin.

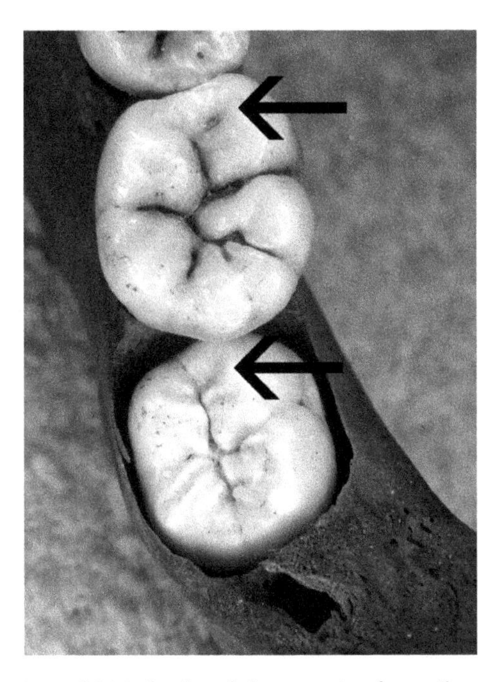

**Figure 28.3** Right LM1 and LM2 both exhibit anterior fovea (between grades 2 and 3).

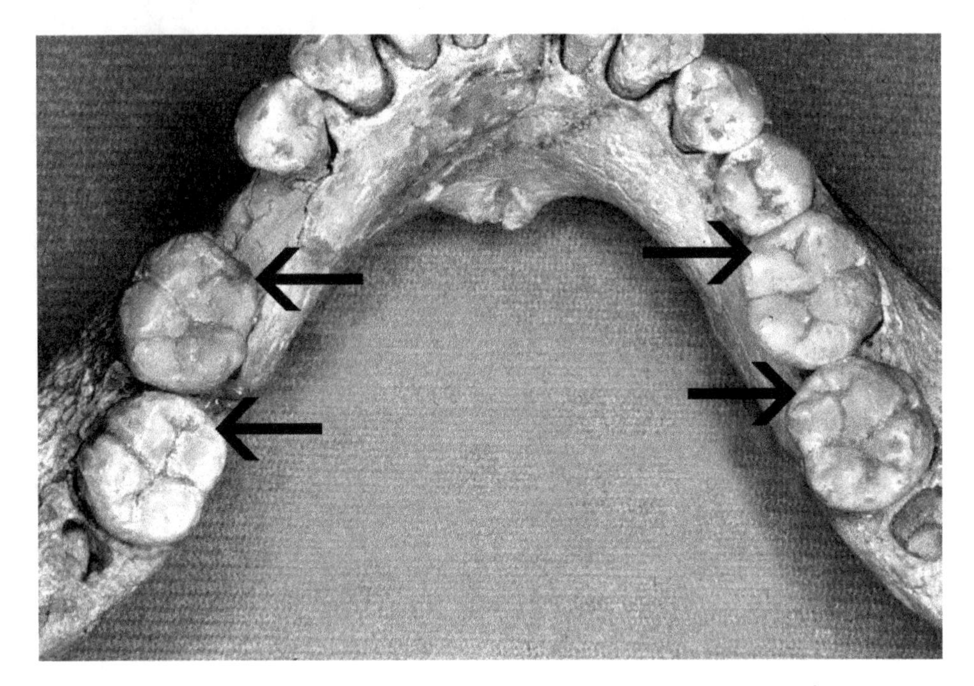

**Figure 28.4** Young individual (LM3 unerupted) exhibiting distinct grade 4 anterior fovea.

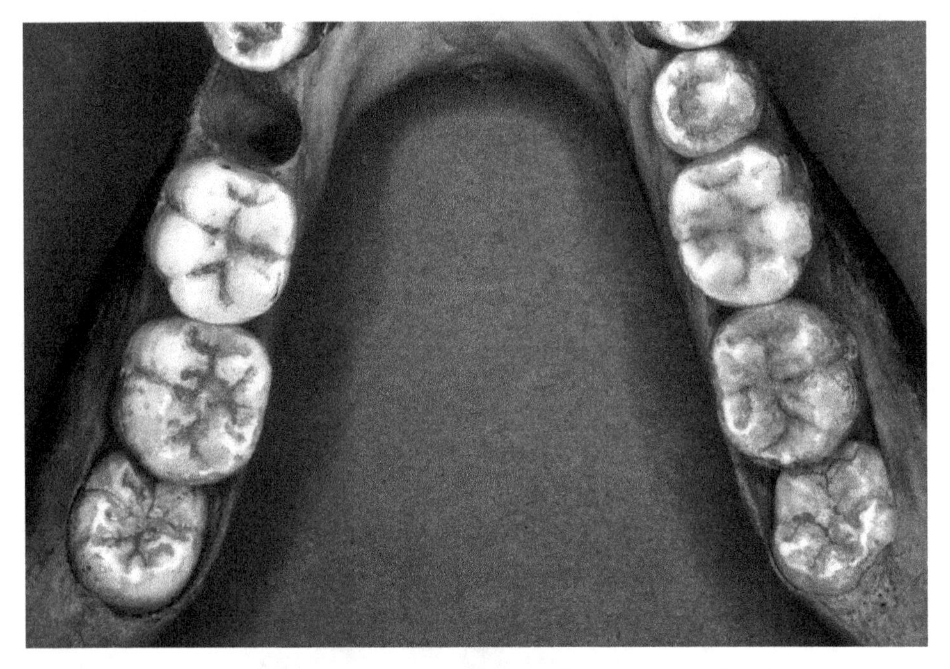

**Figure 28.5** Anterior fovea on LM1 and LM2; it is symmetrical and either grade 3 or 4.

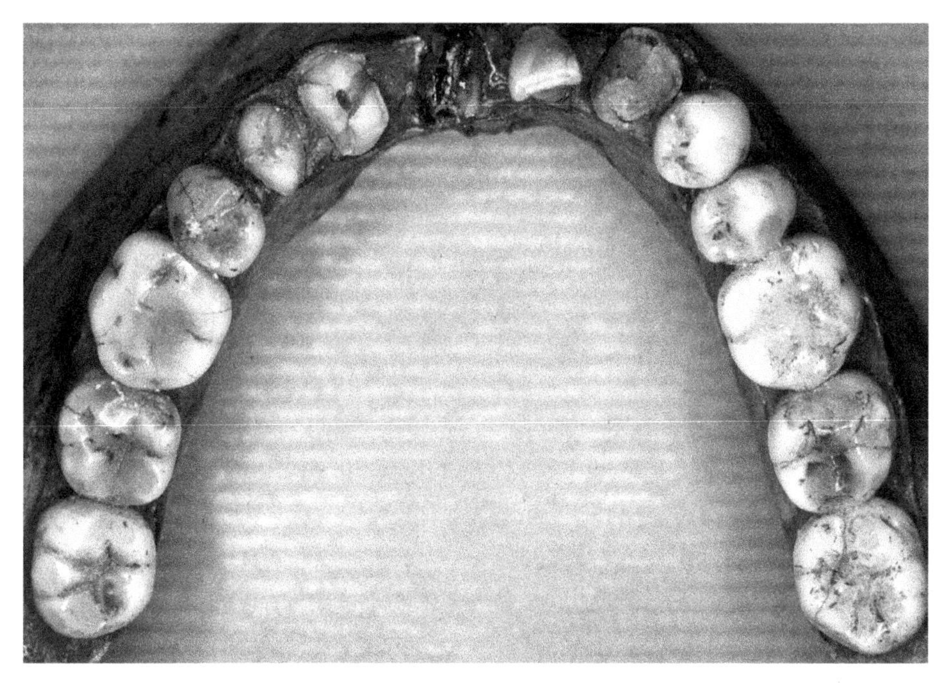

**Figure 28.6** Although this individual is a young adult, crown wear has obliterated the locus of the anterior fovea, so these teeth cannot be scored (Turner recommended scoring only children and teenagers for this reason).

## Select Bibliography

Bailey, S.E. (2002). A closer look at Neanderthal postcanine dental morphology: the mandibular dentition. *Anatomical Record* 269, 148–156.

Begun, D.R., and Kordos, L. (1993). Revision of *Dryopithecus brancoi* Schlosser, 1901 based on the fossil hominoid material from Rudabánya. *Journal of Human Evolution* 25, 271–285.

Broom, R. (1950). The genera and species of the South African fossil ape-men. *American Journal of Physical Anthropology* 8, 1–14.

Coffing, K., Feibel, C., Leakey, M., and Walker, A. (1994). Four-million-year-old hominids from East Lake Turkana, Kenya. *American Journal of Physical Anthropology* 93, 55–65.

Haillie-Selassie, Y., Suwa, G., and White, T.D. (2004). Late Miocene teeth from the middle Awash, Ethiopia and early hominid dental evolution. *Science* 303, 1503–1505.

Hrdlička, A. (1924). New data on the teeth of early man and certain fossil European apes. *American Journal of Physical Anthropology* 7, 109–132.

Martinón-Torres, M., Bermúdez de Castro, J.M., Gómez-Robles, A., *et al.* (2007). Dental evidence on the hominin dispersals during the Pleistocene. *Proceedings of the National Academy of Sciences of the USA* 104, 13279–13282.

McCrossin, M.C. Human molars from later Pleistocene deposits of Witkrans Cave, Gaap Escarpment, Kalahari margin. *Human Evolution* 7, 1–10.

Turner, C.G., II, Nichol, C.R., and Scott, G.R. (1991). Scoring procedures for key morphological traits of the permanent dentition: the Arizona State University dental anthropology system. In *Advances in Dental Anthropology*, ed. M.A. Kelley and C.S. Larsen. New York: Wiley-Liss, pp. 13–31.

# 29

# Mandibular Torus

## Observed on

Lingual surface of mandible

## Key Tooth

N/A (oral torus)

## Synonyms

*Torus mandibularis*, mandibular hyperostosis (Hrdlička 1940)

## Description

Mandibular torus is expressed as one or two (or more) lobes that originate on the lingual surface of the mandible below the canine. The torus varies from a small elevation below the canine and first premolar to a multi-lobed exostosis that extends back as far as the second molar. Scoring mandibular torus involves a combination of sight and touch. Low grades of expression, for example, can be palpated even though they are difficult to observe.

## Classification

Mandibular tori show a wide range of variation. Turner adopted a relatively simple scale, with four categories ranging from absent to marked.

Grade 0: absence of torus (palpation required)
Grade 1: small (slight elevation below LC and LP1)
Grade 2: moderate (larger elevation with more extended coverage, sometimes as two small lobes)
Grade 3: marked (more pronounced expression, extends from LC to LM1)
Grade 4: very marked (extends from LC to LM2, with very little separation of lobes across the mandible)

## Breakpoint

Based on frequencies in the literature, there is some disagreement as to what constitutes the presence of mandibular torus. If an observer palpates the locale where the torus originates, even very low-grade expressions can be scored consistently. For observations on the living, it might be best to use moderate forms of torus (grade 2+) as the breakpoint.

## Potential Problems in Scoring

Two types of exostoses can be manifest on the lingual surface of the mandible. The mandibular torus, as noted above, and pleated hyperostosis. The distinction between the two can be defined as follows:

(1) Mandibular torus originates around LC and LP1 and depending on grade of expression can extend back as far as LM2; it is not associated directly with the alveolar bone.
(2) Pleated hyperostosis involves a bony elevation associated with the alveolar bone that begins in the area of the lower premolars and extends as far back as LM3; vertical grooves separate elevations of varying heights, hence the term pleated (like a curtain).

There are instances where the two forms co-occur and a pleated hyperostosis overlays a mandibular torus.

## Geographic Variation

Mandibular torus and palatine torus do not co-vary in terms of individual correlations or population distributions. Mandibular torus is strongly tied to northern populations in both Europe and Asia. As one moves up in latitude, torus frequencies go up. Moving toward the equator, torus frequencies go down. Although mandibular torus has some genetic component, most agree that it has an environmental component of variance as well. World clinal variation in mandibular torus frequencies is illustrated in Scott *et al.* (2016).

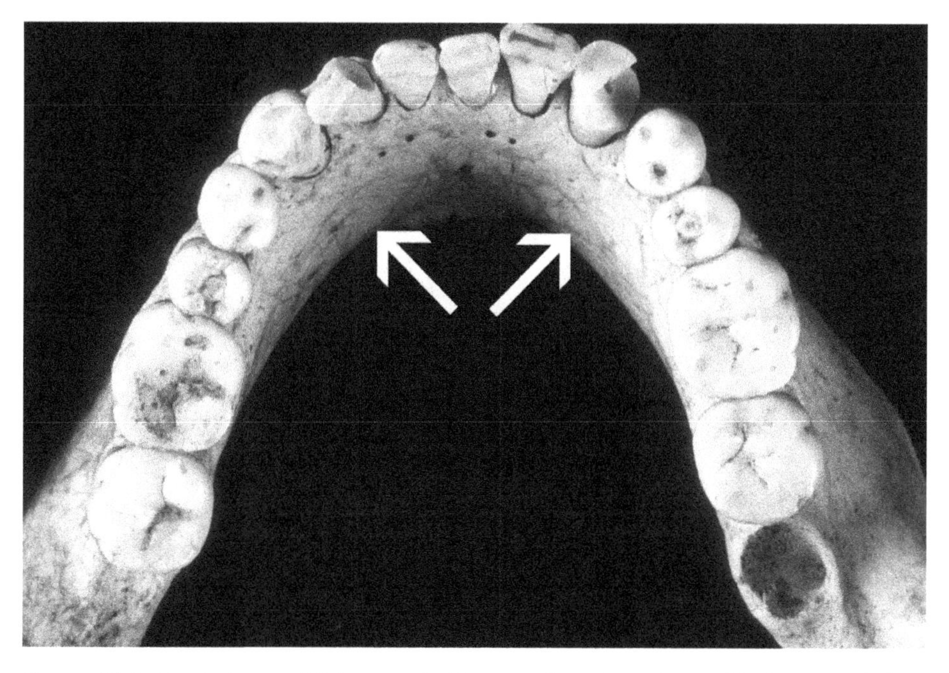

**Figure 29.1** Grade 0: no visible or palpable manifestation of mandibular torus below LC and LP1 (note odontomes on LP2).

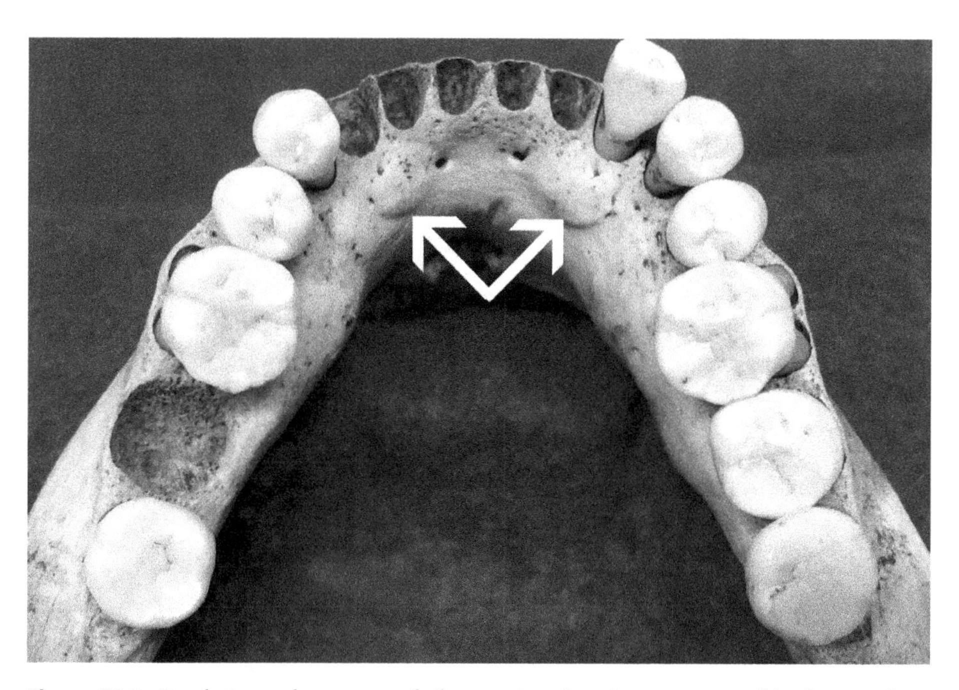

**Figure 29.2** Grade 2: moderate torus below canine that does not extend back to molars.

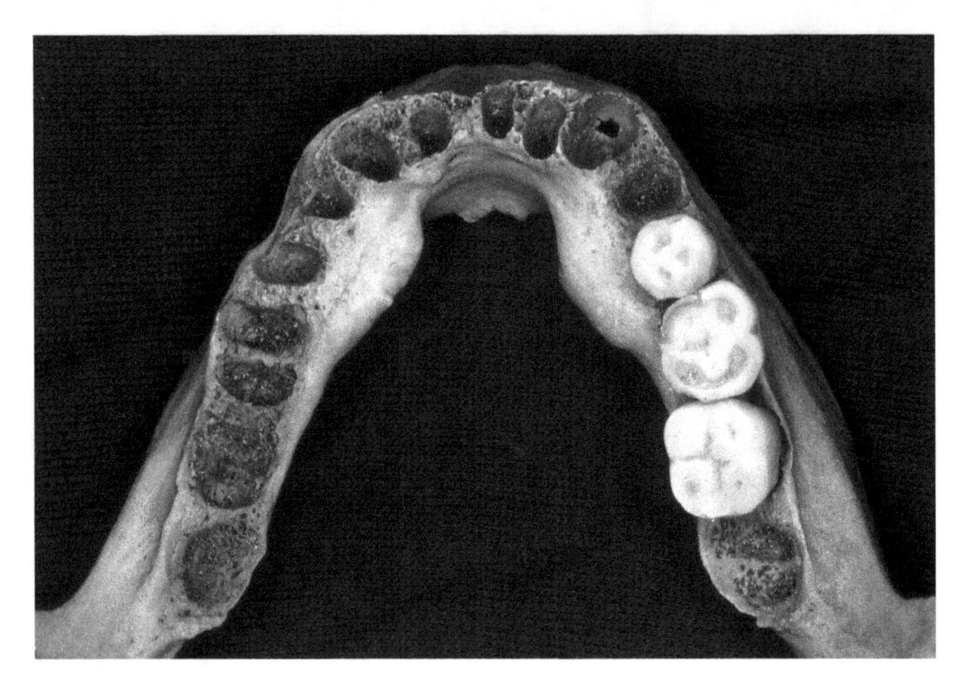

**Figure 29.3** Grade 3: marked torus that extends from LC to midpoint of LM1.

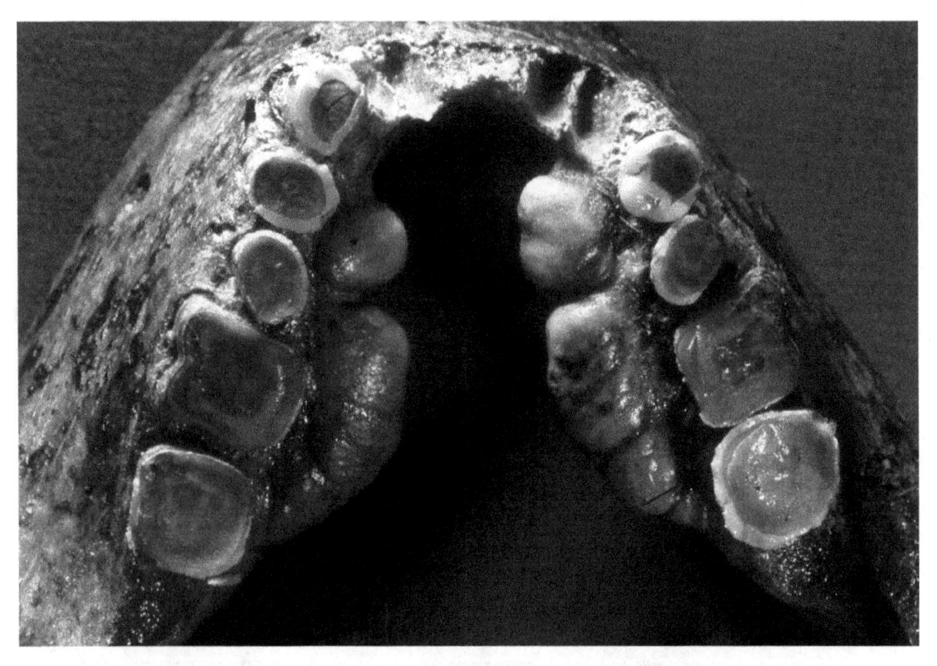

**Figure 29.4** Grade 4: very marked torus divided into two lobes that extends from LC to midpoint on LM2.

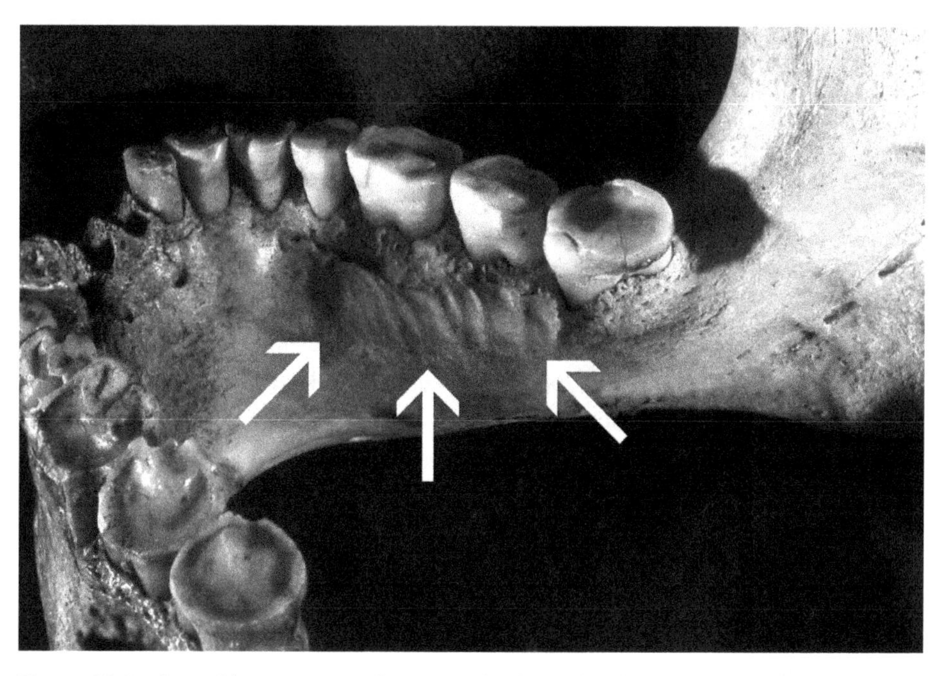

**Figure 29.5** Pleated hyperostosis that extends along the alveolar bone of the cheek teeth; such a manifestation may appear alone or in association with a torus.

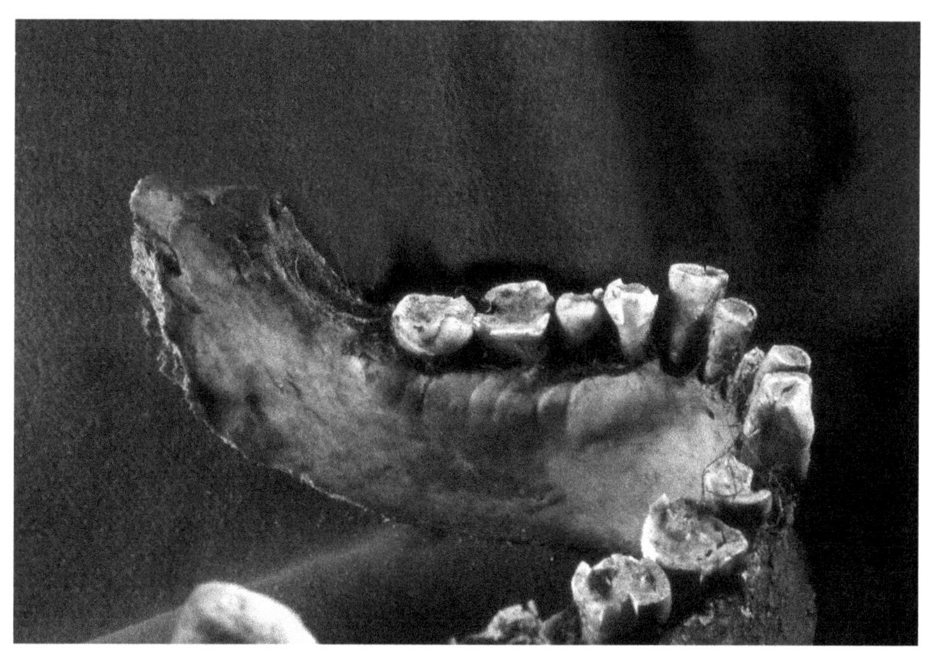

**Figure 29.6** Mandibular torus (premolar region) in association with pleated hyperostosis (molar region).

## Select Bibliography

Axelsson, G., and Hedegard, M. (1981). Torus mandibularis among Icelanders. *American Journal of Physical Anthropology* 54, 383–389.

Hasset, B. (2006). Mandibular torus: etiology and bioarchaeological utility. *Dental Anthropology* 19, 1–14.

Haugen, L.K. (1992). Palatine and mandibular tori: a morphologic study in the current Norwegian population. *Acta Odontologica Scandinavica* 50, 65–77.

Hrdlička, A. (1940). Mandibular and maxillary hypertostoses. *American Journal of Physical Anthropology* 27, 1–67.

Ihunwo, A.O., and Phukubye, P. (2006). The frequency and anatomical features of torus mandibularis in a Black South African population. *Homo – Journal of Comparative Human Biology* 57, 253–262.

Johnson, C.C., Gorlin, R.J., and Anderson, V.E. (1965). Torus mandibularis: a genetic study. *American Journal of Human Genetics* 17, 433–442.

Moorrees, C.F.A. (1957). *The Aleut Dentition.* Cambridge, MA: Harvard University Press.

Ohno, N., Sakai, T., and Mizutani, T. (1988). Prevalence of torus palatinus and torus mandibularis in five Asian populations. Aichi-Gakuin Dental *Science* 1, 1–8.

Ossenberg, N.S. (1978). Mandibular torus: a synthesis of new and previously reported data and a discussion of its causes. In *Contributions to Physical Anthropology*, ed. J.S. Cybulsky. Ottawa: National Museum of Canada, pp. 1–52.

Pechenkina, E.A., and Benfer, R.A., Jr. (2002). The role of occlusal stress and gingival infection in the formation of exostoses on mandible and maxilla from Neolithic China. *Homo – Journal of Comparative Human Biology* 53, 112–130.

Sawyer, D.R., Allison, M.J., Elzay, R.P., and Pezzia, A. (1979). A study of torus palatinus and torus mandibularis in Pre-Columbian Peruvians. *American Journal of Physical Anthropology* 50, 525–526.

Scott, G.R., Schomberg, R., Swenson, V., Adams, D., and Pilloud, M.A. (2016). Northern exposure: mandibular torus in the Greenlandic Norse and the whole wide world. *American Journal of Physical Anthropology* 161, 513–521.

Seah, Y.H. (2009). Torus palatinus and torus mandibularis: a review of the literature. *Australian Dental Journal* 40, 318–321.

Sellevold, B.J. (1980). Mandibular torus morphology. *American Journal of Physical Anthropology* 53, 569–572.

Suzuki, M., and Sakai, T. (1960). A familial study of torus palatinus and torus mandibularis. *American Journal of Physical Anthropology* 18, 262–273.

# 30

# Lower Molar Groove Pattern

## Observed on

LM1, LM2, LM3

## Key Tooth

LM2

## Synonyms

*Dryopithecus* Y pattern (Gregory and Hellman 1926)

## Description

Lower molars have five major cusps. The cusps were originally named and numbered by Osborn and Cope: (1) protoconid (mesiobuccal), (2) metaconid (mesiolingual), (3) hypoconid (distobuccal), (4) entoconid (distolingual), and (5) hypoconulid (distal). Gregory and Hellman noted that hominoids showed a distinct pattern in terms of contact between major cusps of the lower molars. Almost invariably, hominoids have lower molars where contact is between cusps 2 and 3, with part of the groove system resembling a Y. As hominoids typically had all five cusps, this was the origin of the classic Y5 pattern. In hominins, there has been some shift in pattern so that cusps 1 and 4 often come in contact at the central sulcus, forming an X pattern. This is still rare in modern humans on LM1 but is common on LM2.

## Classification

Gregory and Hellman (1926); Jørgensen (1956)

Y pattern: contact between cusps 2 and 3
X pattern: contact between cusps 1 and 4
+ pattern: contact between cusps 1, 2, 3, and 4 at central sulcus

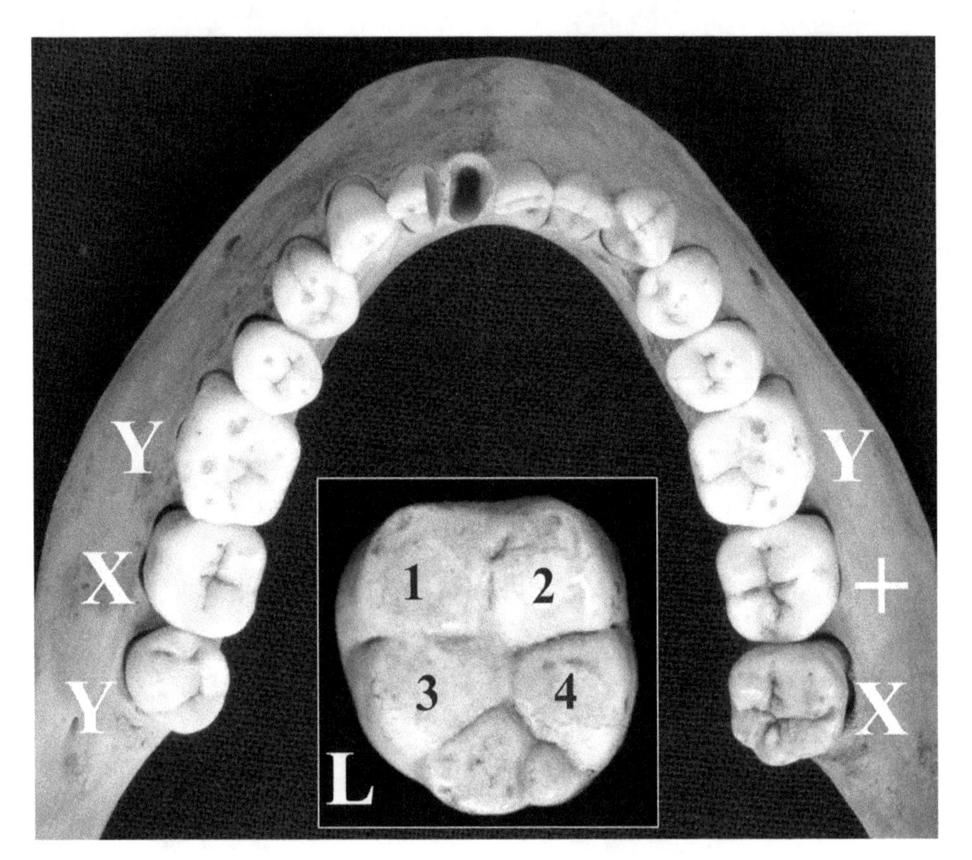


**Figure 30.1** Groove pattern is determined by cusp contact (inset shows five lower molar cusps and their numbers). Both LM1s show Y patterns (2–3 contact) while left LM2 shows an X pattern (1–4 contact) and right LM2 shows a + pattern (1–2–3–4 contact). The abnormal left LM3 has a Y pattern, while its normal antimere has an X pattern.

## Breakpoint

As the Y pattern is close to 100% on LM1 of all human groups, focus is typically on LM2, which shows all three patterns. In Scott and Turner (1997), the frequency of the Y pattern on LM2 is used to compare world populations.

## Potential Problems in Scoring

For lower molars, cusp contact is usually easy to distinguish. However, in cases where it is difficult to discern, researchers vary in terms of how often they use the + category to score groove pattern. Contact between the major cusps of the lower molars can also be obscured by wear, primarily in skeletal samples, and by fillings in living samples. As the central sulcus is often the focus of caries formation, sample size in modern groups can be severely limited. Other morphological variants can also impact cusp contact, notably the deflecting wrinkle (cf. Morris 1970) and distal trigonid crest.

## Geographic Variation

The frequency range of Y pattern on LM2 is not great even on a world scale. Sub-Saharan African and New Guinea samples have the highest frequency of LM2 Y pattern (sometimes above 40%) while most human groups have frequencies between 20% and 30%. Although this trait was one of the earliest morphological variants to be described and classified, its utility is limited in population studies because of a narrow range of variation.

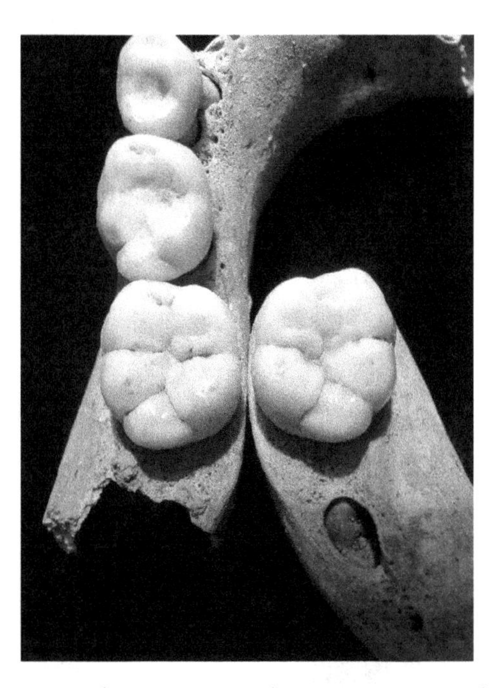

**Figure 30.2** LM1 antimeres showing contact between cusps 2 and 3, forming a Y pattern. A large distal accessory ridge (part of distal trigonid crest?) on the metaconid comes in direct contact with the hypoconid.

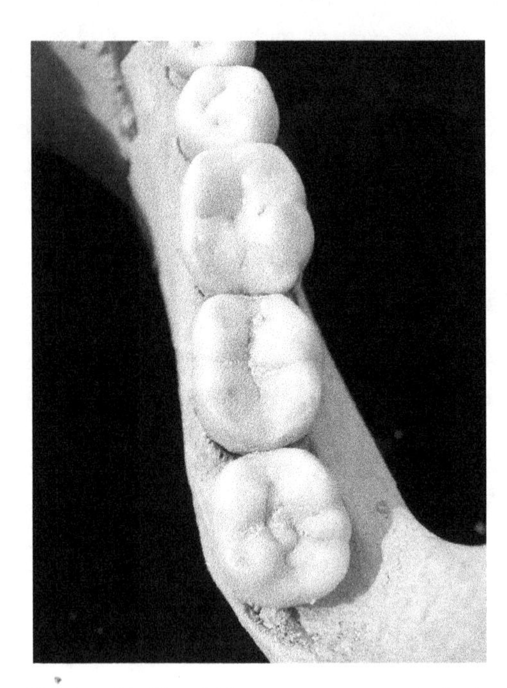

**Figure 30.3** LM1 and LM2 with Y and X patterns, respectively, but note the occlusal tubercle in the central fossa of LM3 that obscures the groove pattern.

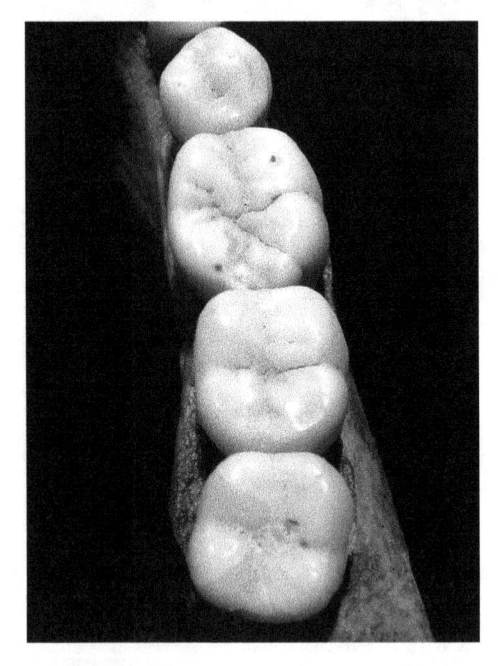

**Figure 30.4** Right LM1 has a deflecting wrinkle on the metaconid that increases the likelihood of cusp 2–3 contact.

## Select Bibliography

Bailey, S.E. (2002). A closer look at Neanderthal postcanine dental morphology: the mandibular dentition. *Anatomical Record* 269, 148–156.

Biggerstaff, R. H. (1970). Morphological variations for the permanent mandibular first molars in human monozygotic and dizygotic twins. *Archives of Oral Biology* 15, 721–730.

Dahlberg, A.A. (1961). Relationship of tooth size to cusp number and groove conformation of occlusal surface patterns of lower molar teeth. *Journal of Dental Research* 40, 34–38.

Erdbrink, D.P. (1965). A quantification of the *Dryopithecus*- and other lower molar patterns in man and some of the apes. *Zeitschrift für Morphologie und Anthropologie* 57, 70–108.

Garn, S.M., Dahlberg, A.A., Lewis, A.B., and Kerewsky, R.S. (1966a). Groove pattern, cusp number, and tooth size. *Journal of Dental Research* 45, 970.

Gregory, W.K., and Hellman, M. (1926). The dentition of *Dryopithecus* and the origin of man. *American Museum of Natural History Anthropological Papers* 28, 1–117.

Hellman, M. (1928). Racial characters in human dentition. Part 1. A racial distribution of the *Dryopithecus* pattern and its modifications in the lower molar teeth of man. *Proceedings of the American Philosophical Society* 67, 157–174.

Jørgensen, K.D. (1955). The *Dryopithecus* pattern in recent Danes and Dutchmen. *Journal of Dental Research* 34, 195–208.

(1956). The deciduous dentition: a descriptive and comparative anatomical study. *Acta Odontologica Scandinavica* 14, 1–202.

Lavelle, C.L.B. (1971). Mandibular molar tooth configurations in different racial groups. *Journal of Dental Research* 50, 1353.

Lavelle, C.L.B., Ashton, E.H., and Flinn, R.M. (1970). Cusp pattern, tooth size and third molar agenesis in the mandibular human dentition. *Archives of Oral Biology* 15, 227–237.

Morris, D.H. (1970). On deflecting wrinkles and the *Dryopithecus* pattern in human mandibular molars. *American Journal of Physical Anthropology* 32, 97–104.

Scott, G.R., and Turner, C.G., II (1997). *The Anthropology of Modern Human Teeth: Dental Morphology and its Variation in Recent Human Populations*. Cambridge: Cambridge University Press.

# NOTES

# 31

# Rocker Jaw

## Observed on

Inferior surface of mandibular body from chin to gonial angle

## Key Tooth

N/A

## Synonyms

Rocker mandible

## Description

A rocker jaw refers, literally, to a mandible that rocks back and forth when placed upon a flat surface and pushed. The left and right inferior horizontal rami are convex, like the two curved rocker components of a rocking chair. Some researchers think the trait develops in response to craniofacial growth, to maintain correct occlusion. Others maintain that diet and chewing forces do not lead to its formation, but rather that it is genetic in origin. Like the palatine and mandibular tori (this volume), it is likely a combination of both environmental and genetic factors; at present, its level of heritability is not known. In any event, rocker jaw only "appears" in adult individuals who have completed craniofacial growth.

## Classification

Martin and Saller (1959)

There are three possible ASUDAS grades:

Grade 0: no expression. Both inferior horizontal rami are flat, or together are tripod-like in appearance; in the latter case, projections of the chin (i.e., gnathion craniometric measurement point) and the two distal-most points of the horizontal rami that transition into the vertical rami (gonion) form the base of the tripod.

Grade 1: near rocker. The inferior horizontal rami are convex enough that the mandible is unstable when laid on a flat surface. The mandible will "rock" for about a second when pushed.

Grade 2: rocker. The horizontal rami are so convex that the mandible will easily rock back and forth on a flat surface for more than a second.

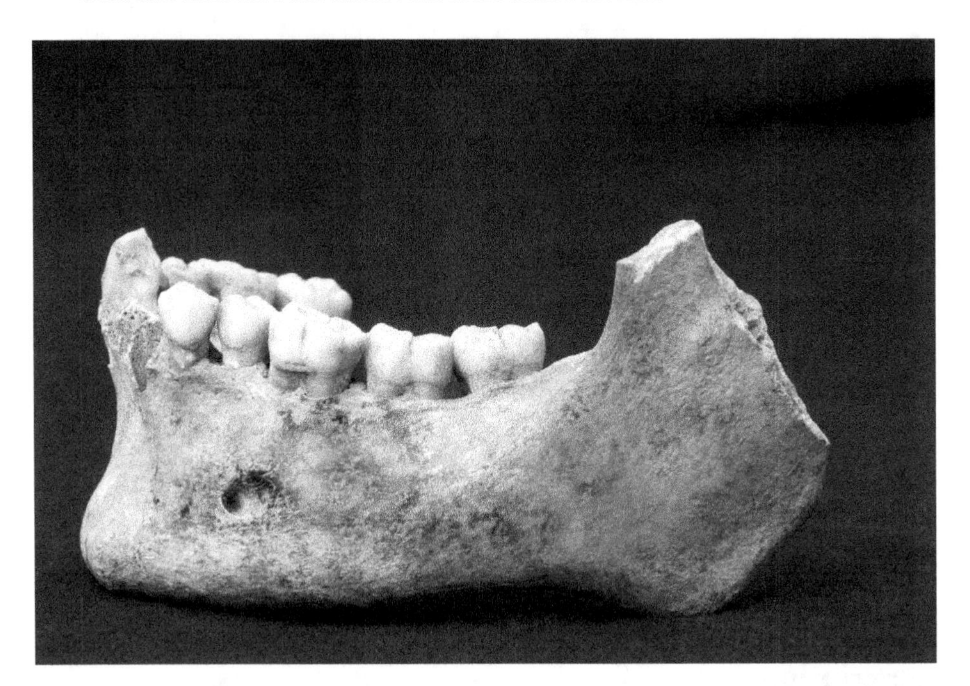

**Figure 31.1** Grade 0 mandible for rocker jaw.

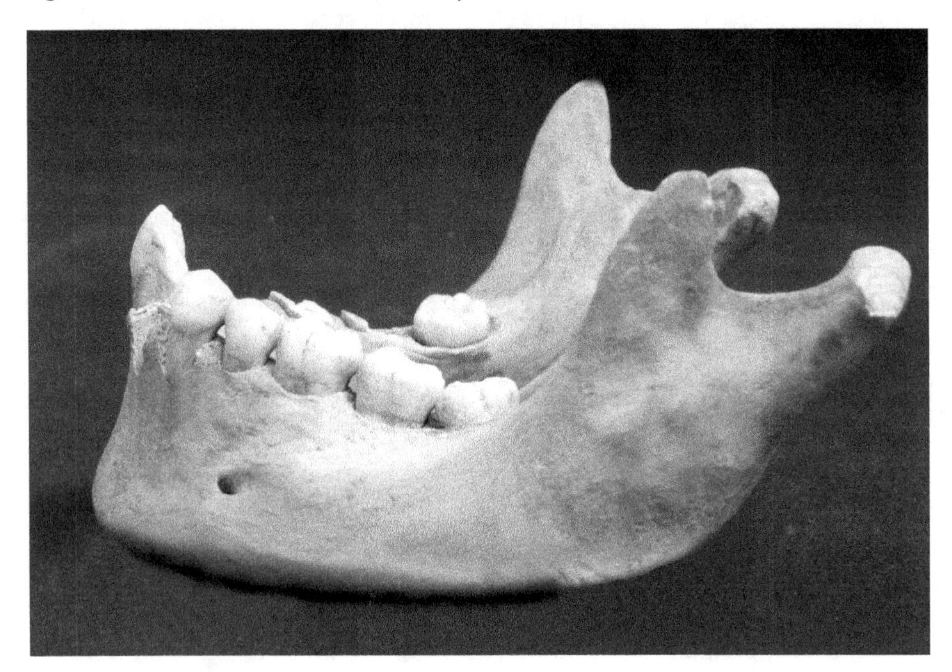

**Figure 31.2** Grade 1 mandible for rocker jaw. Near rocker.

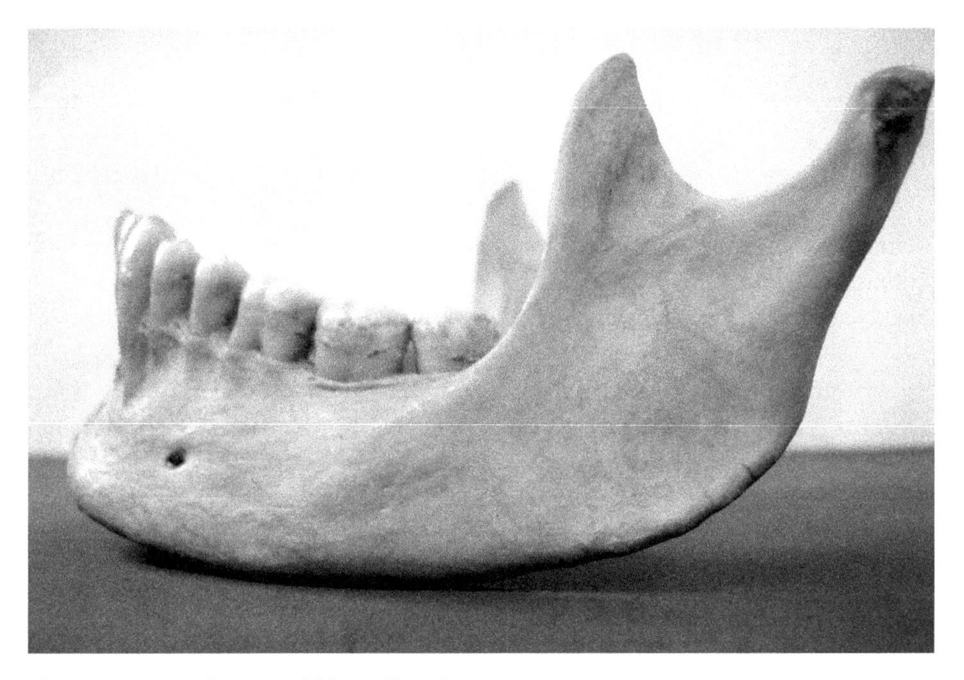

**Figure 31.3** Grade 2 mandible. Full rocker expression.

## Breakpoint

1

## Potential Problems in Scoring

As noted, the trait only manifests in adults after the completion of craniofacial growth. Therefore it should not be recorded in younger individuals. If one half of the mandible is missing, the trait may still be observed based on the convexity of the remaining horizontal ramus.

## Geographic Variation

Rocker jaw has not routinely been recorded in all world populations. However, it is known as a "Polynesian" trait, where frequencies of >80% have been observed. It is also common in Southeast and mainland Asian populations (e.g., >60% in some studies). In unpublished data, the second author (JDI) recorded trait presence (grades 1–2) of up to 20% in several Mediterranean-area European samples; published results reveal a range of 0.0–33.3% in 14 Sub-Saharan samples (mean of 8.7%) and 0.0–41.2% in 14 North African samples (mean 19.3%).

## Select Bibliography

Houghton, P. (1980). *The First New Zealanders*. Auckland: Hodder & Stoughton.

Irish, J.D. (1993). *Biological Affinities of Late Pleistocene through Modern African Aboriginal Populations: The Dental Evidence*. PhD dissertation, Department of Anthropology, Arizona State University, Tempe.

Martin, R., and Saller, K. (1959). *Lehrbuch der Anthropologie*, 3rd edn. Stuttgart: Gustav Fischer.

Pietrusewsky, M., and Douglas, M.T. (2002). *Ban Chiang, a Prehistoric Village Site in Northeast Thailand, Volume 1: The Human Skeletal Remains*. Philadelphia: University of Pennsylvania Museum of Archaeology and Anthropology.

Schendel, S.A., Walker, G., and Kamisugi, A. (1980). Hawaiian craniofacial morphometrics: average Mokapuan skull, artificial cranial deformation, and the "rocker" mandible. *American Journal of Physical Anthropology* 52, 491–500.

Wendell, W.S.K. (1971). *Discriminant Analysis of Rocker Jaws*. MA thesis, McMaster University. https://macsphere.mcmaster.ca/bitstream/11375/10379/1/fulltext.pdf (accessed November 2016).

# 32

# Lower Molar Cusp Number

## Observed on

LM1, LM2, LM3

## Key Tooth

LM2

## Synonyms

Hypoconulid (Osborn 1897), cusp 5 (Gregory and Hellman 1926)

## Description

Lower molar cusp number, contrary to the observations of some early workers, depends entirely on the presence of cusp 5, or the hypoconulid. This cusp, like the hypocone of the upper molars, was the last to be added in hominoid evolution and is the first to be lost in hominin evolution. Apes and early hominins almost invariably exhibit the hypoconulid on all lower molars. At some point in hominin evolution, it became a polymorphic trait where you could have either a five-cusped or four-cusped lower molar. The hypoconulid is typically present on LM1 in most human populations but is highly variable in terms of presence and size on LM2. It is also variable on LM3 but that tooth is, in general, too variable to be useful for population comparisons. Interestingly, cusp number on the lower molars is sometimes 5–4–5, a pattern that would rarely be observed for the upper molar hypocone (i.e., 4–3–4). Although technically a distal cusp of the lower molars, the hypoconulid is integrated more closely with the hypoconid than the entoconid.

Although Turner developed the following classification to score the hypoconulid for strength of expression, the tables in the Appendix focus on lower molar cusp number as 4, 5, or 6.

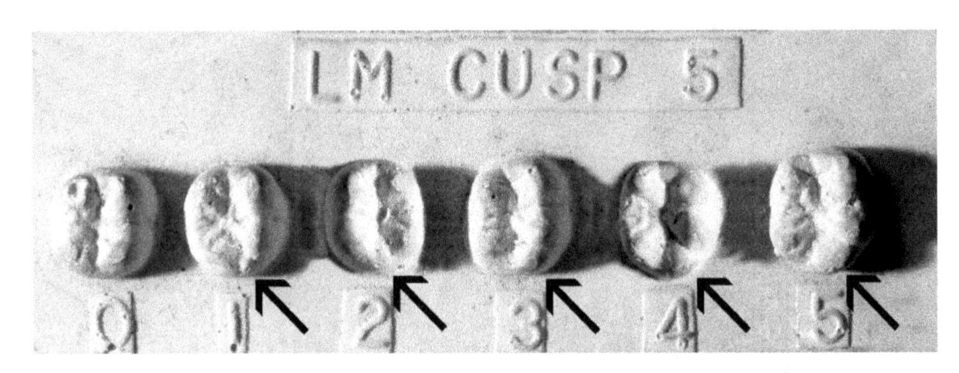


**Figure 32.1** Standard plaque for hypoconulid expression in the lower molars.

## Classification

Turner *et al.* (1991)

Grade 0: hypoconulid is absent (four-cusped tooth)
Grade 1: trace expression
Grade 2: slight
Grade 3: moderate
Grade 4: strong
Grade 5: pronounced

## Breakpoint

Any manifestation of the hypoconulid, from very small to very large, makes a tooth five-cusped. Consistent with their treatment of the hypocone, Scott and Turner (1997) focused on hypoconulid absence rather than presence. That is, their characterization of geographic variation focused on four-cusped LM1 and LM2. This places more emphasis on the divergence of a given group from the ancestral five-cusped lower molar (100% for all three lower molars in hominoids and early hominins).

## Potential Problems in Scoring

As with groove pattern, crown wear and dental fillings are the major impediments to making observations on the hypoconulid. Even when wear is relatively minor, the grooves separating the hypoconulid from the hypoconid and entoconid can be partly or completely obliterated. The key is to distinguish four-cusped from five-cusped teeth, so this is an important consideration. In unworn teeth, another phenotype may cause some confusion in determining whether cusp 5 is present or not. That is, the hypoconid sometimes extends around the distal border of the tooth and shows

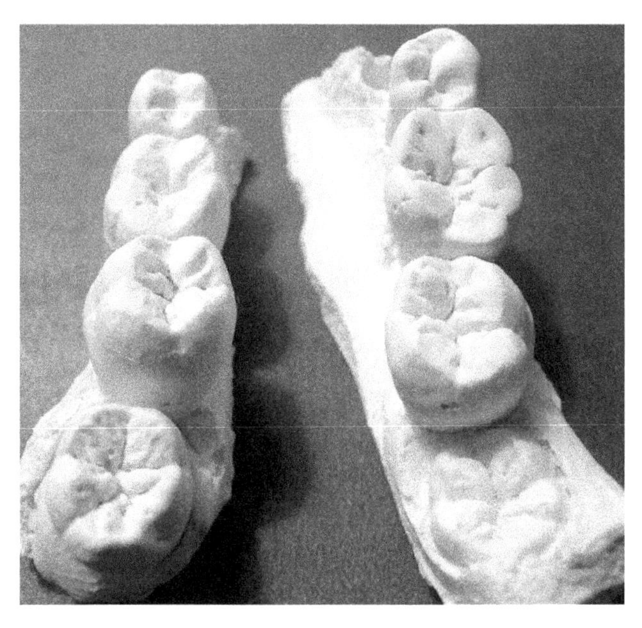

**Figure 32.2** Lower molars showing 5–4–5 cusp number. LM1 hypoconulid is large and distinctive; it is absent on LM2, and present and unusually large on LM3.

an occlusal groove but no vertical groove on the distal surface. Without micro-CT, it is impossible to determine whether or not a hypoconulid is present on the enamel–dentine junction. To be consistent, a hypoconulid should be scored as present when both occlusal and vertical grooves set the cusp apart from the adjoining hypoconid and entoconid.

## Geographic Variation

There are two ways to characterize lower molar cusp number in human populations. One could, for example, use the total frequency of the hypoconulid (i.e., the frequency of five-cusped lower molars). Another way to characterize variation is to focus on the frequency of hypoconulid loss, also noted as four-cusped lower molars. Scott and Turner (1997) adopted the second method in their regional characterizations. They found that Europeans have the most distinctive profile for four-cusped lower molars, with the world's highest frequencies on both LM1 (ca. 10%) and LM2 (70–90%). At the opposite extreme, Sub-Saharan Africans and Australians have the lowest frequencies of four-cusped lower molars. Asian and Asian-derived populations rarely have four-cusped LM1, but frequencies for four-cusped LM2 are highly variable for these groups. One puzzling finding for lower molar cusp number was the relatively high frequency of four-cusped lower molars in New Guinea, which is in contrast to neighboring Australians.

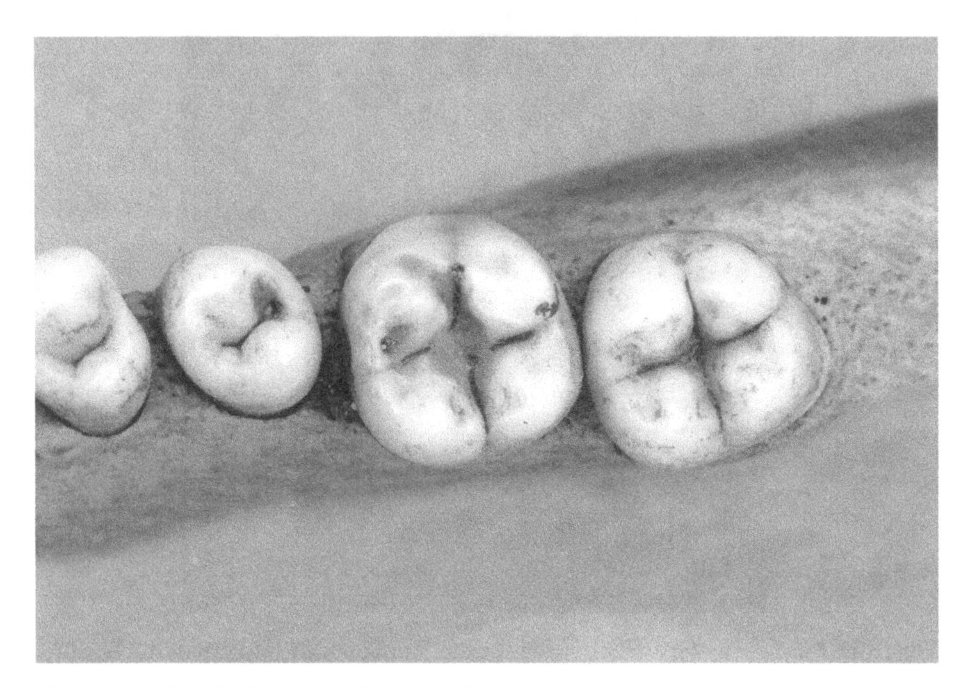

**Figure 32.3** Simple four-cusped LM1 and LM2.

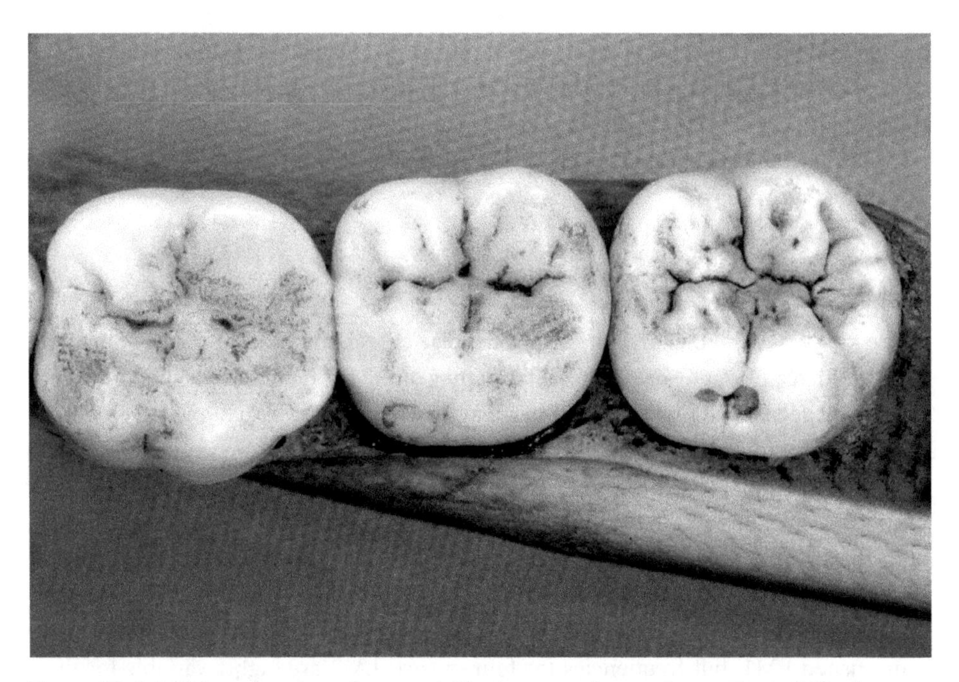

**Figure 32.4** Left lower molars show variable degrees of wear that make it difficult to score the hypoconulid; LM1 probably had a hypoconulid, but the grooves are worn away, making scoring a guess. LM2 has a small hypoconulid and small cusp 6. LM3 has a larger hypoconulid than LM2 (which is not unusual) and also a small distinct cusp 6.

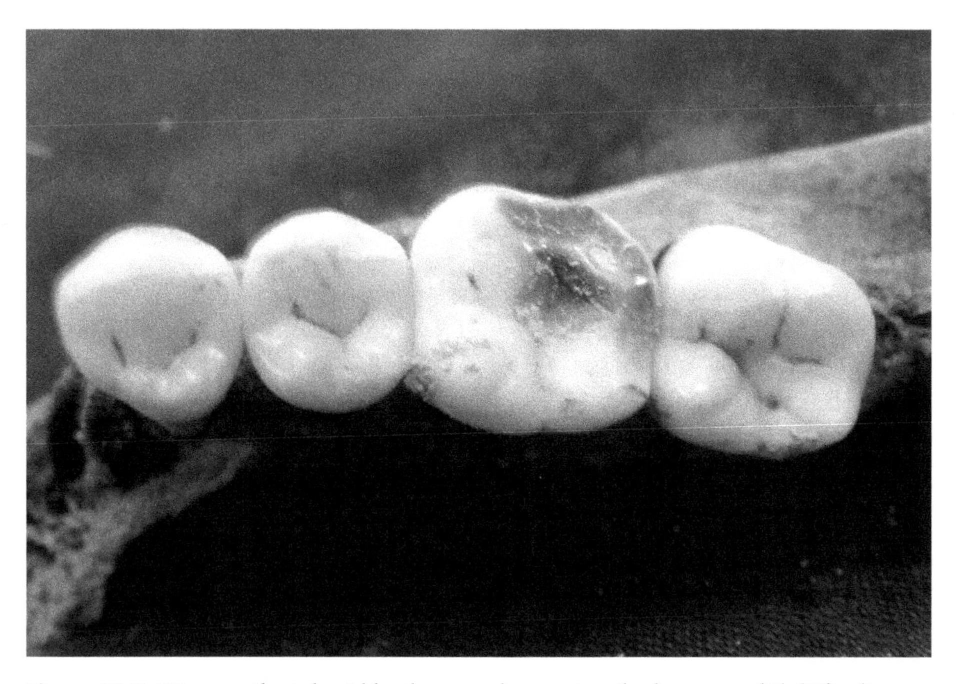

**Figure 32.5** Wear on the talonid heel can make scoring the hypoconulid difficult. Even in this relatively young individual, the distal segment of LM1 is worn to the point where one cannot score the hypoconulid; LM2 also shows wear, but in this case an observer would be safe in scoring the hypoconulid as absent, given the overall quadrate form of the tooth and no hint of a distal projection that would suggest the presence of a hypoconulid.

## Select Bibliography

Cope, E.D. (1874). On the homologies and origin of the types of molar teeth in Mammalia educabilia. *Journal of the Academy of Natural Sciences, Philadelphia* 8, 71–89.

(1888). On the tritubercular molar in human dentition. *Journal of Morphology* 2, 7–26.

Dahlberg, A.A. (1961). Relationship of tooth size to cusp number and groove conformation of occlusal surface patterns of lower molar teeth. *Journal of Dental Research* 40, 34–38.

Davies, P.L. (1968). Relationship of cusp reduction in the permanent mandibular first molar to agenesis of teeth. *Journal of Dental Research* 47, 499.

Gregory, W.K. (1916). Studies on the evolution of the primates. I. The Cope–Osborn "theory of trituberculy" and the ancestral molar patterns of the Primates. *Bulletin of the American Museum of Natural History* 35, 239–257.

(1922). *The Origin and Evolution of the Human Dentition*. Baltimore: Williams and Wilkins.

(1934). A half century of trituberculy: the Cope–Osborn theory of dental evolution with a revised summary of molar evolution from fish to man. *Proceedings of the American Philosophical Society* 73, 169–317.

Gregory, W.K., and Hellman, M. (1926). The dentition of *Dryopithecus* and the origin of man. *American Museum of Natural History Anthropological Papers* 28, 1–117.

Osborn, H.F. (1897). Trituberculy: a review dedicated to the late Professor Cope. *American Naturalist* 31, 993–1016.

Scott, G.R., and Turner, C.G., II (1997). *The Anthropology of Modern Human Teeth: Dental Morphology and its Variation in Recent Human Populations*. Cambridge: Cambridge University Press.

Turner, C.G., II, Nichol, C.R., and Scott, G.R. (1991). Scoring procedures for key morphological traits of the permanent dentition: the Arizona State University dental anthropology system. In *Advances in Dental Anthropology*, ed. M.A. Kelley and C.S. Larsen. New York: Wiley-Liss, pp. 13–31.

# 33

# Deflecting Wrinkle

## Observed on

LM1

## Key Tooth

LM1

## Synonyms

N/A

## Description

The deflecting wrinkle is expressed on the occlusal surface of the mesiolingual cusp (metaconid) of the lower molars, notably LM1. It is so rare on LM2 that scoring this tooth is unnecessary. Basically, this trait is an unusual manifestation of the essential ridge of the metaconid. In most instances, this ridge runs a direct course from the cusp tip of the metaconid to the central occlusal fossa. However, in some instances, the ridge changes course (or deflects) about halfway along its length before it terminates in the central sulcus.

## Classification

Turner *et al.* (1991)

Grade 0: deflecting wrinkle absent; essential ridge of metaconid runs a straight course from cusp tip to central occlusal fossa
Grade 1: essential ridge is straight but with midpoint constriction
Grade 2: essential ridge deflects at halfway point toward central occlusal fossa but does not contact hypoconid
Grade 3: essential ridge shows strong deflection at midpoint and does contact hypoconid

**Figure 33.1** Standard plaque for the deflecting wrinkle of the lower molars.

## Breakpoint

2+. Although the deflecting wrinkle has three grades of presence, only the two distinct forms (grades 2 and 3) are used for calculating population frequencies.

## Potential Problems in Scoring

As the deflecting wrinkle is on the occlusal surface, wear and caries remain the biggest problems in scoring. There is no dentine involvement, so even moderate wear makes it difficult to score this trait.

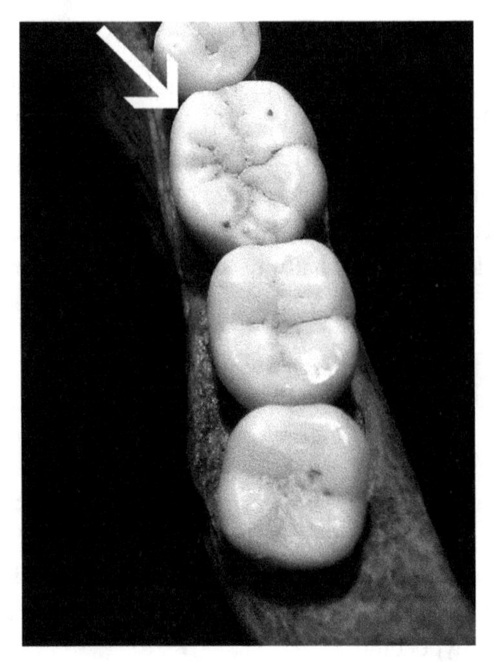

**Figure 33.2** Grade 2 deflecting wrinkle on right LM1; this trait is rarely observed on LM2 or LM3 in recent humans.

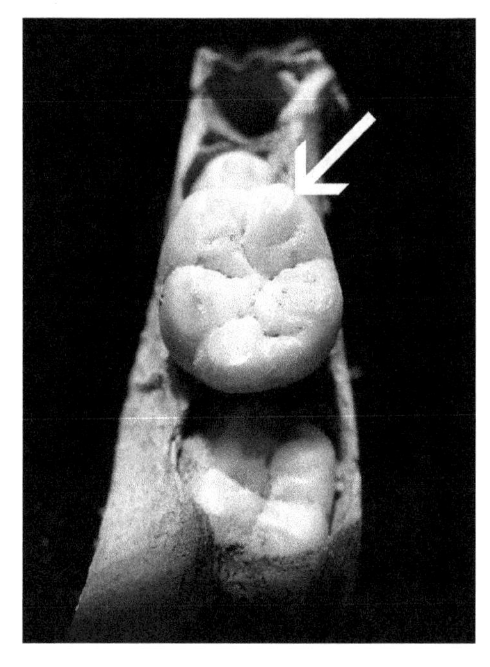

**Figure 33.3** Grade 3 deflecting wrinkle on metaconid of left UM1; unerupted LM2 has no deflecting wrinkle.

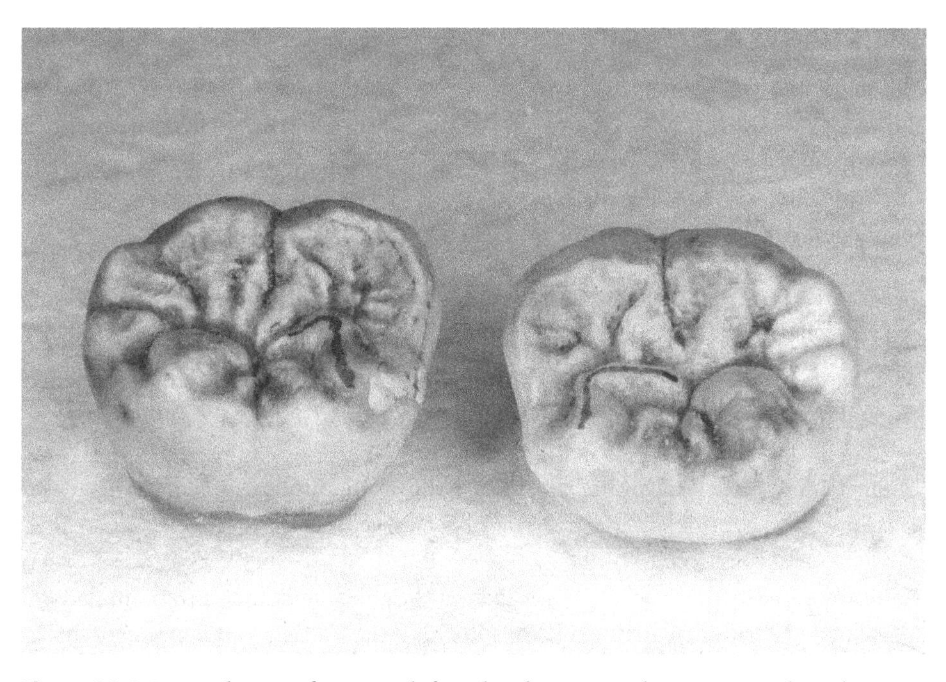

**Figure 33.4** Lingual view of unworn left and right LM1 with symmetrical grade 3 deflecting wrinkles (highlighted).

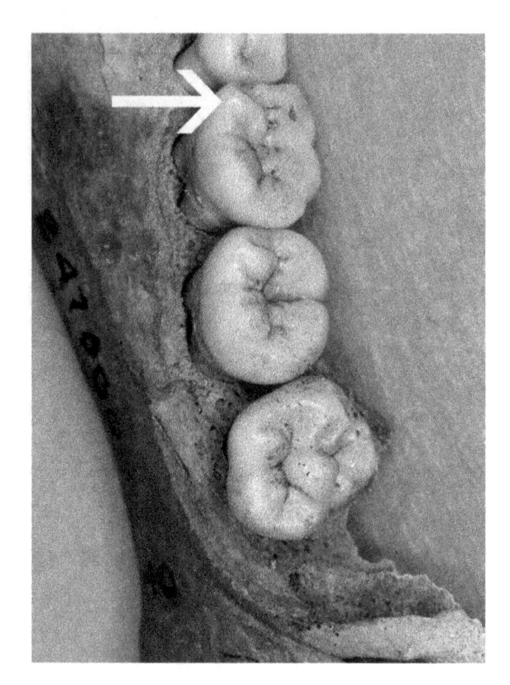

**Figure 33.5** Grade 3 deflecting wrinkle on right LM1; unlike the majority of crown traits, which are expressed to varying degrees by all teeth in a field, this trait is rare on LM2 and LM3.

## Geographic Variation

The deflecting wrinkle shows a wide range of variation on a global level. Western Eurasians (along with New Guinea) have the lowest deflecting wrinkle frequencies, varying generally between 5% and 15 %. It is more common in Africans (20–30%) and East and Southeast Asian populations (30–40%). The highest frequencies (55–70%) are found in northeast Siberian and New World populations.

### Select Bibliography

Axelsson, G., and Kirveskari, P. (1977). The deflecting wrinkle on the teeth of Icelanders and the Mongoloid dental complex. *American Journal of Physical Anthropology* 47, 321–324.

Hanihara, K. (1963). Crown characteristics of the deciduous dentition of the Japanese–American hybrids. In *Dental Anthropology*, ed. D.R. Brothwell. New York: Pergamon Press, pp. 105–124.

Hanihara, T. (1992). Negritos, Australian Aborigines, and the "Proto-Sundadont" dental pattern: the basic populations in East Asia, V. *American Journal of Physical Anthropology* 88, 183–196.

Hanihara, K., Kuwasima, T., and Sakao, N. (1964). The deflecting wrinkle on the lower molars in recent man. *Journal of the Anthropological Society of Nippon* 72, 1–7.

Morris, D.H. (1970). On deflecting wrinkles and the *Dryopithecus* pattern in human mandibular molars. *American Journal of Physical Anthropology* 32, 97–104.

Remane, A. (1960). Zähne und gebiss. In *Primatologia, Handbuch der Primatenkunde III. Teil 2*, ed. H. Hofer, A.H. Schultz, and D. Starck. Basel and New York: S. Karger, pp. 637–846.

Suzuki, M., and Sakai, T. (1956). On the "deflecting wrinkle" in recent Japanese. *Journal of the Anthropological Society of Nippon* 65, 49–53 (in Japanese with English summary).

Swindler, D.R., and Ward, S. (1988). Evolutionary and morphological significance of the deflecting wrinkle in the lower molars of the hominoidea. *American Journal of Physical Anthropology* 75, 405–411.

Turner, C.G., II (1990). Major features of Sundadonty and Sinodonty, including suggestions about East Asian microevolution, population history, and late Pleistocene relationships with Australian Aboriginals. *American Journal of Physical Anthropology* 82, 295–317.

Turner, C.G., II, Nichol, C.R., and Scott, G.R. (1991). Scoring procedures for key morphological traits of the permanent dentition: the Arizona State University dental anthropology system. In *Advances in Dental Anthropology*, ed. M.A. Kelley and C.S. Larsen. New York: Wiley-Liss, pp. 13–31.

Weidenreich, F. (1937). The dentition of *Sinanthropus pekinensis*: a comparative odontography of the hominids. *Palaeontologica Sinica*, Whole series 101, New series D-1, 1–180.

# NOTES

# 34

# Distal Trigonid and Mid-Trigonid Crests

## Observed on

LM1, LM2, LM3

## Key Tooth

LM1, LM2

## Synonyms

C1–C2 contact

## Description

The major mesial cusps of the lower molars form the trigonid. The two major cusps (protoconid and metaconid) can exhibit ridges that are connected. If the location of the ridge runs from one essential cusp to the other, the trait is referred to as a mid-trigonid crest. A distal trigonid crest is present when the distal accessory ridges run a direct course along the distal portion of the cusps and come in contact at a point close to the central occlusal sulcus. Both crests can be continuous or discontinuous. Neither mid-trigonid nor distal trigonid crests are common in modern humans (<10%), but they are very common in earlier hominins, especially *Homo heidelbergensis* and Neanderthals. These groups have been studied through micro-CT scans, and the contact is apparent at the enamel–dentine junction, even with crown wear.

Although Turner *et al.* (1991) specified LM1 as the key tooth for the distal trigonid crest, we feel this position should be examined more closely. Both mid-trigonid and distal trigonid crests (DTC) are more common and easier to evaluate on LM2. One issue with scoring DTC on LM1 is the manner in which the essential ridge of the metaconid is typically directed toward the central occlusal sulcus and hypoconid (forming the Y-groove pattern that is almost invariant on LM1). Moreover, the authors are not certain why the standard plaque shows DTC on the deciduous second molar rather than on the permanent molars (LM1 and LM2). Developing a new standard plaque using permanent lower molars and determining the key tooth for the two trigonid crests are separate issues that should be rectified.

## Classification

Hanihara (1961)

### Distal Trigonid Crest

Grade 0: distal trigonid crest absent
Grade 1: distal trigonid crest present

**Figure 34.1** Distal trigonid crest; this crest is formed when the distal accessory ridges of protoconid and metaconid meet at the central occlusal sulcus (can be continuous or discontinuous).

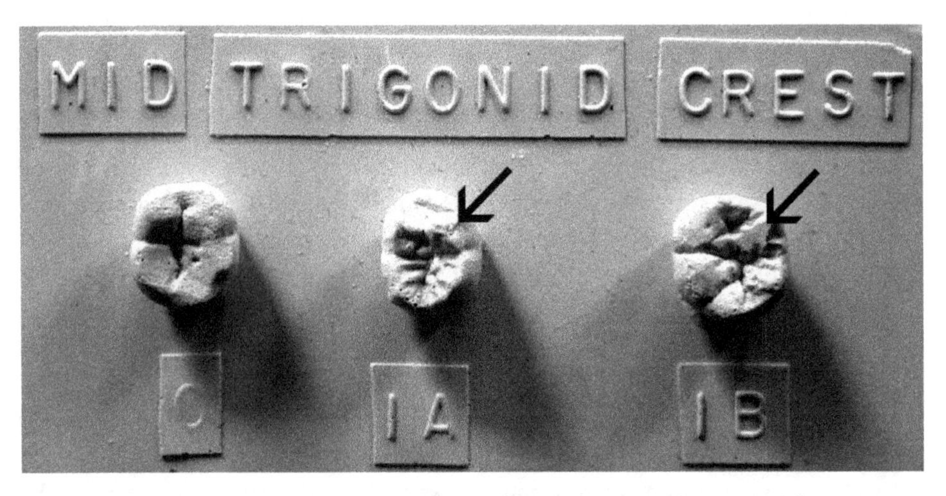


**Figure 34.2** Mid-trigonid crest; this crest runs from cusp tip of protoconid to cusp tip of metaconid (can be continuous or discontinuous).

## Mid-Trigonid Crest

Grade 0: mid-trigonid crest absent
Grade 1: mid-trigonid crest present

## Breakpoint

Trait presence

## Potential Problems in Scoring

In unworn teeth, it is possible to follow the course of the ridges that connect the essential cusps of the protoconid and metaconid and also the distal accessory ridges. When present, the mid-trigonid and distal trigonid crests can be continuous or discontinuous. The outer enamel surface exhibits only a limited part of the range of trigonid crests that have been revealed by micro-CT. The reader is referred to the studies of Bailey *et al.* (2011) and Martínez de Pinillos *et al.* (2014, 2015) for the more complex trigonid crest patterns evident at the enamel–dentine junction. The latter authors define 14 different combinations of mid-trigonid and distal trigonid crest configurations, most of which cannot be readily observed on the external crown surface. This point is aptly illustrated by Bailey *et al.* (2011).

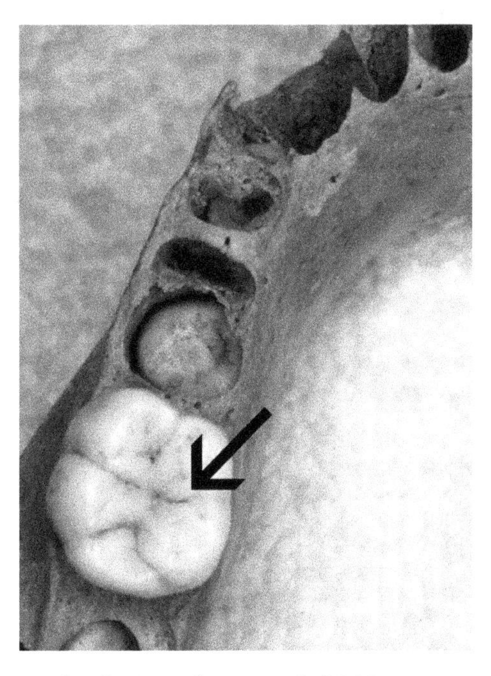

**Figure 34.3** Continuous distal trigonid crest on left LM1.

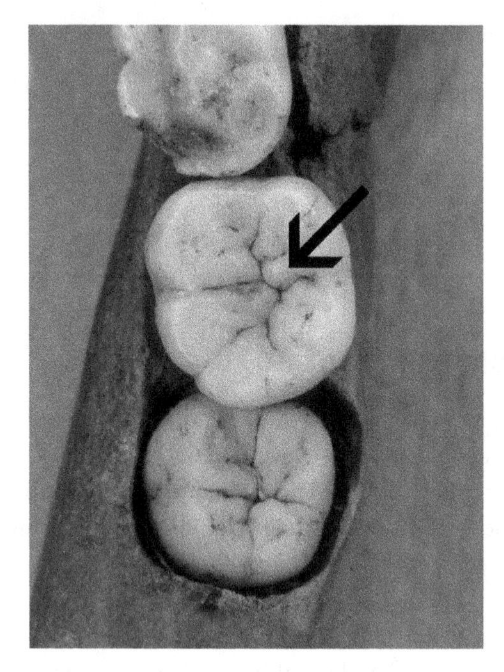

**Figure 34.4** Discontinuous distal trigonid crest on left LM1.

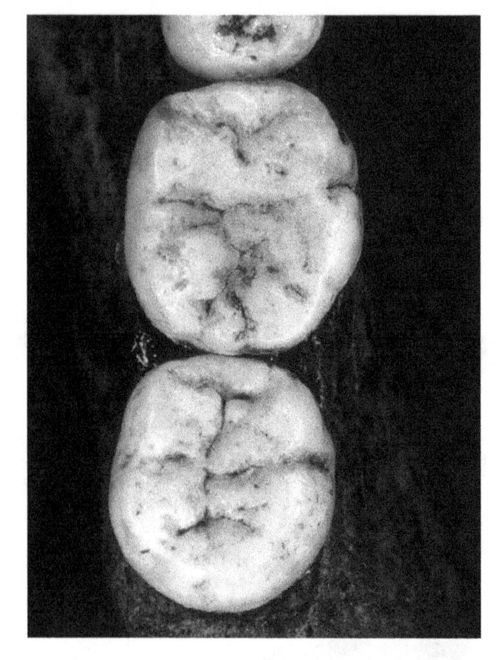

**Figure 34.5** Continuous distal trigonid crest on right LM1 that is visible despite moderate wear.

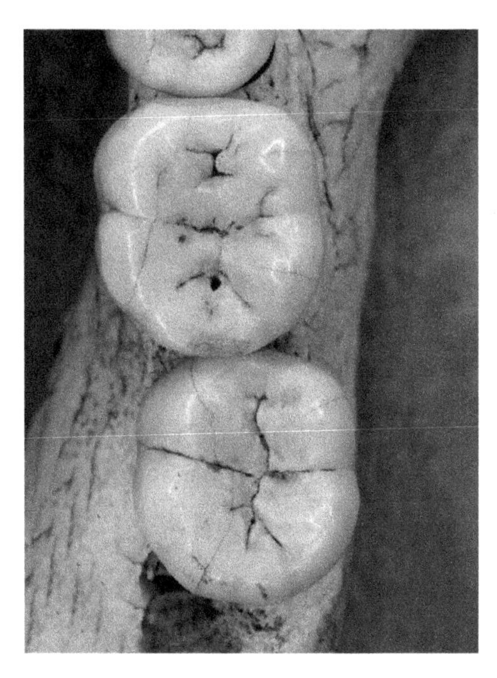

**Figure 34.6** A left LM1 that exhibits a pronounced and continuous distal trigonid crest and a smaller discontinuous mid-trigonid crest.

## Geographic Variation

The geographic variation in the mid-trigonid crest is unknown, but frequencies are available for the distal trigonid crest because it was one of Turner's 29 key traits. The DTC is rare in Africans and Austral-Melanesians (0–2%), with a slightly higher frequency in Western Eurasians (3–7%). It is most common in Asian and Asian-derived populations, where the frequency range is 7–15%. The marked reduction in trigonid crest occurrence is one hallmark of the modern human dentition.

### Select Bibliography

Bailey, S.E. (2002). *Neanderthal Dental Morphology: Implications for Modern Human Origins.* PhD dissertation, Department of Anthropology, Arizona State University, Tempe.

Bailey, S.E., Skinner, M.M., and Hublin, J.J. (2011). What lies beneath? An evaluation of lower molar trigonid crest patterns based on both dentine and enamel expression. *American Journal of Physical Anthropology* 145, 505–518.

Hanihara, K. (1961). Criteria for classification of crown characters of the human deciduous dentition. *Zinriugaku Zassi* 69, 27–45.

Hrdlička, A. (1924). New data on the teeth of early man and certain fossil European apes. *American Journal of Physical Anthropology* 3, 429–465.

Irish, J.D. (1998). Diachronic and synchronic dental trait affinities of Late and Post-Pleistocene peoples from North Africa. *Homo – Journal of Comparative Human Biology* 49, 138–155.

Korenhof, C.A.W. (1978). Remnants of the trigonid crests in medieval molars of man of Java. In *Development, Function and Evolution of Teeth*, ed. P.M. Butler and K. Joysey. New York: Academic Press, pp. 157–169.

(1982). Evolutionary trends of the inner enamel anatomy of deciduous molars from Sangiran (Java, Indonesia). In *Teeth: Form, Function, and Evolution*, ed. B. Kurtén. New York: Columbia University Press, pp. 350–365.

Martínez de Pinillos, M., Martinón-Torres, M., Skinner, M.M., *et al.* (2014). Trigonid crest expression in Atapuerca-Sima de los Huesos lower molars: internal and external morphological expression and evolutionary inferences. *Comptes Rendus Palevol* 13, 205–221.

Martínez de Pinillos, M., Martinón-Torres, M., Martín-Francés, L., Arsuaga, J.L., and Bermúdez de Castro, J.M. (2015). Comparative analysis of the trigonid crests patterns in *Homo antecessor* molars at the enamel and dentine surfaces. *Quaternary International*, in press. http://dx.doi.org/10.1016/j.quaint.2015.08.050 (accessed November 2016).

Turner, C.G., II, Nichol, C.R., and Scott, G.R. (1991). Scoring procedures for key morphological traits of the permanent dentition: the Arizona State University dental anthropology system. In *Advances in Dental Anthropology*, ed. M.A. Kelley and C.S. Larsen. New York: Wiley-Liss, pp. 13–31.

Wu, L., and Turner, C.G., II (1993). Variation in the frequency and form of the lower permanent molar middle trigonid crest. *American Journal of Physical Anthropology* 91, 245–248.

# 35

# Protostylid

## Observed on

LM1, LM2, LM3

## Key Tooth

LM1 (less frequent but often more pronounced on LM2 and LM3)

## Synonyms

Protoconidal cingulum (Robinson 1956)

## Description

Dahlberg (1956) set up the original classification for the protostylid. Like Carabelli's trait, the protostylid is a cingular derivative. In some regards it is a mirror image in that it is expressed on the mesiobuccal cusp of the lower molars rather than the mesiolingual cusp of the upper molars. The two traits are linked to some extent (Scott 1978), although the inter-trait correlations are relatively low. Grade 1, or the buccal pit, may or may not be part of the protostylid complex. It may have been included originally because there is also a pit form of Carabelli's trait. All other expressions of the protostylid are positive. Grade 2 shows a slight mesial deviation of the groove separating the protoconid and hypoconid. Grades 3 and higher involve increasingly pronounced forms of this cingular trait. Despite some parallels, the protostylid never attains the status of a pronounced Carabelli's cusp in modern humans, although it was more pronounced in earlier hominins (Hlusko 2004).

## Classification

Dahlberg (1956)

Grade 0: no pit or positive expression on buccal surface of lower molar
Grade 1: buccal pit (a pit of varying sizes, situated around the midpoint of the crown in the protoconid–hypoconid inter-lobal groove)

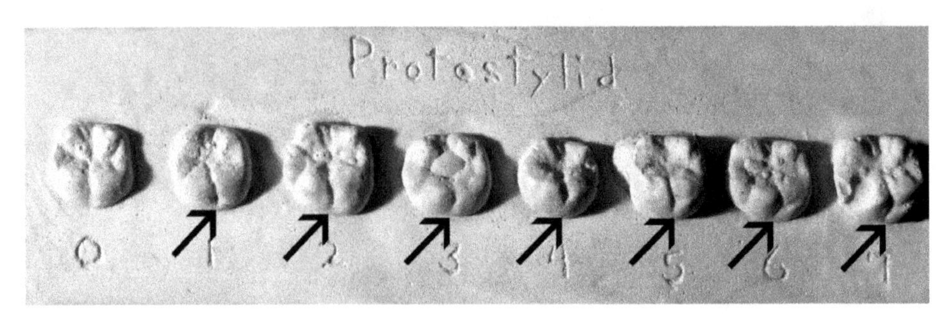


**Figure 35.1** Standard plaque for protostylid expression.

Grade 2: a very slight swelling and associated groove coursing mesially from buccal groove
Grade 3: slight positive expression on mesiobuccal cusp
Grade 4: moderate positive expression
Grade 5: strong positive expression
Grade 6: pronounced positive expression
Grade 7: most distinctive form of protostylid, expressed as tubercle

## Breakpoint

The relationship between the buccal pit (grade 1) and positive forms of the protostylid is unclear. This classification is useful, however, as it allows buccal pits and positive expressions to be considered separately. The breakpoint for positive expressions is grade 2. All manifestations above grade 2 should be scored as protostylids. Buccal pit frequencies can also be directly compared between groups.

## Potential Problems in Scoring

There are pros and cons to scoring cingular traits versus occlusal traits. Cingular traits are less impacted by wear, but in earlier populations the buccal cusps of the lower molars wore down more quickly than the lingual cusps. For scoring cingular traits, lighting is very important, as lower grades of the protostylid are subtle and difficult to score. Unlike a number of other traits, the protostylid is often more pronounced on LM2 and LM3. On occasion, there are slight ridges with associated grooves on the buccal surface of the hypoconid. These are not part of the protostylid complex.

One of us (GRS) considers it likely that positive expressions of the protostylid are more difficult to score on real teeth as opposed to casts because of the marked difference in the findings of C.G. Turner (many publications), who observed thousands of skulls, and G.R. Scott (1973, Scott and Dahlberg 1982, Scott *et al.* 1983), who observed thousand of casts. The frequency of positive protostylid expressions on dental casts is significantly higher than comparable frequencies based on real teeth.

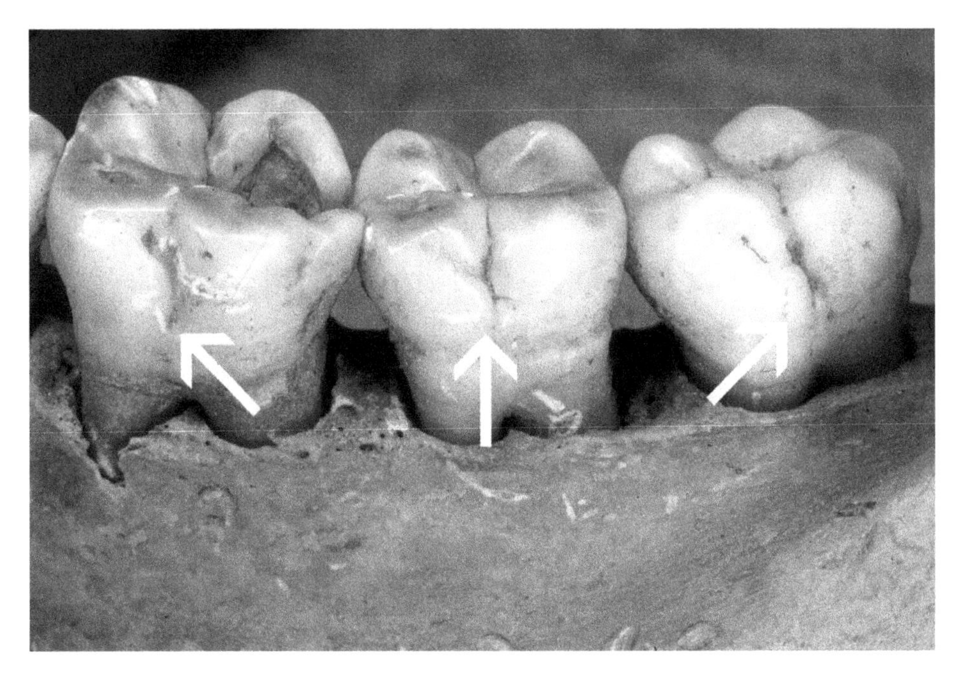

**Figure 35.2** Positive expressions of the protostylid on LM1, LM2, and LM3. For LM1 and LM2, there is a close connection to the groove separating the protoconid and hypoconid. For LM3, this is not the case.

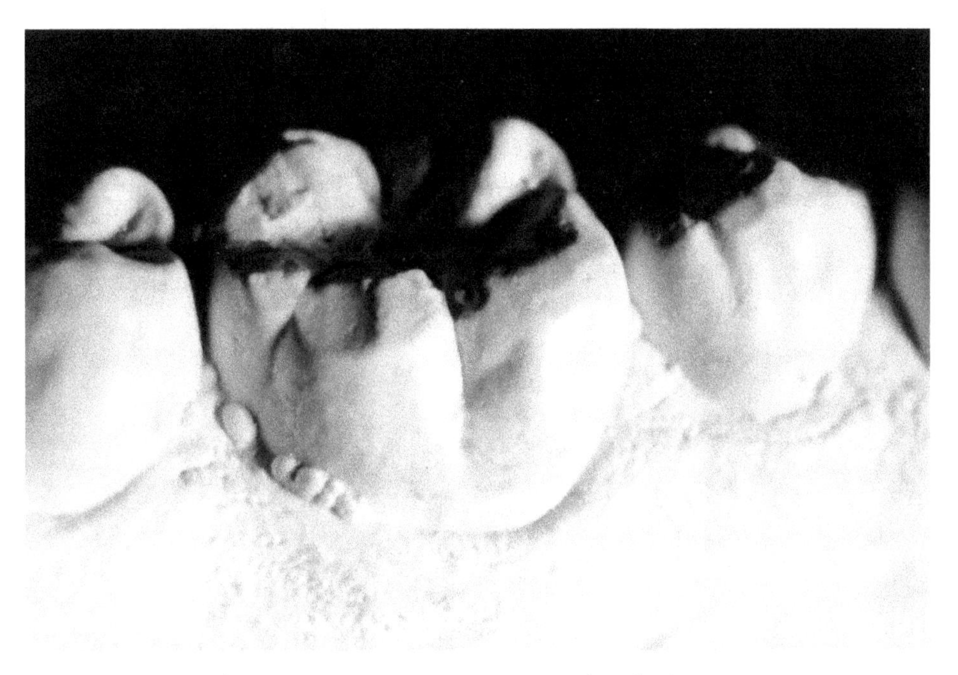

**Figure 35.3** A moderate positive expression on LM1 (grade 4).

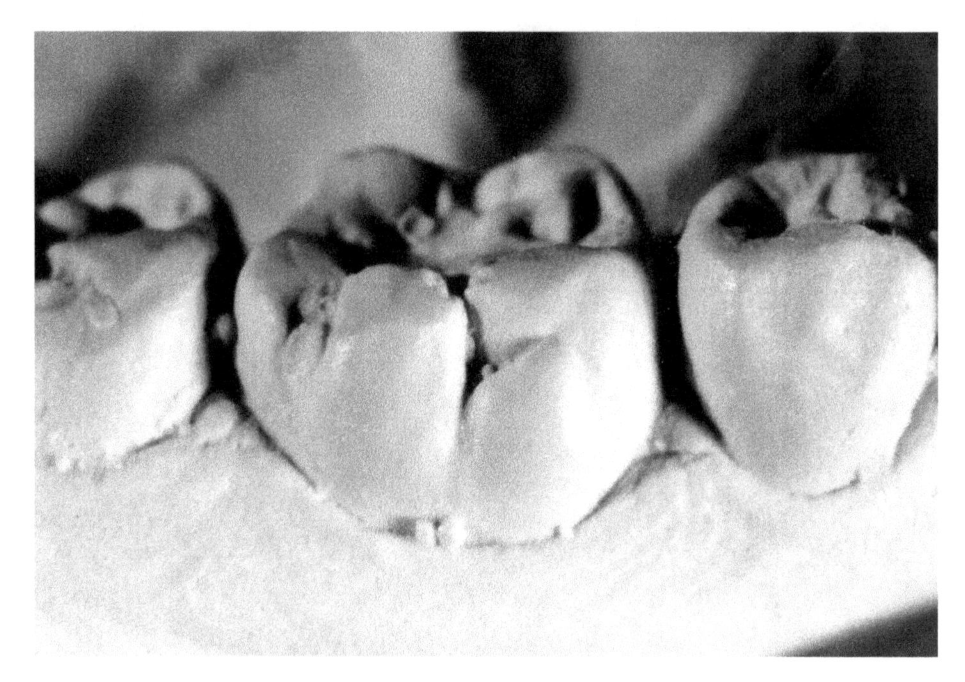

**Figure 35.4** A well-developed positive expression on LM1 (grade 6). In modern humans, the protostylid rarely exhibits a free apex comparable to a large Carabelli's cusp.

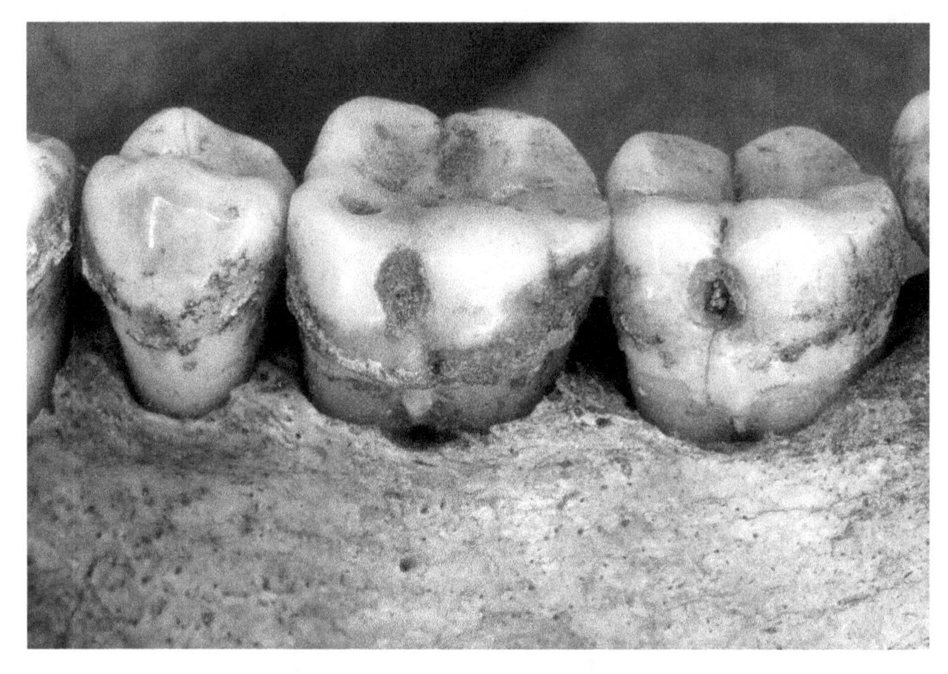

**Figure 35.5** LM1 and LM2 both exhibit buccal pits (grade 1).

## Geographic Variation

The protostylid was far more common in Australopithecines than in modern humans, with expression centered on LM2 rather than LM1 (Hlusko 2004). In modern populations, it is rare in Western Eurasians. Asian and Asian-derived groups have the highest frequencies, but when buccal pits are excluded the trait is not exceptionally common in any group (see above for possible caveat).

## Select Bibliography

Dahlberg, A.A. (1950). The evolutionary significance of the protostylid. *American Journal of Physical Anthropology* 8, 15–25.

(1956). Materials for the establishment of standards for classification of tooth characters, attributes, and techniques in morphological studies of the dentition. Zollar Laboratory of Dental Anthropology, University of Chicago (mimeo).

Hlusko, L. (2004). Protostylid variation in *Australopithecus*. *Journal of Human Evolution* 46, 579–594.

Robinson, J.T. (1956). *The Dentition of the Australopithecinae*. Pretoria: Transvaal Museum Memoir, Number 9.

Scott, G.R. (1973). *Dental Morphology: A Genetic Study of American White Families and Variation in Living Southwest Indians*. PhD dissertation, Department of Anthropology, Arizona State University, Tempe.

(1978). The relationship between Carabelli's trait and the protostylid. *Journal of Dental Research* 57, 570.

Scott, G.R., and Dahlberg, A.A. (1982). Microdifferentiation in tooth crown morphology among Indians of the American Southwest. In *Teeth: Form, Function, and Evolution*, ed. B. Kurten. New York: Columbia University Press, pp. 259–291.

Scott, G.R., Potter, R.H.Y., Noss, J.F., Dahlberg, A.A., and Dahlberg, T. (1983). The dental morphology of Pima Indians. *American Journal of Physical Anthropology*, 61, 13–31.

Skinner, M.M., Wood, B.A., and Hublin, J.-J. (2009). Protostylid expression at the enamel–dentine junction and enamel surface of mandibular molars of Paranthropus robustus and *Australopithecus africanus*. *Journal of Human Evolution* 56, 76–85.

Smith, P., Propopec, M., and Pretty, G. (1988). Dentition of a prehistoric population from Roonka Flat, South Australia. *Archaeology in Oceania* 23, 31–36.

Turner, C.G., II (1976). Dental evidence on the origins of the Ainu and Japanese. *Science* 193, 911–913.

Wood, B.A., Abbott, S.A., and Graham, S.H. (1983). Analysis of the dental morphology of Plio-Pleistocene hominids. II. Mandibular molars – study of cusp areas, fissure pattern and cross sectional shape of the crown. *Journal of Anatomy* 137, 287–314.

# NOTES

# 36

# Cusp 6

## Observed on

LM1, LM2, LM3

## Key Tooth

LM1

## Synonyms

*Tuberculum sextum*, entoconulid (Turner *et al.* 1969)

## Description

Like the hypoconulid, cusp 6 is expressed on the distal portion of the lower molars. While the hypoconulid is associated with the hypoconid (cusp 3), cusp 6 is associated with the entoconid (cusp 4), hence the term entoconulid. In this classification, cusp 5 has to be present for cusp 6 to be scored.

## Classification

Turner *et al.* (1969)

Grade 0: absence of cusp 6
Grade 1: cusp 5 is more than twice the size of cusp 6
Grade 2: cusp 5 is about twice as large as cusp 6
Grade 3: cusps 5 and 6 are about equal in size
Grade 4: cusp 6 is slightly larger than cusp 5
Grade 5: cusp 6 is markedly larger than cusp 5

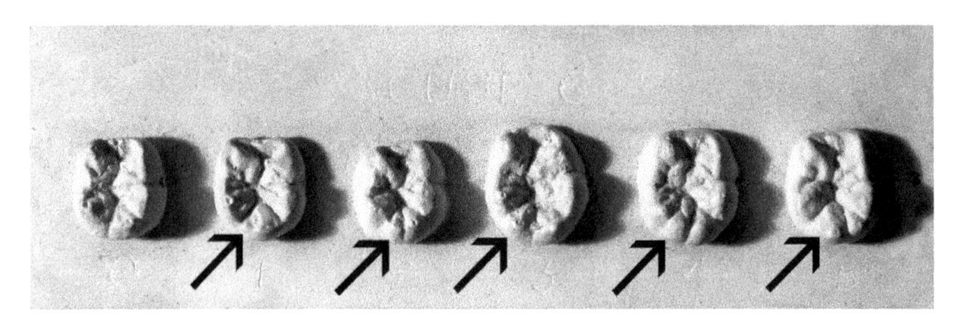

**Figure 36.1** Standard plaque for cusp 6 of lower molars.

## Breakpoint

1+ (presence)

## Potential Problems in Scoring

Some authors have raised the issue of whether or not cusp 6 can be present in the absence of cusp 5. In the Turner classification, this possibility is precluded. In those cases where the distal cusp is clearly in association with the entoconid rather than the hypoconid, this might indicate the presence of cusp 6 and absence of cusp 5. Until this problem is resolved, one should be judicious in assigning the presence of cusp 6 in the absence of cusp 5.

## Geographic Variation

Cusp 6 shows a wide range of geographic variation, making it especially useful for biodistance and forensic studies. The trait is relatively rare in Europeans (5–15%) but is extremely common in Asian and Asian-derived groups (40–50%), sometimes reaching frequencies in excess of 70%. The trait is moderate in African populations (20–40%) and quite high in Australian and Melanesian populations, where the frequencies fall between 50 and 60%.

**Figure 36.2** Even with some wear, a grade 2 cusp 6 is clearly evident on the right LM1.

**Figure 36.3** For this right LM1, cusp 5 and cusp 6 are of about equal size (grade 3).

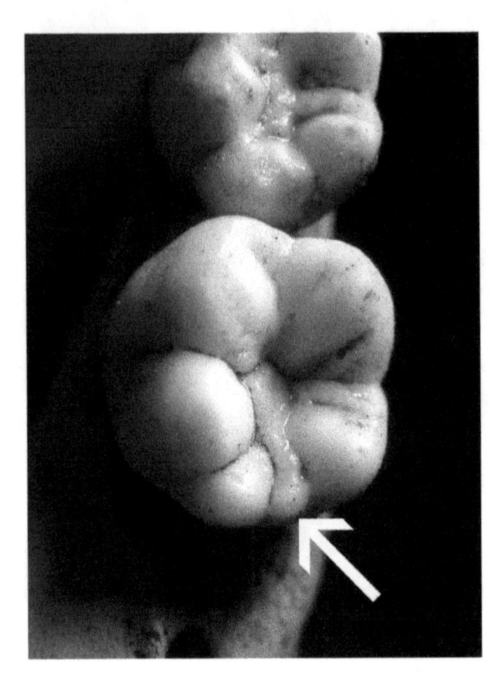

**Figure 36.4** This unworn molar exhibits an unusual enamel formation that extends from the central occlusal fossa to the distal border of the tooth between the hypoconulid and entoconid. If the tooth were worn, it would appear to be a grade 2 cusp 6. Without wear, it is an idiosyncratic variant that is independent of cusp 6.

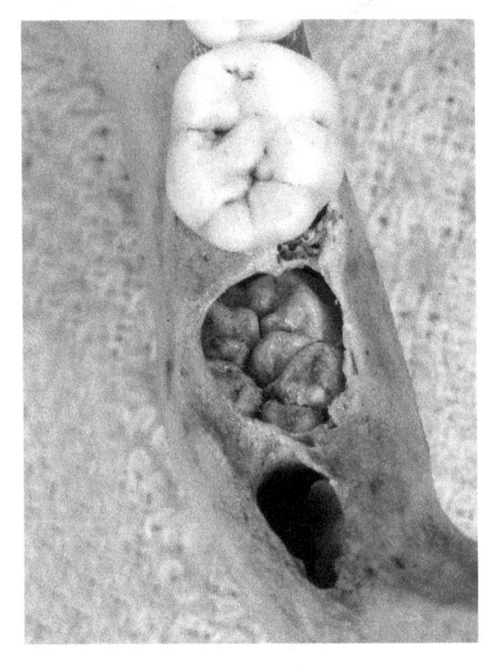

**Figure 36.5** A right dm2 with a distinct grade 2 cusp 6 (presence of cusp 6 on unerupted LM1 likely but not possible to discern in photo).

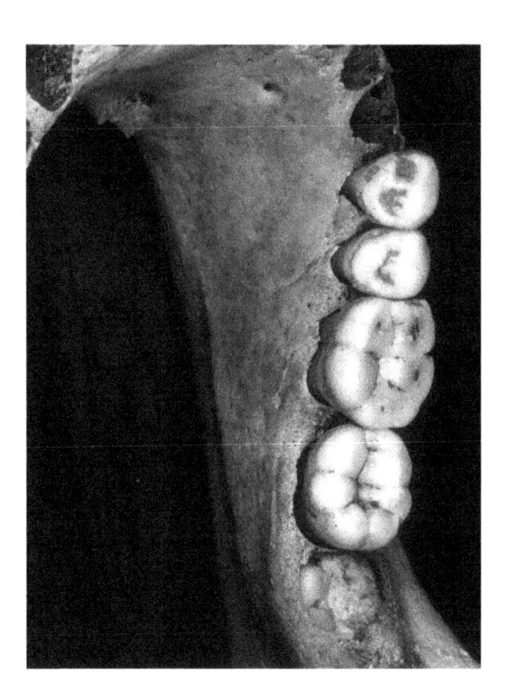

**Figure 36.6** A mandible with a very large cusp 6 on the right LM2. Although cusp 6 is less common on LM2 than LM1, it is frequently expressed as a higher grade on LM2.

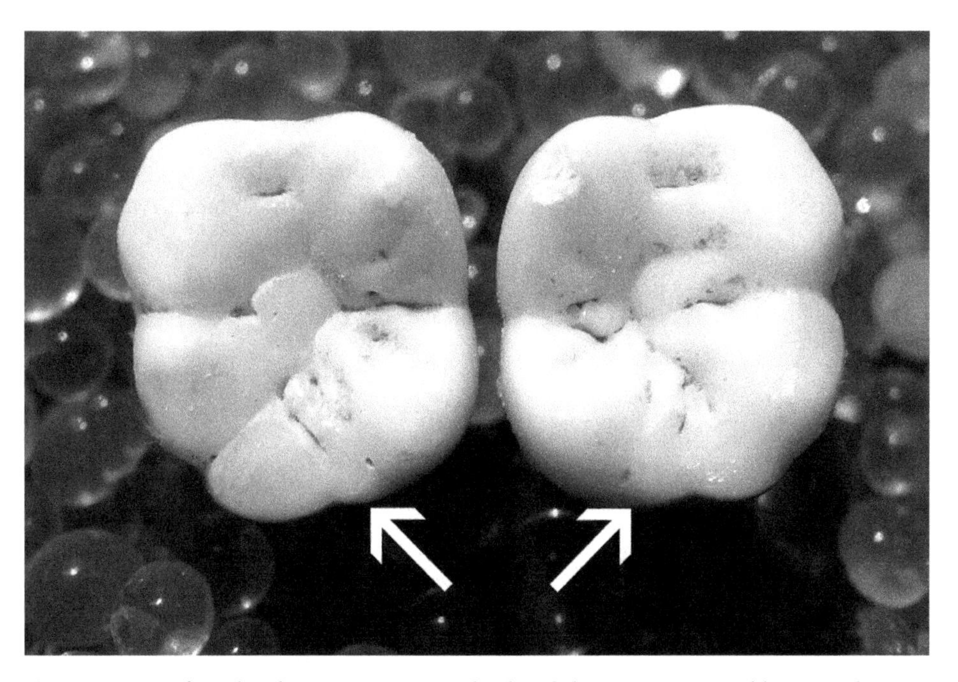

**Figure 36.7** Left and right LM1 antimeres both exhibiting an unusual but mostly symmetrical form of cusp 6.

## Select Bibliography

Axelsson, G., and Kirveskari, P. (1979). Sixth and seventh cusp on lower molar teeth of Icelanders. *American Journal of Physical Anthropology* 51, 79–82.

Dahlberg, A.A. (1961). Relationship of tooth size to cusp number and groove conformation of occlusal surface patterns of lower molar teeth. *Journal of Dental Research* 40, 34–38.

Irish, J.D. (1997). Ancestral dental traits in recent Sub-Saharan Africans and the origins of modern humans. *Journal of Human Evolution* 34, 81–98.

Keene, H.J. (1994). On the classification of C6 (tuberculum sextum) of the mandibular molars. *Human Evolution* 9, 231–247.

Mayhall, J.T., and Saunders, S.R. (1986). Dimensional and discrete dental trait asymmetry relationships. *American Journal of Physical Anthropology* 69, 403–411.

Skinner, M.M., Gunz, P., Wood, B.A., Boesch, C., and Hublin, J.-J. (2010). Discrimination of extant *Pan* species and subspecies using the enamel–dentine junction morphology of lower molars. *American Journal of Physical Anthropology* 140, 234–243.

Townsend, G.C., Hiroyuki, Y., and Smith, P. (1990). Expression of the entoconulid (sixth cusp) on mandibular molar teeth of an Australian aboriginal population. *American Journal of Physical Anthropology* 82, 267–274.

Turner, C.G., II (1985). Expression count: A method for calculating morphological dental trait frequencies by using adjustable weighting coefficients with standard ranked scales. *American Journal of Physical Anthropology* 68, 263–267.

Turner, C.G., II, and Hanihara, K. (1977). Additional features of Ainu dentition. *American Journal of Physical Anthropology* 46, 13–24.

Turner, C.G., II, Scott, G.R., and Rose, T.A. (1969). Mandibular molar cusp 6 plaque and definitions of variation. Department of Anthropology, Arizona State University, Tempe.

Wood, B.A., and Abbott, S.A. (1983). Analysis of the dental morphology of Plio-Pleistocene hominids. I. Mandibular molars: crown area measurements and morphological traits. *Journal of Anatomy* 136, 197–219.

# 37

# Cusp 7

## Observed on

LM1, LM2, LM3

## Key Tooth

LM1

## Synonyms

*Tuberculum intermedium*, metaconulid (Turner *et al.* 1970)

## Description

Cusp 7 is a wedge-shaped accessory cusp expressed between cusps 2 (metaconid) and 4 (entoconid); hence the synonym *tuberculum intermedium*. In the original classification, Turner added a 1A grade that assumes a form similar to that expressed on lower second deciduous molars. However, there are cases where this expression is coupled with a cusp expression, which raises the issue as to whether or not this expression should be included with other forms of cusp 7. Scoring 1A as cusp 7 obscures a distinctive pattern of geographic variation. We recommend that when 1A is scored, it should not be included in deriving the total frequency of cusp 7.

## Classification

Turner *et al.* (1970)

Grade 0: no accessory cusp between cusps 2 and 4
Grade 1: small, wedge-shaped cusp between cusps 2 and 4
Grade 1A: this expression does not assume the typical wedge-shaped form of a cusp 7 but is marked by a groove on the lingual surface of the metaconid
Grade 2: distinct but small cusp
Grade 3: moderate cusp
Grade 4: large cusp

## Breakpoint

1+(total frequency of presence, minus category 1A)

## Potential Problems in Scoring

Although cusp 7 is an occlusal trait, it can withstand a moderate amount of wear and still be scored. The key is to recognize two vertical grooves on the lingual surface of LM1, with one on the metaconid and one on the entoconid. In most instances, the

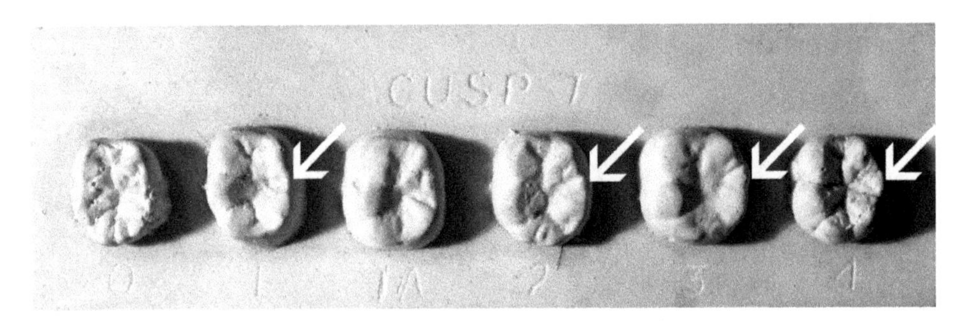


**Figure 37.1** Standard plaque for cusp 7 of the lower molars.

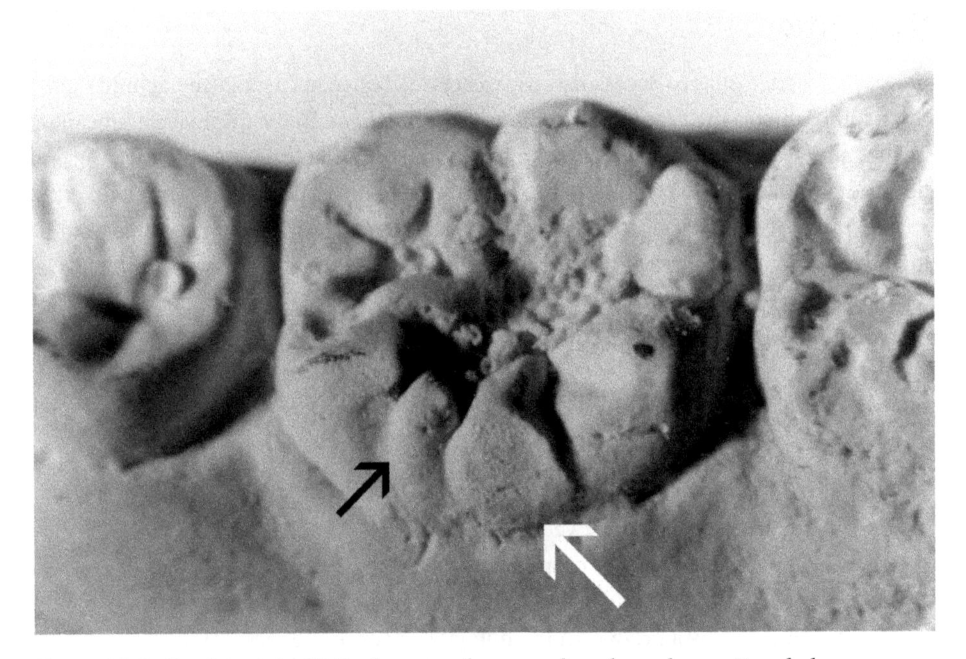

**Figure 37.2** On this right LM1, there is a large wedge-shaped cusp 7 and also a post-metaconulid that Turner referred to as grade 1A. Although 1A can be scored, it should not be added to the frequency for conventional cusp 7 expressions.

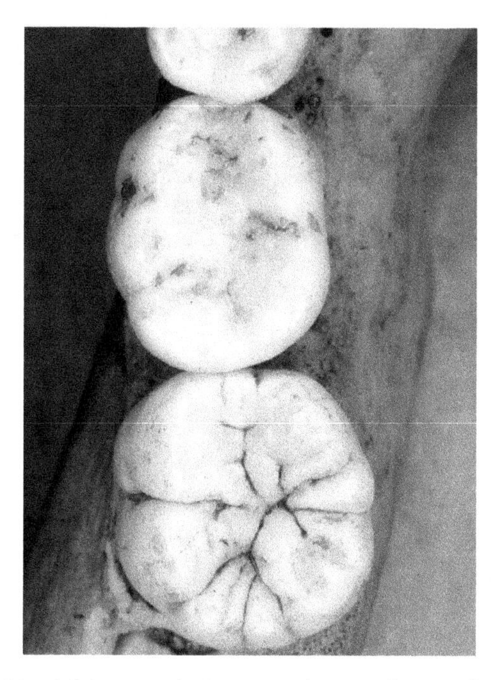

**Figure 37.3** Left LM1 exhibits a grade 3 cusp 7 along with a grade 2 cusp 6. Although these traits sometimes occur together, they do not show a significant inter-trait correlation.

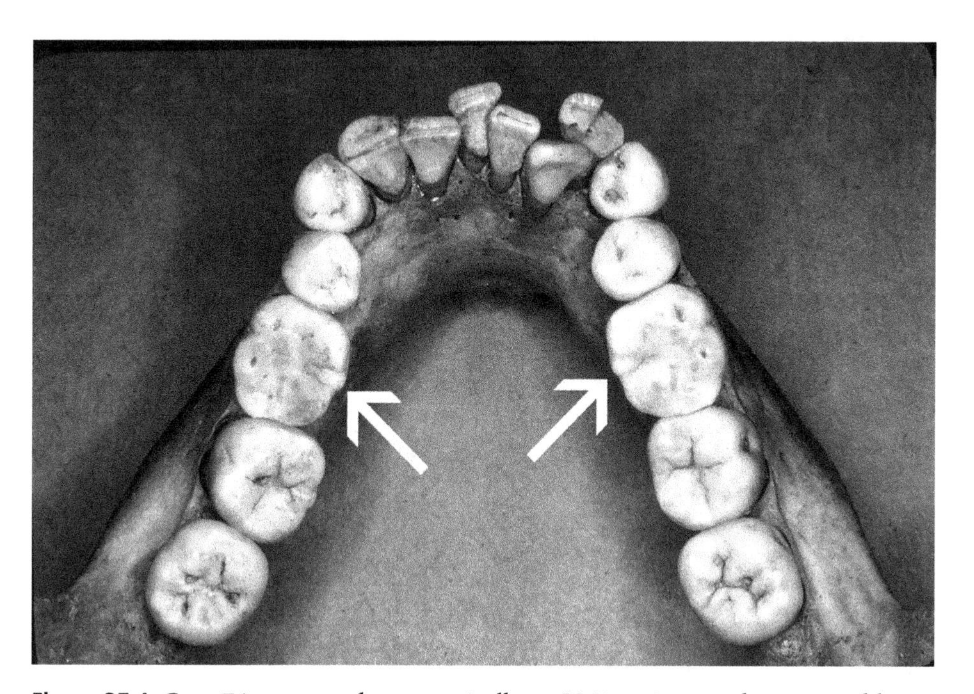

**Figure 37.4** Cusp 7 is expressed symmetrically on LM1 antimeres; this trait, unlike cusp 6, is rarely present on LM2 or LM3, although 1A forms sometimes occur on these teeth.

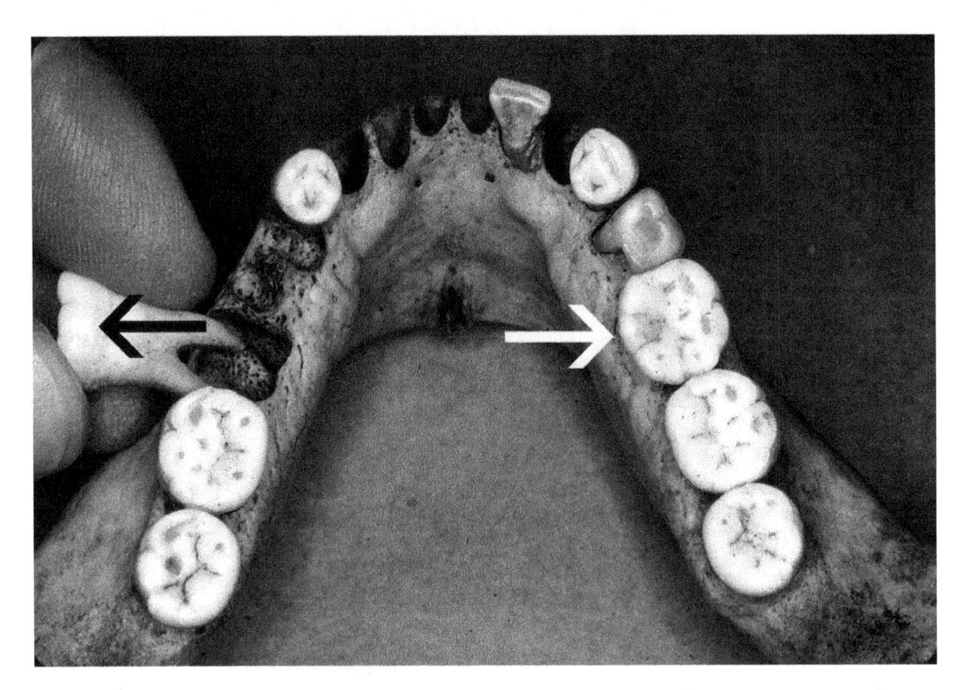

**Figure 37.5** An extraordinarily large cusp 7 on right LM1; in this instance, it is almost as large as the major cusps of the lower molars.

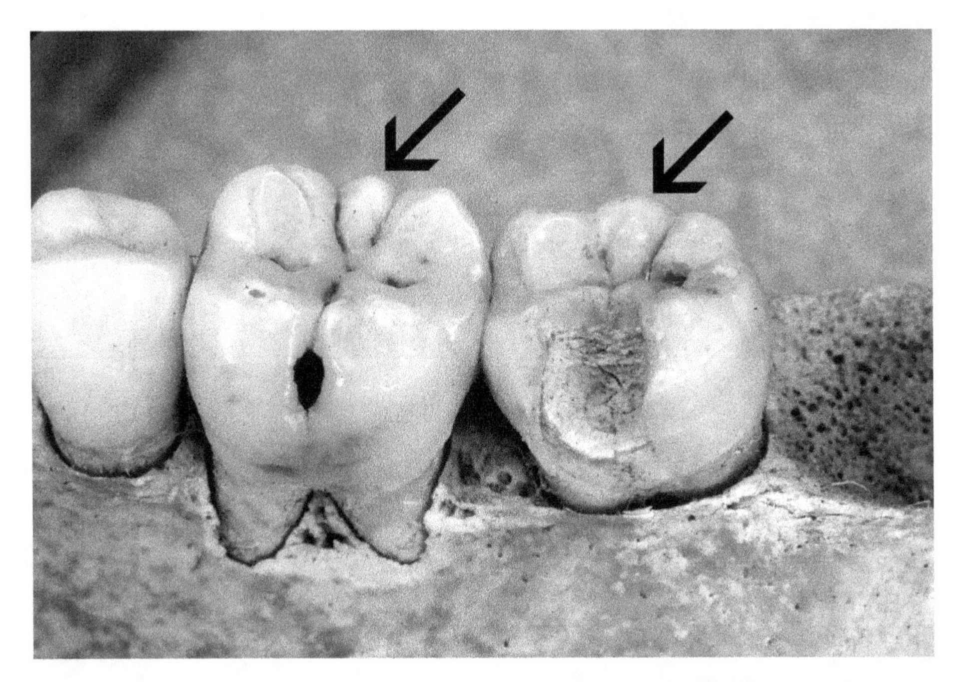

**Figure 37.6** Cusp 7 from a different angle; in this buccal view of left lower molars, a distinct cusp 7 of grade 2 is present between the metaconid and entoconid of LM1, with an even greater expression on LM2 (grade 4). (This photo was taken for the buccal pits, including the one on LM2 that was the focus of a large caries.)

wedge form of cusp 7 should be distinguishable. One has to be careful not to score the distal accessory ridge as a cusp 7, as this has led to some highly inflated cusp 7 frequencies.



## Geographic Variation

Cusp 7 shows a distinctive pattern of geographic variation. In most world populations, it is relatively rare, falling between 3% and 8%. The only exception is for African populations, where the frequencies range from 25% to 40%. For this reason, it is one of the most useful variables in distinguishing African and African-derived groups from other populations.

## Select Bibliography

Axelsson, G., and Kirveskari, P. (1979). Sixth and seventh cusp on lower molar teeth of Icelanders. *American Journal of Physical Anthropology* 51, 79–82.

Biggerstaff, R.H. (1970). Morphological variations for the permanent mandibular first molars in human monozygotic and dizygotic twins. *Archives of Oral Biology* 15, 721–730.

Gupta, S.K., and Saxena, P. (2013). Prevalence of cusp 7 in permanent mandibular first molars in an Indian population: a comparative study of variations in occlusal morphology. *Journal of Investigative and Clinical Dentistry* 4, 240–246.

Irish, J.D., and Guatelli-Steinberg, D. (2003). Ancient teeth and modern human origins: an expanded comparison of African Plio-Pleistocene and recent world dental samples. *Journal of Human Evolution* 45, 113–144.

Irish, J.D., and Koningsberg, L.W. (2006). The ancient inhabitants of Jebel Moya Redux: Measures of population affinity based on dental morphology. *International Journal of Osteoarchaeology* 17, 138–156.

Skinner, M.M., Wood, B.A., Boesch, C., Olejniczak, A.J., Rosas, A., Smith, T.M., and Hublin, J.-J. (2008). Dental trait expression at the enamel–dentine junction of lower molars in extant and fossil hominoids. *Journal of Human Evolution* 54, 173–186.

Turner, C.G., II, Scott, G.R., and Larsen, M. (1970). Mandibular molar cusp 7 plaque and definitions of variation. Department of Anthropology, Arizona State University, Tempe.

# NOTES

# 38

# Lower First Premolar Root Number

## Observed on

LP1

## Key Tooth

LP1

## Synonyms

Tomes' root (Tomes 1923)

## Description

Lower premolars do not have two prominent cusps like upper premolars, and this difference is reflected in their respective roots. In the upper premolars, each cusp, buccal and lingual, is associated with its own independent root or distinct root cone. In the lower premolars, the buccal cusp is more prominent than the lingual cusp. The lingual cusp is not only smaller than the buccal cusp, but it is commonly off-center. This reflects the condition of the roots. There are not clearly distinct buccal and lingual root cones, but a buccal root cone and one or more lingual cones, with the most prominent on the mesial boundary of the tooth. An LP1 may exhibit one, two, three, or four root radicals. There may or may not be inter-radicular projections separating these radicals. Tomes' root constitutes that instance where a mesiolingual root cone exhibits an inter-radicular projection, producing an independent root. This is rarely as pronounced as the root bifurcation observed in the upper premolars.

## Classification

Turner *et al.* (1991)

In the six-grade scale for Tomes' root, grades 1–3 exhibit root grooves that are presumed to anticipate a Tomes' root. Only at grade 5 is there a distinct extra root.

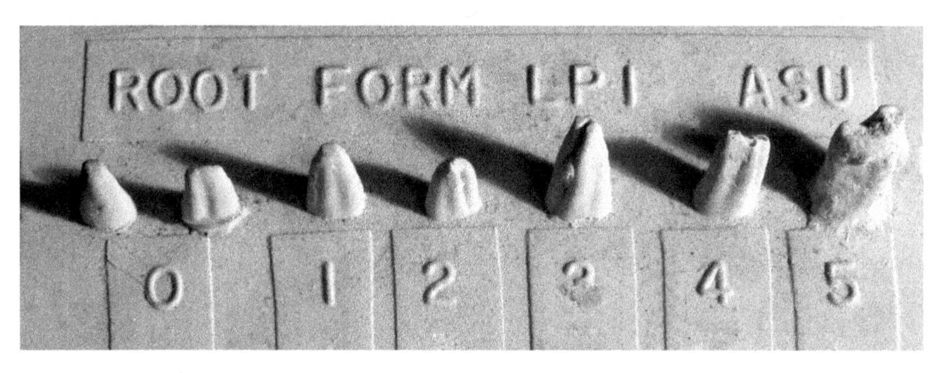


**Figure 38.1** Standard plaque for Tomes' root of LP1.

Grade 0: slight or no groove separating cones on mesial surface of LP1 root
Grade 1: slight V-shaped groove separating cones
Grade 2: deeper V-shaped groove separating cones
Grade 3: deep developmental groove separating root cones along at least ⅓ of root
Grade 4: deep grooving on both mesial and distal surfaces of root
Grade 5: inter-radicular projection present so LP1 has two roots, a large buccal root and a smaller mesial/lingual root

## Breakpoint

Grades 4 and 5 are considered the only clear manifestations of a Tomes' root. Turner later added two higher grades (6 and 7) that involve more pronounced inter-radicular projections setting off distinct LP1 Tomes' roots. Although we do not have an illustration of these higher grades, they are included in the tables in the Appendix. For the eight-grade scale (0–7), Turner used grade 4+ as the breakpoint for tabulating Tomes' root frequencies.

## Potential Problems in Scoring

Most root traits can be scored based on the socket, even in the absence of the tooth. Since Tomes' root is not as distinct as other major root variants, scoring presence or absence based on the socket alone should be done with extreme care.

## Geographic Variation

Sub-Saharan Africans exhibit the highest frequency of Tomes' root (>25%). It is moderately common in Australia and Southeast Asia (15–25%). Europeans and East Asians have relatively low frequencies of a distinct Tomes' root (5–15%).

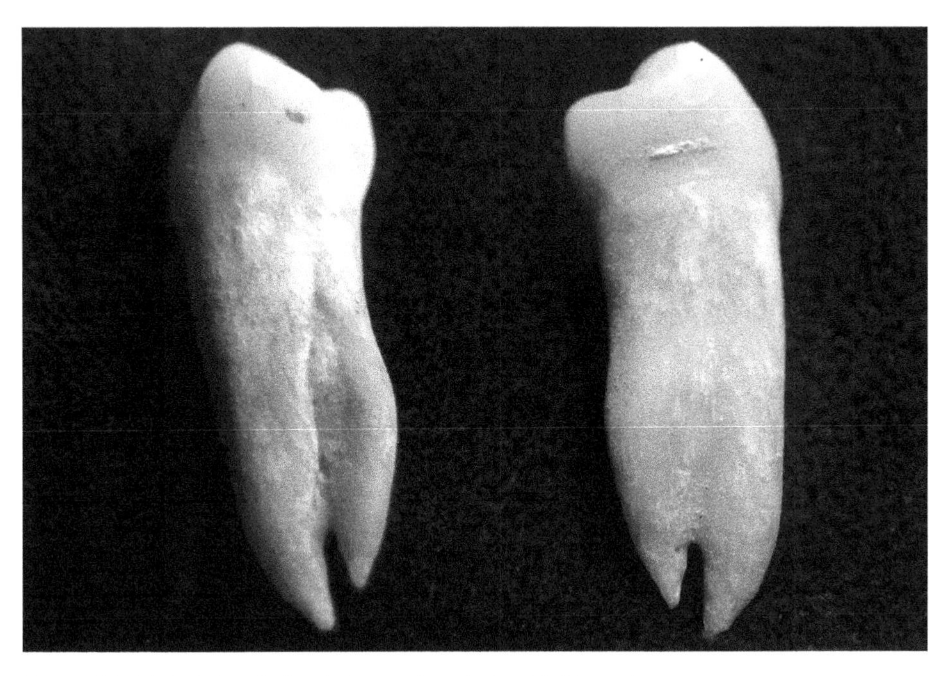

**Figure 38.2** Mesial and distal views of an LP1 with Tomes' root.

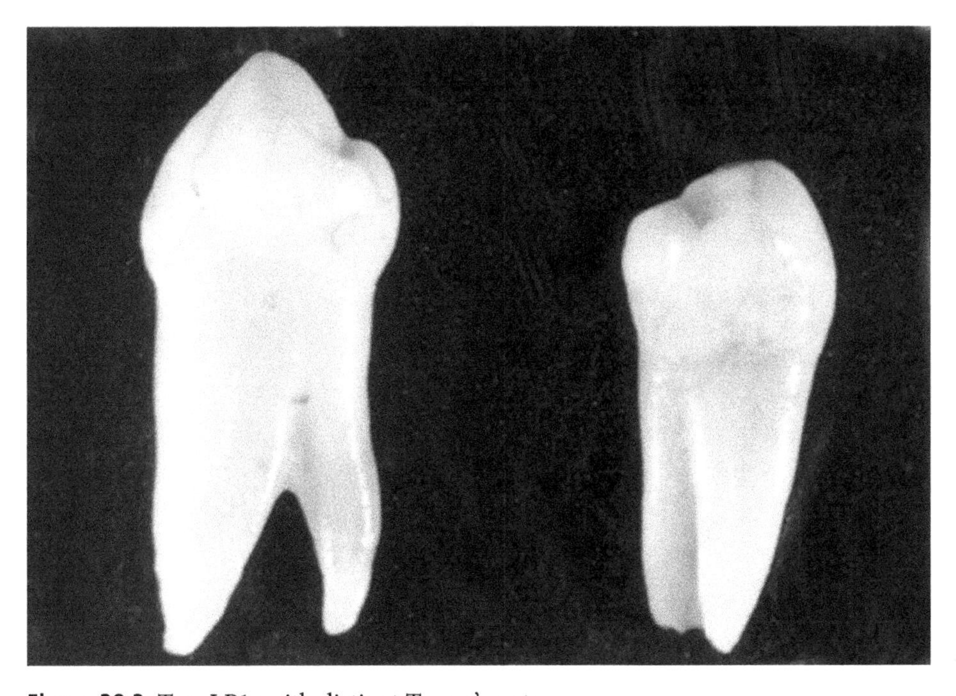

**Figure 38.3** Two LP1s with distinct Tomes' roots.

## Select Bibliography

Abbott, S.A. (1984). *A Comparative Study of Tooth Root Morphology in the Great Apes, Modern Man and Early Hominids.* PhD dissertation, Department of Anatomy, University of London.

Cleghorn, B.M., Christie, W.H., and Dong, C.C. (2007). The root and canal morphology of the mandibular first premolar: a literature review. *Journal of Endodontics* 33, 509–516.

Gu, Y., Zhang, Y., and Liao, Z. (2013). Root and canal morphology of mandibular first premolars with radicular grooves. *Archives of Oral Biology* 58, 1609–1617.

Peiris, R. (2008). Root and canal morphology of human permanent teeth in a Sri Lankan and Japanese population. *Anthropological Science* 116, 123–133.

Shields, E.D. (2005). Mandibular premolar and second molar root morphological variation in modern humans: what root number can tell us about tooth morphogenesis. *American Journal of Physical Anthropology* 128, 299–311.

Tomes, C.S. (1923). *A Manual of Dental Anatomy: Human and Comparative*, 8th edn. New York: MacMillan.

Turner, C.G., II, Nichol, C.R., and Scott, G.R. (1991). Scoring procedures for key morphological traits of the permanent dentition: the Arizona State University dental anthropology system. In *Advances in Dental Anthropology*, ed. M.A. Kelley and C.S. Larsen. New York: Wiley-Liss, pp. 13–31.

Varrela, J. (1992). Effect of 45,X/46,XX mosaicism on root morphology of mandibular premolars. *Journal of Dental Research* 71, 1604–1606.

Wheeler, R.C. (1965). *A Textbook of Dental Anatomy and Physiology*, 4th edn. Philadelphia and London: W.B. Saunders.

Wood, B.A., Abbott, S.A., and Uytterschaut, H. (1988). Analysis of the dental morphology of Plio-Pleistocene hominids. IV. Mandibular postcanine root morphology. *Journal of Anatomy* 156, 107–139.

# 39

# Lower Canine Root Number

## Observed on

LC

## Key Tooth

LC

## Synonyms

Two-rooted lower canines

## Description

Upper and lower canines are typically single-rooted teeth. In some instances, however, the lower canines exhibit two roots. Although canines are single-cusped teeth, they are associated with two root cones. These cones ordinarily take the form of radicals separated by root grooves rather than independent roots separated by an inter-radicular projection. Although relatively uncommon, two-rooted lower canines can reach frequencies of around 10% in some populations, and the variation shows distinct geographic patterning.

## Classification

Grade 0: one-rooted LC, with or without root grooves separating buccal and lingual cones

Grade 1: two-rooted LC, with inter-radicular projection separating buccal and lingual cones by at least ¼ to ⅓ of total root length

## Breakpoint

No scale has been established for two-rooted lower canines, although they vary in terms of the location of the inter-radicular projection. Any form of two distinct roots constitutes a two-rooted LC.

## Potential Problems in Scoring

A researcher has to use judgment as to when to classify a lower canine as two-rooted. In populations where the trait is fairly common, this principle can be violated in several ways, and the presumption is that the genes for a two-rooted tooth are in play (see Figures 39.1–39.4). This is especially true when an unusual root form is symmetrical. This trait can be scored even in the absence of the lower canines, but caution is urged: a thin trabecular bone in the empty socket has to indicate clearly an inter-radicular projection.

## Geographic Variation

Two-rooted lower canines are one of the few traits that distinguish Western Eurasians from other populations. The trait is most common in Europeans, often falling in a range of 5–10%. East Asians sometimes exhibit the trait, but it is usually in a frequency around or below 1%. It may be a useful indicator of gene flow between European and Asian populations where admixed groups show intermediate trait frequencies (Lee and Scott 2011). In African, Pacific, and New World populations, the trait is rarely observed.

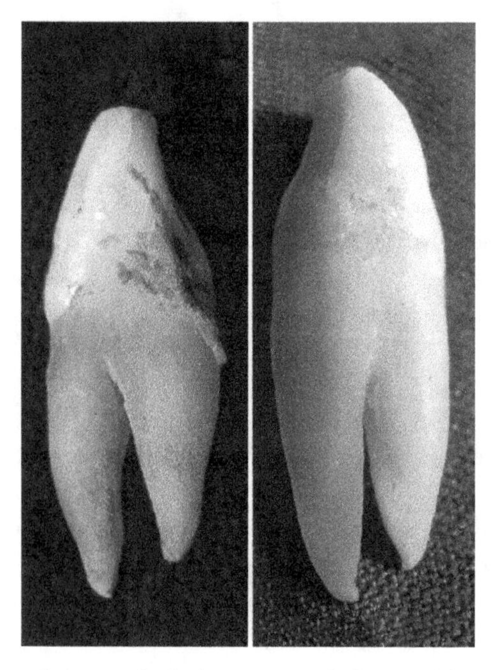

**Figure 39.1** Two-rooted LC in which the roots are bifurcated for more than half of total root length.

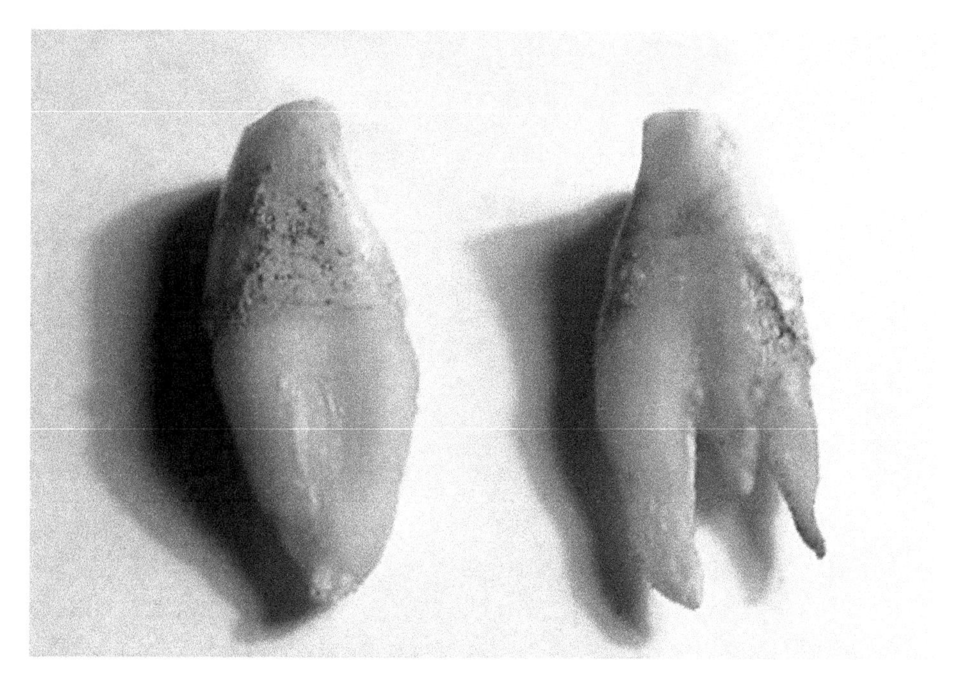

**Figure 39.2** An extremely rare and asymmetrical three-rooted LC.

**Figure 39.3** Distal aspect of two LC antimeres; even though the roots are not bifurcated, these teeth came from a population with a high (10%) frequency of two-rooted LC. Given that, these teeth could be scored as two-rooted.

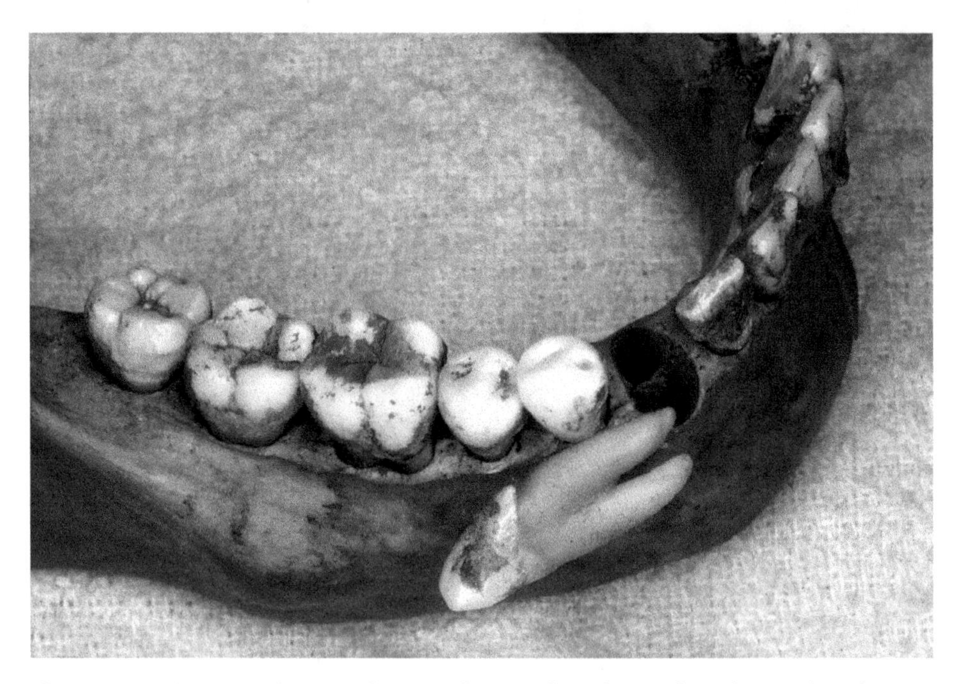

**Figure 39.4** Two-rooted LC can be scored even when the tooth is absent; there has to be a distinct plate of alveolar bone separating the two roots.

## Select Bibliography

Alexandersen, V. (1962). Root conditions in human lower canines with special regard to double-rooted canines. II. Occurrence of double-rooted lower canines in *Homo sapiens* and other primates. *Sætryk af Tandlægebladet* 66, 729–760.

(1963). Double-rooted human lower canine teeth. In *Dental Anthropology*, ed. D.R. Brothwell. New York: Pergamon Press, pp. 235–244.

Harborow, C. (1934). The two-rooted mandibular canine. *British Dental Journal* 56, 244–246.

Lee, C., and Scott, G.R. (2011). Brief communication: two-rooted lower canines – a European trait and sensitive indicator of admixture across Eurasia. *American Journal of Physical Anthropology* 146, 481–485.

Scott, G.R., Anta, A., Schomberg, R., and de la Rua, C. (2013). Basque dental morphology and the "Eurodont" dental pattern. In *Anthropological Perspectives on Tooth Morphology: Genetics, Evolution, Variation*, ed. G.R. Scott and J.D. Irish. Cambridge: Cambridge University Press, pp. 296–318.

# 40

# Three-Rooted Lower Molar

## Observed on

LM1, LM2, LM3

## Key Tooth

LM1

## Synonyms

3RM1, lower molar distolingual root, *radix entomolaris* (Carlsen 1987), supernumerary radicular structure

## Description

The standard root number for lower molars is two. There are two large roots, one mesial and one distal. The mesial root is associated with the trigonid (now missing the paraconid) and the distal root is associated with the talonid. Both roots are narrow mesiodistally and broad buccolingually. Often, a groove separates two root cones for these two major roots. An interesting variant of lower molar root number takes the form of an accessory distolingual root. This root is notably smaller than the associated distobuccal root. As this accessory root is often aligned with the distobuccal root, it is frequently overlooked in lateral x-rays. This can lead to complications in tooth extraction or endodontic treatment.

## Classification

Turner (1971)

Grade 1: one-rooted lower first molar (no inter-radicular projection separating roots)
Grade 2: two-rooted lower first molar (distinct mesial and distal roots)
Grade 3: three-rooted lower first molar (3RM1) (distinct distolingual accessory root)

## Breakpoint

Along with other root traits, 3RM1 is scored as present or absent. This root takes a distinctive form quite apart from a simple division of the distal root.

## Potential Problems in Scoring

A slight separation of the distal root about halfway along its breadth should not be confused with 3RM1. This trait requires a distinct and relatively small accessory root to qualify as present.

## Geographic Variation

Three-rooted LM1 constitutes the best-known human root polymorphism because it shows a distinctive pattern of geographic variation. North and East Asian populations have 3RM1 frequencies around 30%, a number mirrored by Siberian populations and North American Eskimos. Their relatives, the Aleuts, have the world's highest frequencies of 3RM1, sometimes reaching 50%. Europeans and Sub-Saharan Africans rarely exhibit 3RM1 (ca. 1%). It occurs at a moderate frequency in Southeast Asian and Pacific populations (5–15%).

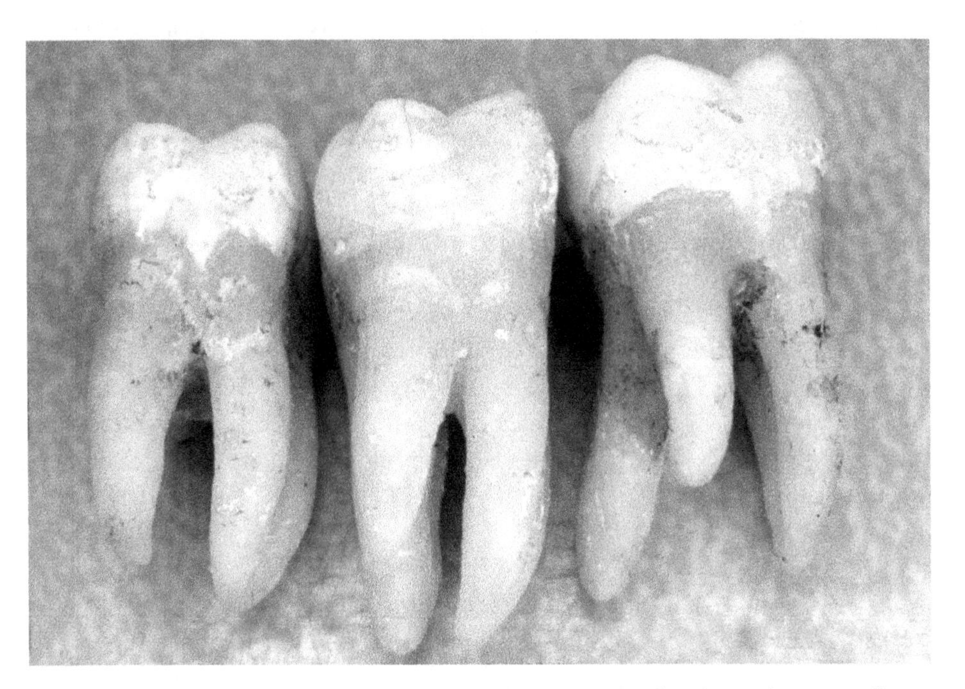

**Figure 40.1** The tooth on the right shows 3RM1; accessory distolingual root smaller than distobuccal root.

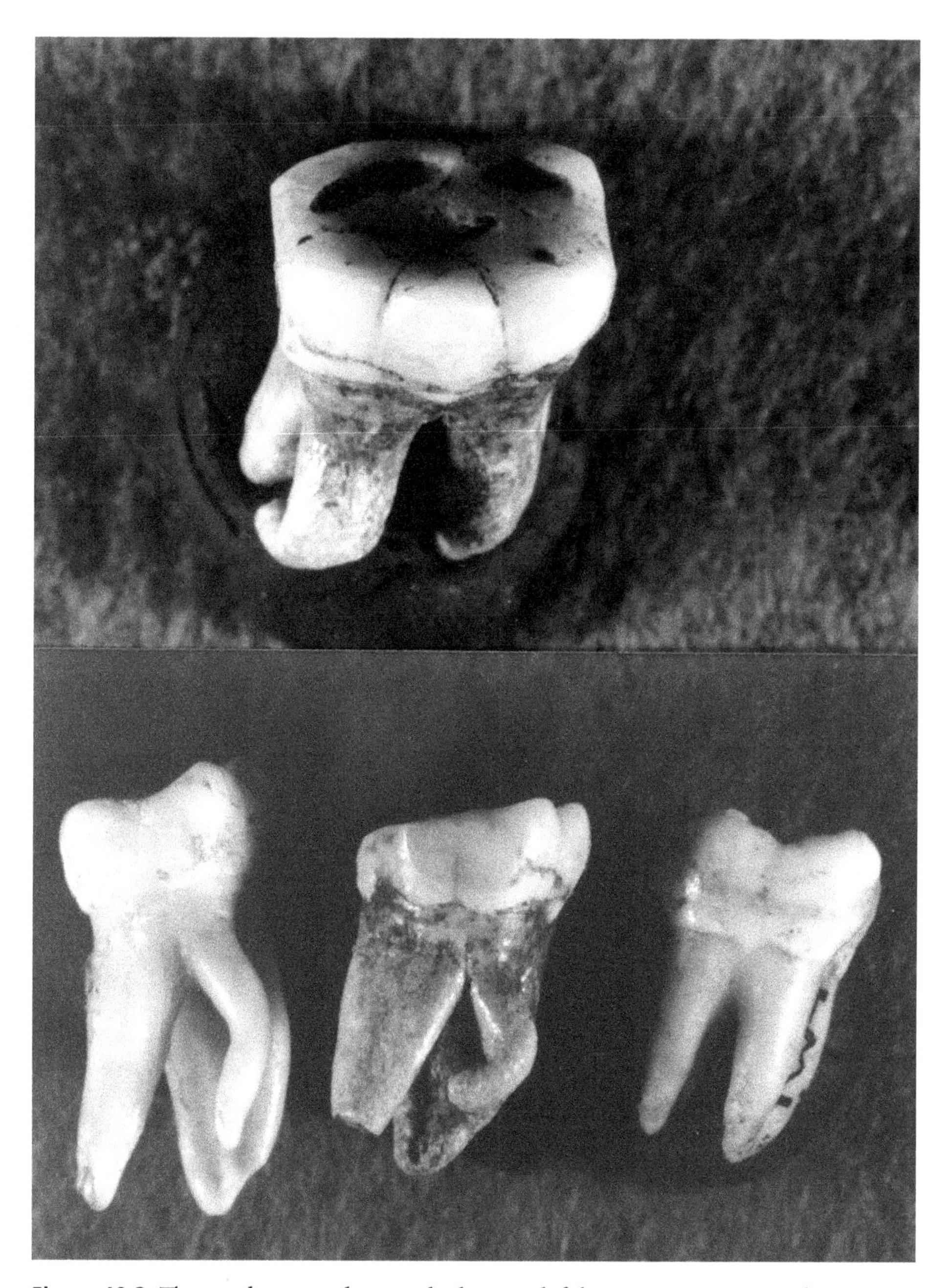

**Figure 40.2** The two lower molars on the bottom left have accessory roots that are bent around the midway point of total root length; such bending would make extraction difficult.

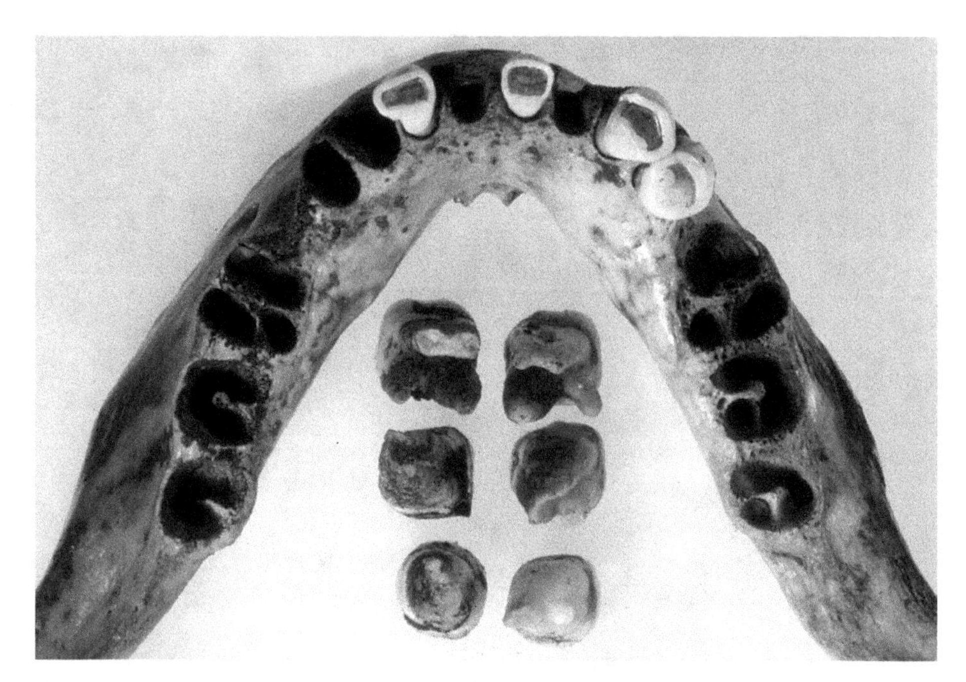

**Figure 40.3** Very rare case of three-rooted LM1, LM2, and LM3.

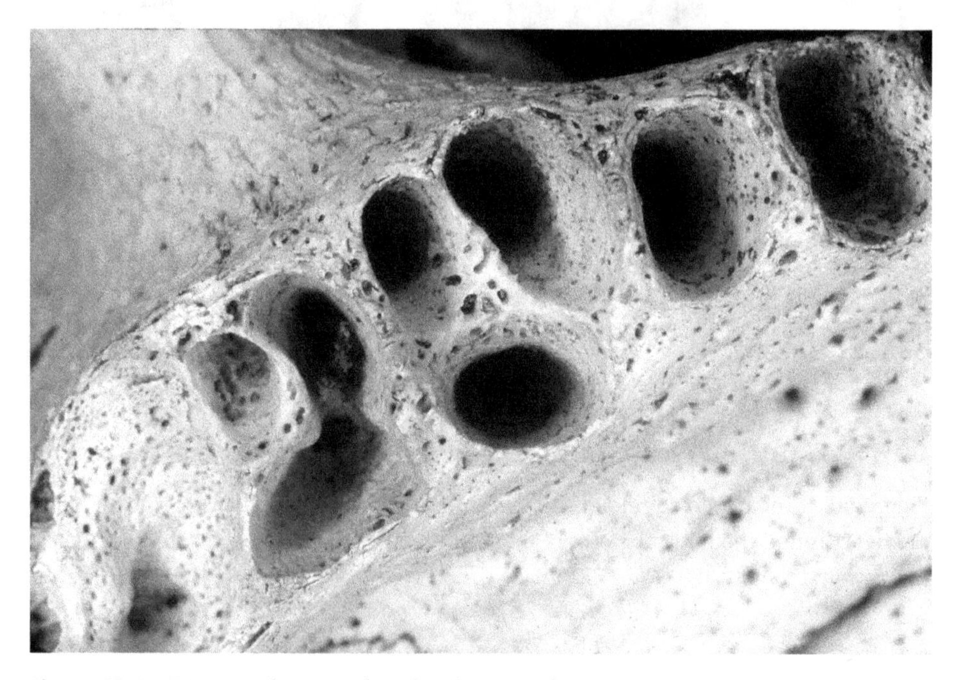

**Figure 40.4** 3RM1 can be scored in the absence of LM1.

## Select Bibliography

Alexandersen, V., and Carlsen, O. (1998). Supernumerary roots of mandibular molar teeth. In *Human Dental Development, Morphology, and Pathology: A Tribute to Albert A. Dahlberg*, ed. J.R. Lukacs. Eugene: University of Oregon Anthropological Papers, Number 54, pp. 201–214.

Carlsen, O. (1987). *Dental Morphology*. Copenhagen: Munksgaard.

Curzon, M.E.J. (1974). Three-rooted mandibular first permanent molars in Greenland Eskimo skulls. *Arctic* 27, 150–153.

Schafer, E., Breuer, D., and Janzen, S. (2008). The prevalence of three-rooted mandibular permanent first molars in a German population. *Journal of Endodontics* 35, 202–205.

Tasa, G.L. (1998). Three-rooted mandibular molars in Northwest Coast populations: implications for Oregon prehistory and peopling of the New World. In *Human Dental Development, Morphology, and Pathology: A Tribute to Albert A. Dahlberg*, ed. J.R. Lukacs. Eugene: University of Oregon Anthropological Papers, Number 54, pp. 215–244.

Tratman, E.K. (1938). Three-rooted lower molars in man and their racial distribution. *British Dental Journal* 64, 264–274.

(1950). A comparison of the teeth of people: Indo-European racial stock with the Mongoloid racial stock. *Dental Record* 70, 31–53.

Turner, C.G., II (1971). Three-rooted mandibular first permanent molars and the question of American Indian origins. *American Journal of Physical Anthropology* 34, 229–241.

Walker, R.T., and Quackenbush, L.E. (1985). Three-rooted lower first permanent molars in Hong Kong Chinese. *British Dental Journal* 159, 298–299.

Younes, S.A., Al-Shammery, A.R., and El-Angbawi, M.F. (1990). Three-rooted permanent mandibular first molars of Asian and black groups in the Middle East. *Oral Surgery, Oral Medicine, Oral Pathology* 69, 102–105.

# NOTES

# 41

# Lower Molar Root Number

## Observed on

LM1, LM2, LM3

## Key Tooth

LM2

## Synonyms

C-shaped canal, one-rooted LM2

## Description

Lower molars generally have two roots, a mesial root associated with the trigonid and a distal root associated with the talonid. As the most conservative tooth in the lower molar district, LM1 retains the ancestral two-rooted phenotype most of the time (not counting a distolingual accessory root, or 3RM1). Given its late development and the restricted space in the jaw of recent humans, LM3 often exhibits fused and shortened roots, resulting in a one-rooted lower molar. As the tooth intermediate between the evolutionarily conservative LM1 and the environmentally labile LM3, LM2 exhibits the polymorphic condition useful in population comparisons. That is, it may exhibit two complete roots, or the two roots may be fused in one of three ways: the buccal aspect of both roots fuses, with the lingual aspect open; the lingual aspect fuses, with the buccal aspect open; or both the buccal and lingual aspects fuse. The C-form of LM2 roots typically involves fusion of the roots at their buccal aspect. By convention, all three phenotypes constitute a one-rooted LM2.

## Classification

Turner (1967)

Grade 1: one-rooted lower molar (mesial and distal roots of lower molars can be fused on either buccal or lingual aspect or both)

Grade 2: inter-radicular structure produces clear separation of mesial and distal roots for at least ¼ to ⅓ of total root length

## Breakpoint

Although the phenotypic expressions could be ranked, at present any manifestation of root fusion down to the apex is considered a one-rooted LM2.

## Potential Problems in Scoring

The Turner convention that bifurcations must be ¼ to ⅓ of total tooth length is sometimes difficult to determine.

## Geographic Variation

One-rooted LM2s are least common in African and Australian populations (<10%). East Asians and Arctic populations exhibit the greatest amount of root fusion (30–40%). Europeans and Southeast Asians fall between these extremes for one-rooted LM2 (20–30%).

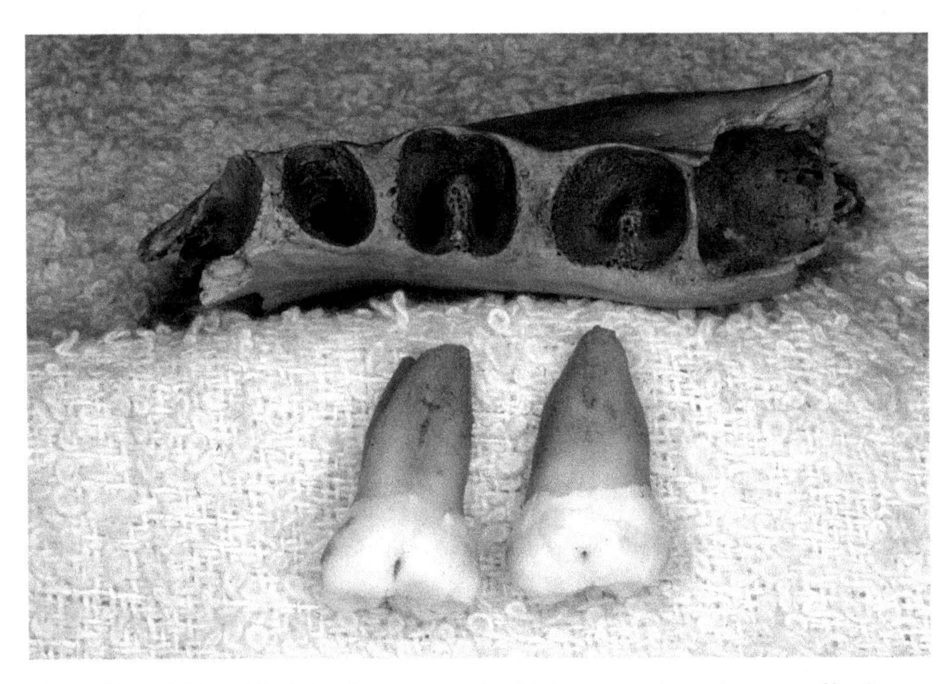

**Figure 41.1** LM1 and LM2 both one-rooted, with fusion on buccal aspect of both teeth (i.e., C-shaped root).

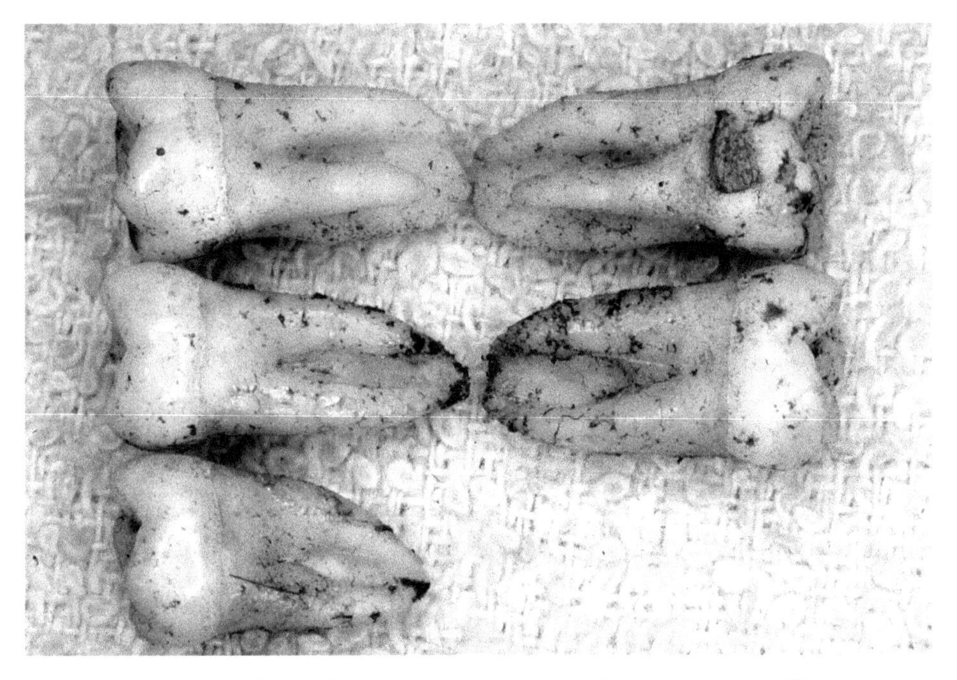

**Figure 41.2** LM2 and LM3 all show buccal fusion, producing one-rooted lower molars.

## Select Bibliography

Gulabivala, K, Opasanon, A., Ng, Y.L., and Alavi, A. (2002). Root and canal morphology of Thai mandibular molars. *International Endodontic Journal* 35, 56–62.

Manning, S.A. (1990). Root canal anatomy of mandibular second molars. Part II. C-shaped canals. *International Endodontics Journal* 23, 40–45.

Peiris, R., Takahashi, M., Sasaki, K., and Kanazawa, E. (2007). Root and canal morphology of permanent mandibular molars in a Sri Lankan population. *Odontology* 95, 16–23.

Shields, E.D. (2005). Mandibular premolar and second molar root morphological variation in modern humans: what root number can tell us about tooth morphogenesis. *American Journal of Physical Anthropology* 128, 299–311.

Turner, C.G., II (1967). *The Dentition of Arctic Peoples.* PhD dissertation, Department of Anthropology, University of Wisconsin, Madison.

Wheeler, R.C. (1965). *A Textbook of Dental Anatomy and Physiology*, 4th edn. Philadelphia and London: W.B. Saunders.

# NOTES

# 42

# Torsomolar Angle

## Observed on

LM3

## Key Tooth

LM3

## Synonyms

Torsoclusion, torsiversion (of mandibular third molars)

## Description

Affecting specifically the LM3s, one or both teeth may be rotated either buccally or lingually relative to a line running through the center of LM1 and LM2. In clinical terms, the torsomolar angle trait is considered a malocclusion, and is referred to most commonly as torsiversion, where a tooth is rotated on its long axis. Rotation can and does occur for any teeth, e.g., premolars (remember the Flores "hobbits"?). In 1978, Neiberger noted LM3 rotation in Native Americans and suggested it is genetically determined at the infracrypt level of development. Thus, the trait was added to the ASUDAS. It is independent of dental crowding; if the latter is present in a dentition, torsomolar angle should not be scored (see below).

## Classification

Neiberger (1978)

To record the feature, observe the long axis of LM3 in relation to those of LM1 and LM2. If it is in line with the two anterior teeth, then the trait is considered absent. If one or both LM3s are rotated along their long axes by ≥10° off this line – either buccally or lingually – the trait is considered present. Neiberger reported that, when present, the rotation ranges between 20° and 60°, with the most common position at 30° in the buccal direction. Scoring can be broken down as follows:

Grade 0: absent (torsomolar angle <10°)
Grade 1: present (torsomolar angle ≥10° in either buccal or lingual direction)

## Breakpoint

Trait presence

## Potential Problems in Scoring

The trait should not be recorded unless the LM3 has completed alveolar eruption with full root formation. If the tooth is partially impacted, if it has shifted position (e.g., mesial drift from antemortem loss of the anterior molars), and/or if general dental crowding is evident, the trait should not be scored.

## Geographic Variation

Unlike the majority of ASUDAS traits, this trait has not been routinely recorded. Thus, its utility for discriminating among world populations remains a work in progress. The second author (JDI) does routinely record the trait, however, and has found it to vary considerably, showing a frequency of 0–20.6% in 14 Sub-Saharan samples (mean 15.7%) and 5.3–53.8% in 13 North African samples (mean 21.6%).

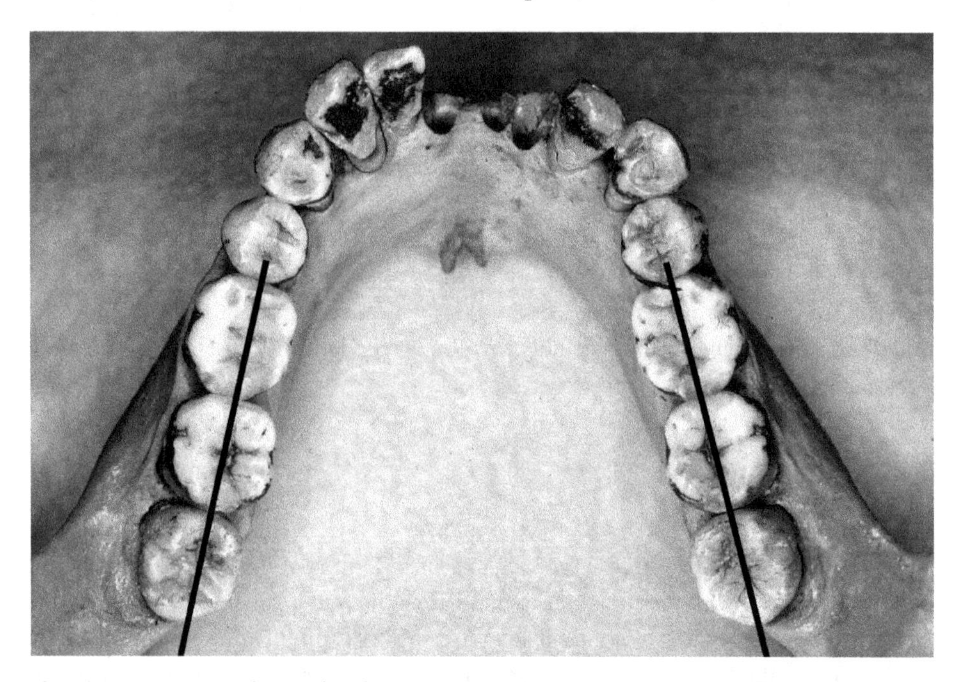

**Figure 42.1** Torsomolar angle of 0° rotation on both sides of this mandible, i.e., scored as absent. The long axes of both LM3s are in line with those of the anterior molars.

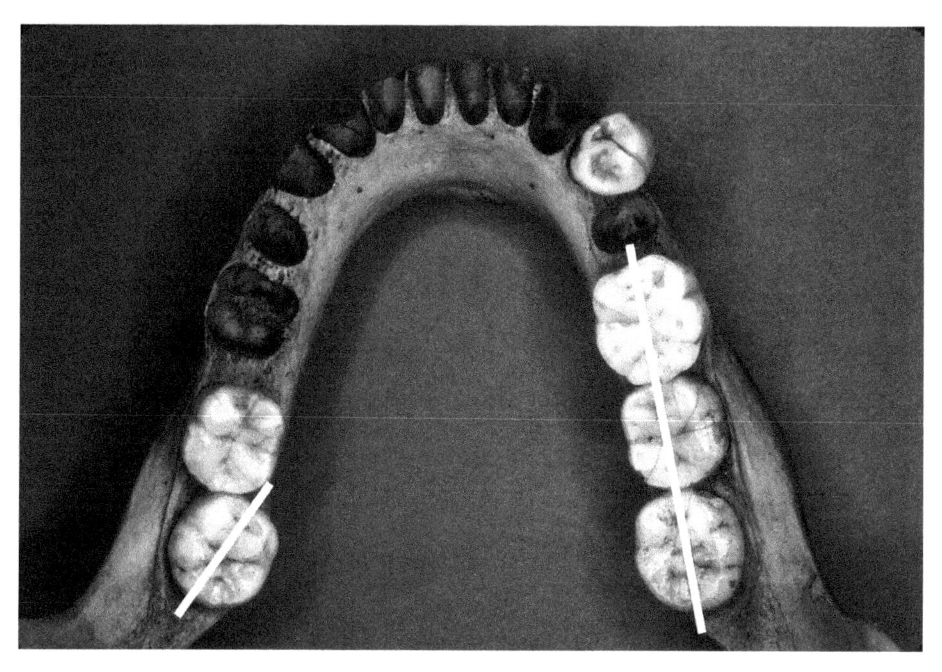

**Figure 42.2** Unilateral torsomolar angle expression. Absent on the right side, but present on the left; the latter is rotated lingually beyond 10°.

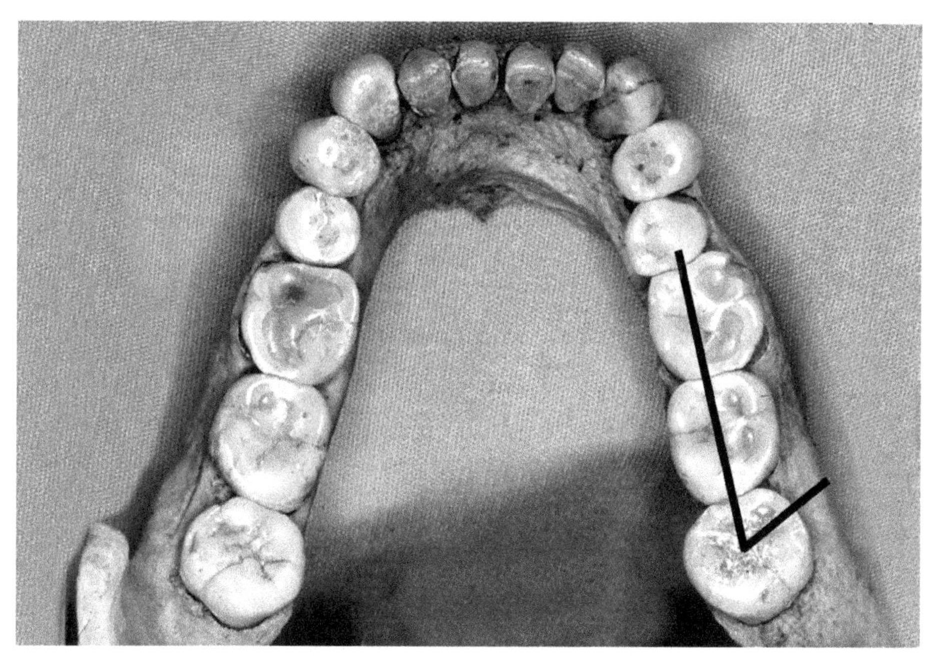

**Figure 42.3** Bilateral torsomolar angle presence. In both cases, the LM3s are rotated buccally to a similar extent, well beyond 10°.

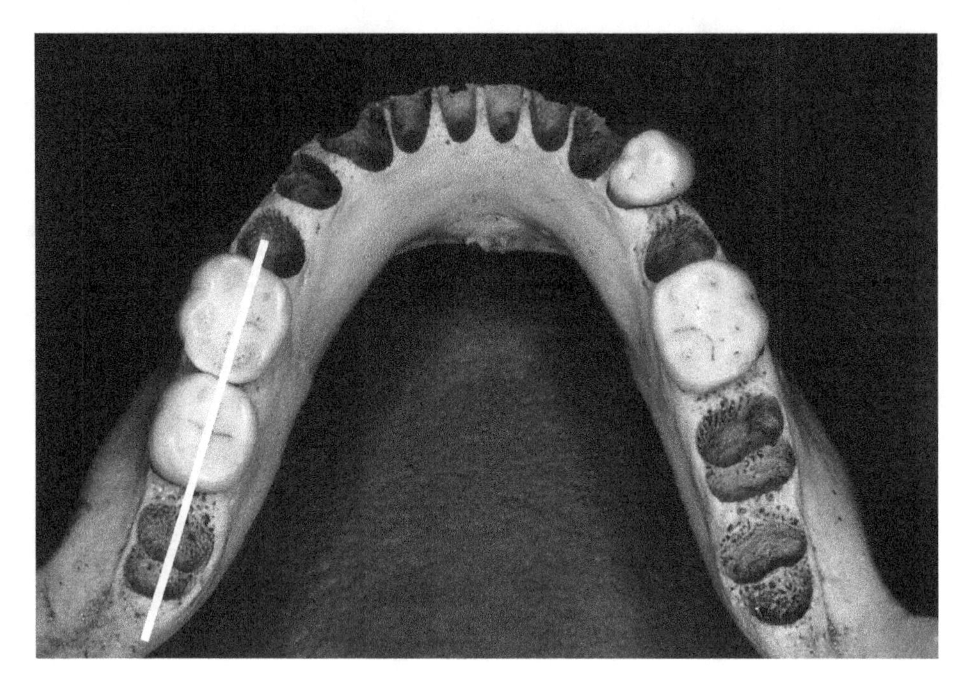

**Figure 42.4** Torsomolar angle can be recorded even if the LM3 is missing post-mortem, by checking the alignment of the empty tooth sockets. Here, the LM3 sockets do not exhibit rotation off the long axis.

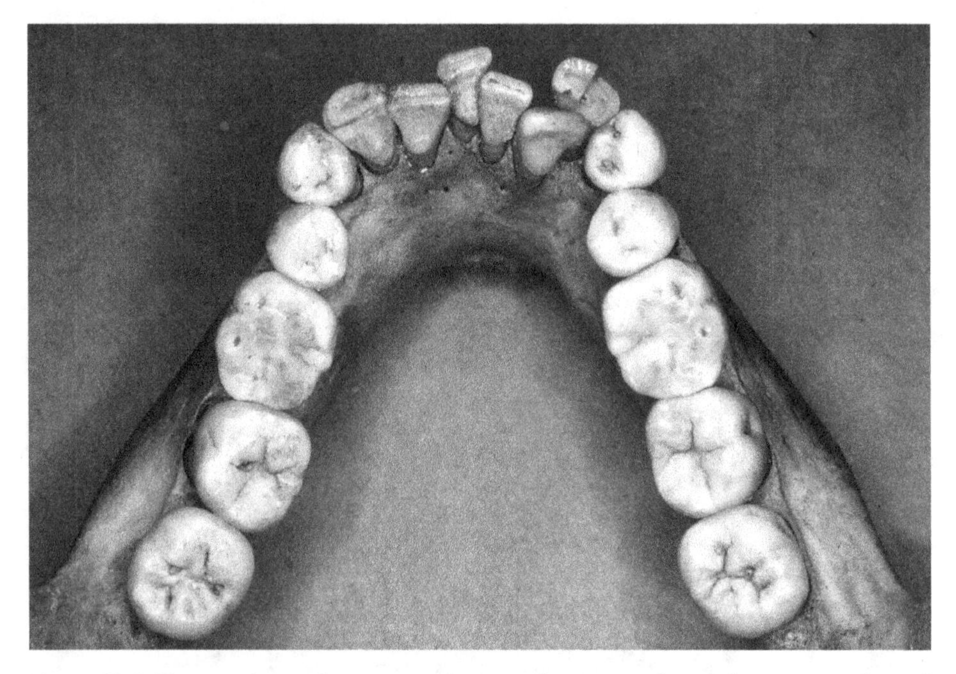

**Figure 42.5** Torsomolar angle must not be scored in cases of partial impaction, loss of anterior molars, or, as in this case, extreme dental crowding that may "falsely" affect alignment of the LM3s.

## Select Bibliography


Goodwin, L. (1964). Age change in torsiversion of the maxillary permanent first molar teeth. *Scholar Archive*. Paper 1379. http://digitalcommons.ohsu.edu/etd/1379/ (accessed November 2016).

Irish, J.D. (1993). *Biological Affinities of Late Pleistocene through Modern African Aboriginal Populations: The Dental Evidence*. PhD dissertation, Department of Anthropology, Arizona State University, Tempe.

Jacob, T., Indriati, E., Soejono, R.P., *et al.* (2006). Pygmoid Australomelanesian *Homo sapiens* skeletal remains from Liang Bua, Flores: Population affinities and pathological abnormalities. *Proceedings of the National Academy of Sciences of the USA* 103, 13421–13426.

Neiberger, E.J. (1978). Incidence of torsiversion in mandibular third molars. *Journal of Dental Research* 57, 209–212.

Saimbi, C.S., Jain, A., and Verma, S. (2004). Bilateral rotation. *British Dental Journal* 196, 378.

Scheid, R.C. (2012). *Woelfel's Dental Anatomy*. New York: Lippincott Williams & Wilkins.

# NOTES

# Part III

# Conclusions

Conducting research on dental morphology involves a number of issues that should be addressed prior to initiating a project. Tooth crown size, for example, shows consistent differences between males and females. Is this also true for crown and root traits? Regarding inter-trait association, are these variables expressed independently of one another or do they co-vary on an individual basis or at the population level? Depending to some degree on the questions asked and the populations studied, which traits are most useful to address a particular problem? As the teeth in the two sides of the jaw are mirror images to a large extent, how should scores be tabulated to arrive at population frequencies? Regarding frequencies, should comparisons be based on total trait frequencies, or are there advantages in using breakpoints along the scale of presence expressions that yield more consistent results between observers? Once frequencies are tabulated for a set of populations, what types of statistical methods, especially those involving biodistance, are appropriate for dental morphological traits? As teeth are prone to wear, and this process impacts or completely conceals trait expression, how does an observer take crown wear into account when making observations? To address a particular problem (e.g., the relationship between crown size and morphology), should a researcher make observations on dental casts or skulls? In laying the foundation for dental morphological research in anthropology, paleontology, genetics, and related fields, many of these questions have been addressed and answered to a point. Advances in technology (e.g., 3D scans and micro-CT) and analytical methods (e.g., geometric morphometrics) should ultimately refine current knowledge on these questions.

## Specific Concerns

### Sex Dimorphism

Dimensional variables (anthropometric, craniometric, odontometric) consistently show a difference between males and females, necessitating intra- and inter-group comparisons that are broken down by sex (cf. Howells 1973, 1989). For many measurements, males exceed females by about 10%, although this is less pronounced for tooth size (2–6%, depending on the tooth). Combining data on males and females is not a good practice for measurement data, especially when the sexes are not evenly represented in a series, as is often the case.

During the twentieth century, it was not uncommon for researchers to compare males and females for dental trait expressions and frequencies. Some researchers would claim a particular trait showed a significant dimorphism, while an equal number would report no difference. Given the tens of thousands of dental casts and skulls that have been scored for crown and root traits, the overall consensus is that these traits are generally non-dimorphic. The only exception may be the canine distal accessory ridge, which was found to be consistently more common and pronounced in males than in females (Scott 1977a, Scott *et al.* 1983). This difference likely reflects evidence that the canines are the most sexually dimorphic teeth in the modern human dentition. Where premolars and molars show male–female differences of 2–4%, the canine dimorphism is about 6%, not high but presumably enough to impact the expression of the distal accessory ridge. This possibility was supported by Noss *et al.* (1983), who reported a significant correlation between lower canine mesiodistal diameters and distal accessory ridge expression. Despite that exception, nonmetric traits of the cranium and dentition are far less sexually dimorphic than metric traits. For that reason, data on males and females can ordinarily be combined, expanding sample size without bias.

## Inter-Trait Association

Metric traits of the skull and dentition are not only sexually dimorphic, they also show significant inter-trait correlations. The length of a skull is not independent of bizygomatic diameter, and the mesiodistal diameter of UM1 is not independent of the buccolingual diameter of LM2, etc. Product moment correlations often fall in the range of 0.30–0.60 for craniometric and odontometric variables. Some early biological distance formulae, such as Pearson's "coefficient of racial likeness," did not take these correlations into account. The generalized $D^2$ developed by Mahalanobis (1936) corrected this deficiency and came to dominate biodistance analyses when the emphasis was on metric traits. Now, multivariate statistics (e.g., principal components and factor analysis) isolate components/factors that are strongly interrelated.

Although the dentition is a highly integrated system, nonmetric traits are usually expressed independently of other nonmetric traits. There is, however, a caveat. The same trait expressed on different members of a morphogenetic field show significant correlations with one another. For example, hypocone expressions on UM1, UM2, and UM3 are not independent. They do not show perfect linear correlations of 1.0, but they show correlations comparable to those of odontometrics variables (e.g., 0.30–0.40). This is also true of shovel-shaped incisors, which show significant inter-trait correlations between members of the same field and among all members of the upper and lower incisor fields (Scott 1977b). For this reason, Turner *et al.* (1991) recommended that researchers use a single tooth to represent a particular trait in a sample. The hypocone, which is relatively invariant on UM1, should be assessed on UM2. For shoveling, UI1 was recommended as the key tooth despite the fact that all incisors can potentially exhibit this trait.

In addition to within-field correlations, there are instances where different traits in the same field or in different fields show significant correlations. For example, the expression of the hypocone and Carabelli's trait shows a low but significant correlation in the upper molar field (Keene 1965, 1968, Scott 1979). This also applies to two cingular traits in different morphogenetic fields, Carabelli's trait of the upper molars and the protostylid of the lower molars (Scott 1978). As pegged/reduced/missing UM3 is one of the variables described in this guidebook, it should be noted that Brook (1984) found significant relationships between tooth size and tooth number: microdontia (small teeth) was associated with hypodontia (missing teeth), and megadontia (large teeth) was associated with hyperodontia (supernumerary teeth).

Even when there are significant correlations between different traits, sample frequencies are not necessarily biased. For example, Carabelli's trait and the protostylid show correlations of around 0.30, but the pattern of population variation does not show this. Carabelli's trait is very common and often pronounced in Western Eurasians, but the protostylid is rare in this group. The protostylid is common in Native Americans, who rarely show pronounced expressions of Carabelli's trait. Although correlation coefficients have not been calculated, shoveling (lingual marginal ridges) and double-shoveling (labial marginal ridges) co-vary on a population level. Groups with high frequencies of shoveling also have high frequencies of double-shoveling (Scott and Turner 1997). This relationship requires more analysis. Except for intra-field correlations, most of the traits described in this guidebook are expressed independently of one another.

## Trait List

In pre-ASUDAS times, it was not uncommon for researchers to describe a single trait in a single population (e.g., Bang and Hasund 1971, 1973, Aas and Risnes 1979) or multiple traits in a single population (e.g., Nelson 1938, Goldstein 1948, Goaz and Miller 1966, Rosenzweig and Zilberman 1967, 1969, Barnes 1969). Authors describing multiple traits would typically consider shoveling, Carabelli's trait, upper molar cusp number, and lower molar groove pattern and cusp number; beyond that, the list of traits was idiosyncratic. In some ways, this parallels early blood group studies that focused on alleles in the ABO, MN, and Rh systems. Although one can gain insights from studying a few variables, results for biodistance or ancestry estimation are far more powerful when based on many variables (Livingstone 1991). One only has to compare the findings of Boyd (1950), who studied world variation based on three blood group systems, to those of Cavalli-Sforza *et al.* (1994), who analyzed over 100 alleles.

Although dental morphologists cannot match the number of variables available to geneticists, the number of independent crown and root traits exceeds 30 (see Figures vii and viii at the end of this chapter for the ASUDAS data sheets). Of this number,

**Arizona State University Dental Anthropology System**
**Standard Score Sheet**

Date:__________ Facility: ____________
File Name & No : __________________________________ Age: ________ Sex: ____

| Maxilla | I1R | I1L | I2R | I2L | CR | CL | P1R | P1L | P2R | P2L | M1R | M1L | M2R | M2L | M3R | M3L | |
|---|---|---|---|---|---|---|---|---|---|---|---|---|---|---|---|---|---|
| Status/Wear | | | | | | | | | | | | | | | | | |
| Caries | | | | | | | | | | | | | | | | | |
| Winging | | | | | | | | | | | | | | | | | |
| Labial Curve | | | | | | | | | | | | | | | | | |
| Shovel | | | | | | | | | | | | | | | | | |
| Double Shovel | | | | | | | | | | | | | | | | | |
| Inter Groove | | | | | | | | | | | | | | | | | |
| I & C TD | | | | | | | | | | | | | | | | | |
| Bushman C | | | | | | | | | | | | | | | | | |
| C DAR | | | | | | | | | | | | | | | | | |
| MxPAR | | | | | | | | | | | | | | | | | |
| P M&D Cusps | | | | | | | | | | | | | | | | | |
| Metacone | | | | | UA-P | | | | | | | | | | | | M |
| Hypocone | | | | | | | | | | | | | | | | | H |
| Cusp 5 | | | | | | | | | | | | | | | | | 5 |
| Carabelli. | | | | | | | | | | | | | | | | | C |
| C2 Parastyle | | | | | | | | | | | | | | | | | P |
| Enamel Ext. | | | | | | | | | | | | | | | | | X |
| Root No. | | | | | | | | | | | | | | | | | R |
| Peg/Reduce | | | | | | | | | | | | | | | | | p |
| Odontome | | | | | | | | | | | | | | | | | O |
| Cong Absent | | | | | | | | | | | | | | | | | c |
| MLD | | | | | | | | | | | | | | | | | m |

Extra Teeth: ____________________
Torus: None___Tr___Med___Mark___
Abscess: ____________________
Perio: G1___ G2___ G3___ Pkts _____
Chipping: ____________________
Cult Treat: ____________________
TMJ Damage: R __________ L __________

Vers. JDI_3.02

**Figure vii** ASUDAS scoring sheet for maxillary dentition.

Turner zeroed in on 29 key traits. For the maxilla, he focused on 14 crown traits and two root traits. Key traits in the mandible include nine crown traits and four root traits. Some variables on the ASUDAS data sheet are subject to wear early in life, so they did not make the list. To avoid problems introduced by inter-trait association, 28 traits are observed on single teeth (e.g., UI1 shoveling), although the odontome, a rare trait in all populations, is scored on all premolars. In the Appendix, we include tables for

| Mandible | I1L | I1R | I2L | I2R | CL | CR | P1L | P1R | P2L | P2R | M1L | M1R | M2L | M2R | M3L | M3R | |
|---|---|---|---|---|---|---|---|---|---|---|---|---|---|---|---|---|---|
| Status/Wear | | | | | | | | | | | | | | | | | |
| Caries | | | | | | | | | | | | | | | | | |
| Shovel | | | | | | | | | | | | | | | | | |
| C DAR | | | | | | | | | | | | | | | | | |
| P Ling Cusps | | | | | | | | | | | | | | | | | |
| MxPAR | | | | | | | | | | | | | ← Anterior Fovea | | | | |
| Groove Pat | | | | | | | | | | | | | | | | | G |
| M Cusp No | | | | | | | | | | | | | | | | | C |
| Def Wrinkle | | | | | | | | | | | | | | | | | D |
| C1—C2 Crest | | | | | | | | | | | | | | | | | T |
| Protostylid | | | | | | | | | | | | | | | | | P |
| Cusp 5 | | | | | | | | | | | | | | | | | 5 |
| Cusp 6 | | | | | | | | | | | | | | | | | 6 |
| Cusp 7 | | | | | | | | | | | | | | | | | 7 |
| Enamel Ext. | | | | | | Tomes → | | | | | | | | | | | X |
| Root No. | | | | | | | | | | | | | | | | | R |
| Odontome | | | | | | | | | | | Torsomolar Ang | | | | | | |
| Cong Abs | | | | | | | | | | | | | | | | | c |

Torus: None____Tr____Med____Mark____
Rocker: None____ Near ____ Rocker ____
Abscess: ____
Perio: G1____ G2____ G3____ Pkts ____
Chipping: ____
Cult Treat: ____
Extra Teeth: ____

Comments:

**Figure viii** ASUDAS scoring sheet for mandibular dentition.

60 samples representing major geographic regions of the world that have the full class frequency distributions for Turner's 29 key traits.

## Counting Method

Because morphological traits are expressed on both right and left teeth (antimeres), there are several possible counting methods that can be followed to derive trait frequencies. A total tooth count involves adding trait expressions for all left and right antimeres together. The major problem is that trait expression is symmetrical over 80% of the time so this method introduces redundancy and artificially increases sample size (which has a big impact on some statistics, e.g., chi-square). There is one circumstance where a total tooth count is justified: when most or all available teeth from a series are not associated with maxillae or mandibles (i.e., loose teeth; this can happen because teeth preserve better than bone). A second method, and one often followed by odontometricians, is the side count. That is, a researcher scores trait expression on only the left or only the right antimeres, often with the provision that if one is missing, the remaining antimere is scored. This method avoids the problem of artificially inflating sample size, but is it biologically sound? That brings us to the third method, the individual count (Turner and Scott 1977). In this instance, an observer scores both antimeres for trait expression, but when data are tabulated the tooth with the most pronounced degree of expression is tallied. The rationale is that individuals who exhibit, say, grade 3 Carabelli's on the left UM1 and grade 6 Carabelli's on the right UM1 have the genotypic potential for grade 6 expression. If an individual is represented on the basis of one expression only, it should be on the score that best reflects the underlying genetic potential.

Many years ago, a reviewer questioned the use of the individual count method in tabulating frequencies for Carabelli's trait. It was a valid point, so all data sheets were reevaluated to provide tallies based on the tooth count, side count, and individual count methods. As it turned out, the counting methods generated almost identical results (Scott 1980). C.G. Turner and both authors have always used the individual count method, to avoid artificial sample size inflation and provide what we feel to be the best representation of an individual's underlying genotype. Despite this, we score trait expression on both left and right teeth. If research on epigenetics suggests there is a better way to summarize data, we will be in a position to make the adjustment.

## Breakpoints

Population comparisons for crown and root traits often focus on frequencies. For many traits, total frequencies are used. For other traits, breakpoints have been adopted. Technically, either is appropriate, assuming that dental traits are quasi-continuous characteristics with polygenic modes of inheritance (Grüneberg 1952, Scott 1973, Harris 1977). Such variables are also referred to as threshold traits (Falconer 1960). The assumption is that these traits are normally distributed like a polygenic trait (e.g., stature) but, unlike a polygenic trait, there are underlying and visible scales separated by a threshold. Individuals that fail to exhibit a trait may nonetheless differ genetically from other individuals who lack the same trait as they fall on different sections of the underlying scale. Some are close to while others are at some remove from the threshold. The visible scale is easier to conceptualize, because one can actually observe different phenotypic expressions that fall along this scale. Individuals with slight trait expressions are assumed to fall just above the threshold, while pronounced expressions fall well above the threshold.

To characterize a population for a trait frequency, all one needs for a threshold trait is a single number. If you have a total frequency and sufficient sample size, you know the trait's distribution. A trait with a total frequency of 50% (which is not uncommon) would have half the population below the threshold (trait absence) and half the population above the threshold (varying degrees of trait presence). A population with a frequency of 75% would have more individuals above the threshold and, in all likelihood, more pronounced trait expressions than you would find for the sample with a 50% frequency.

Grüneberg (1952) noted there was a correlation between total trait frequency and degree of expression. For dental traits, there is no better illustration of this than shovel-shaped incisors. Shoveling is so common in Native Americans that it is almost invariant, and these groups show by far the most pronounced expressions of the trait. For a large Southwest Native American sample, shoveling basically transitioned from a quasi-continuous character to a continuous character, as trait expression, from slight to pronounced, is expressed along a normal distribution with essentially no underlying scale (Scott and Turner 1997).



## Quantitative Analyses

Once data have been collected they need to be analyzed, based on what the researcher wants to know, e.g., exploratory analysis versus testing a specific hypothesis. Whichever the case, the analyses generally proceed as follows (as detailed in Irish 2010).

First, the trait data should be summarized to allow initial characterization and comparisons, if more than one sample has been recorded. To do so, the ASUDAS data are dichotomized as described above. This step is also required for most multivariate applications. These data can be presented in tabular format (e.g., Irish 1993, 1997). Although useful in a qualitative sense, it is difficult to fully interpret a large amount of summary data.

Second, the data may need to be "cleaned up," or edited, prior to conducting multivariate quantitative analyses (see below), depending upon their completeness in samples and the focus of the researcher. Individual trait frequencies based on only a few observations (e.g., <10) should be deleted from analyses, because they are likely not representative, particularly if multiple samples are affected. Traits that do not vary in expression across the samples under study may be deleted as well because they provide no discriminatory value (Harris and Sjøvold 2004). In some cases, invariance is obvious (e.g., 0.0 or 100.0%) across samples; otherwise, the ones that are least or, on the other hand, most likely to drive inter-sample variation can be identified using correspondence analysis (Clausen 1988, Benzécri 1992, Phillips 1995) or principal components analysis (PCA) (Irish 2010). If the mean measure of divergence (MMD) distance statistic (below) is used, any strongly correlated trait pairs (i.e., >0.5) must be deleted, or differential weighting of the underlying dimensions can produce erroneous distances (Sjøvold 1977). To test for correlation, a tetrachoric correlation statistic can be used with dichotomized data, or the non-dichotomized rank-scale data can be submitted to a nonparametric test, such as Kendall's tau-b correlation coefficient. As noted, ASUDAS traits are often minimally correlated, but it is best to make sure before proceeding.

The third step of analysis, if multiple samples are under study, is to assess overall among-sample similarities (i.e., phenetic affinities) based on the final set of edited traits. Of the many distance statistics available, dental researchers have used two with equivalent success: the Mahalanobis $D^2$ statistic for nonmetric traits (Konigsberg 1990) and the MMD (Sjøvold 1977). Both have pros and cons, but yield highly comparable distances (Irish 2010). Both methods provide a dissimilarity measure among sample pairs, where lower values are indicative of greater phenetic similarity, and vice versa. Beyond a $D^2$ or MMD matrix of inter-sample distances, affinities can be visualized using one of several illustrative approaches, such as cluster analysis and multidimensional scaling (MDS). The former depicts samples in a dendrogram, where degrees of relationships are identified via branching points in the display (Romesburg 1984). Another effective technique to graphically portray relationships is through MDS. It provides a spatial representation of 1 to $n$ dimensions consisting of a geometric configuration of points (the samples), like on a map (Kruskal and Wish 1978). Plotting samples into groups illustrates degrees of similarity.

## The Impact of Wear on Scoring Traits

Ongoing research by Burnett (1998, 2016, Burnett *et al.* 2013) revealed two forms of bias that crown wear, or attrition, can have on recording nonmetric crown traits. He refers to the first bias as "grade shift"; it occurs when a particular trait is recorded at either a lower or higher grade than it actually should be. The potential result is that the "true" trait frequency is either artificially decreased or increased, respectively. Second, a sampling bias may occur by not recording teeth that are subjectively deemed to be "too worn." In doing so, the observer assumes that the missing data are missing completely at random (MCAR). However, the "occurrence of a particular trait, even in the presence of heavy wear, may influence an observer to include the tooth for study, though a similarly worn tooth with absence of the trait is not. In the latter case, concern that the trait was worn away results in exclusion of the tooth (scored as no data), when the trait was in fact absent and should be scored as grade 0" (Burnett 2016: 428). When grading higher than one should (above), the frequency of an affected trait is artificially increased through violation of the MCAR assumption. A number of traits have been assessed for these two possible biases in the above references.

What can be done in terms of scoring morphology on worn teeth? One solution is to only compare samples that have similar levels of wear. Finding such similarity among samples, particularly in archaeological contexts, may not be possible, greatly limiting comparisons. The alternative is to acknowledge that there are major differences in wear among the samples being compared; a cautionary note should be provided in the write-up of results stating that some trait frequencies may be affected. Another possibility is to only select specific traits that are minimally or not affected by wear (e.g., root traits). That is, traits near the occlusal surface, such as UC distal accessory ridge, are more affected at early wear stages than those located lower on the crown (e.g., LM groove pattern). Lastly, and the easiest of solutions, is to shift the breakpoints upward. For example, a large Carabelli's cusp affects crown shape, so it can be identified even with moderate wear. Shifting some standard breakpoints upward (e.g., Irish 1993, 2006, Irish *et al.* 2014) can reduce or eliminate certain wear biases. Doing so permits recording of moderately worn dentitions for the purpose of maximizing all-important sample size; as well, existing data can still be used (Burnett 1998, 2016, Burnett *et al.* 2013).

Whatever the case, crown wear can affect how a researcher scores nonmetric dental traits. A paper by Stojanowski and Johnson (2015) provides an excellent illustration of what happens when two researchers, observing the same skeletal series, approach the issue of scoring worn teeth very differently. Most, if not all, crown traits cannot be observed on heavily worn teeth. In cases of lesser wear, it is up to the researcher to judge "how much is too much." However, in conjunction with the above possible solutions, it is best to err on the side of caution.

## Real Teeth versus Casts, Including Intra-Oral Observations

The four pillars that help researchers justify their time studying teeth are variability, heritability, preservability, and observability. The final two relate to how well teeth preserve in the fossil and archaeological (and forensic) record and how they are the only hard part of the skeleton directly observable in the living. Given the enormous skeletal collections curated at hundreds of museums throughout the world, the potential for making observations directly on teeth is almost unlimited (and more remains are constantly being unearthed). Of course, this is an avenue that many dental researchers (professionals and graduate students alike) follow. But what about the living? How can researchers take advantage of the fact that morphological variables can be studied in modern populations?

There are two major avenues for studying the living, one direct and one indirect. The direct study involves intra-oral observations. Although we do not recommend this method, it has been done (e.g., Sofaer *et al.* 1972). To examine the subtleties of morphological traits, one has to observe expression from multiple angles with good sources of light. This is not easy to do when you tell someone to open wide! If one is only concerned with scoring large tubercle forms of Carabelli's cusp, this is feasible. To score all the traits in this guidebook, impossible.

An indirect method for studying the living involves dental impressions. In earlier times, wax bite impressions were made, but these have serious limitations. Zubov (1968, 1977) and his Russian collaborators studied the negative impressions made by wax bites, resulting in a field of study called odontoglyphics (to parallel dermatoglyphics). Focus was on the patterns produced by grooves and furrows. Although occlusal traits could be studied in children, crown wear made the method impractical in the study of adults. The method also had difficulty capturing the expression of cingular traits on lingual or facial surfaces. In his earliest work among the Pima of Arizona, Dahlberg took wax bite impressions and used them to make positive plaster casts. One of us (GRS) examined many of these casts, and the limitations of this method were all too apparent.

Clinical dentistry places demands on the dental materials industry, and the result is products that improve on a regular basis. Long ago, wax bites were replaced by alginate gels or similar materials to produce highly accurate negative impressions. When done properly, a positive plaster cast can be produced from negative impressions with a minimum of distortion, allowing researchers to study both tooth size and morphology in living populations. Christy Turner and both of the authors have studied both skeletons and casts. Before embarking on a project, we encourage researchers to evaluate the pros and cons of each, because they both have advantages and disadvantages.

## Teeth in Skeletons: Pros and Cons

Pro:

(1) Teeth are directly observable so there is no potential for distortion through casting error.
(2) One can include observations on root traits.

Con:

(1) Single-rooted teeth are often lost during excavation, preservation, or curation (in skeletal collections, it is frequent to see a jaw with all the multi-rooted posterior teeth but no anterior teeth). The result is highly variable sample sizes for say shoveling versus Carabelli's trait. This is apparent in the data tables in the Appendix, which are based on skeletal collections.
(2) The age profile of skeletal assemblages includes many infants (lacking permanent teeth) and many adults (with worn or diseased teeth). Individuals with the best teeth for scoring morphology are children (6–12 years) and teens (13–19), and these cohorts are typically the smallest in a skeletal sample.
(3) The enamel of real teeth, unless properly maintained, is prone to flaking off at the enamel–dentine junction.
(4) The surfaces of real teeth reflect light, making it difficult to score low grades of expression for some traits, especially cingular derivatives (e.g., scoring protostylid on real teeth versus dental casts).

## Teeth in Casts: Pros and Cons

Pro:

(1) Anterior tooth loss is not an issue with casts, so sample sizes for anterior and posterior crown traits are typically similar.
(2) In living populations, you can focus on children with teeth little impacted by wear or disease. In his work among the Pima, Al Dahlberg would go to schools on the reservation and take dental impressions of children.
(3) Casts are less reflective than real teeth, allowing observations on subtle grooves and ridges that are more difficult to observe on real teeth.

Con:

(1) One cannot study roots in dental casts, no matter how hard you try!
(2) Casting error can result in distortion or artifacts (e.g., from bubbles in the plaster that have not been properly removed by vibration).
(3) While real teeth can flake, casts can get chipped through improper handling or storage.

# Final Cautionary Notes

## The Two "Faces" of Crown Morphology

Over the past 25 years, the use of dental morphology has advanced significantly in studies of fossil hominin dentitions and biodistance among recent populations, with additional strides made in research on nonhuman primates and forensic anthropology. One should, however, never lose sight of the fact that morphology is expressed on two surfaces: the outer enamel surface (OES) and the enamel–dentine junction (EDJ).

In 1960, Korenhof pioneered the study of the "ins and outs" of human teeth. He found teeth in Java where the roots had been destroyed but the enamel caps had been perfectly preserved. From these caps, he made replicas of the inside surfaces of the crowns, thus providing a positive expression of the EDJ that did not involve tooth destruction. His illustrations that compare the same tooth for the OES and the EDJ are telling. Although the two surfaces show close correspondence, enamel deposition can obscure fine details more clearly seen on the EDJ. Given the success of dental morphology in unraveling a variety of anthropological problems, this qualification should be noted, but it is not enough to make one "abandon ship." One should be aware of the situation, however, as it may be responsible for generating antimeric asymmetry or differences (albeit subtle) between monozygotic twins.

Recent technological advances in micro-CT make it possible to study the EDJ of complete teeth (no root destruction necessary) (Braga *et al.* 2010, Braga 2016). This method has been widely adopted by researchers who focus on the hominin fossil record. Fossil teeth are invaluable and number far fewer than the teeth available in archaeological collections. Researchers have gained many valuable insights into the evolution of hominin dental morphology using micro-CT (e.g., Skinner *et al.* 2008, Bailey *et al.* 2011, Macchiarelli *et al.* 2013, Martínez de Pinillos *et al.* 2014, Martinón-Torres *et al.* 2014), but is it practical at the present time for widespread adoption? Any real tooth, of course, could be subjected to micro-CT analysis, but the time and money involved make it impractical for most bioarchaeological studies. We are reluctant to say this will always be the case, because technology evolves rapidly. If this guidebook generates enough interest to warrant a second edition, this observation can be revisited.

## Observer Error

The goal of ranked standards as established by Dahlberg (1956) and Turner *et al.* (1991) was to elevate the accuracy and replicability of dental morphological observations. To a significant extent, this goal was achieved. Workers throughout the world following ASUDAS provide data on comparable sets of traits that have been scored using the same set of standards. We have noted that in graduate theses and journal articles over the past 25 years, the quality of morphological data has increased dramatically. However, like any other effort that requires experience and decision making, this guidebook should not be confused with a cookbook. A good point to remember is that some traits are easy to score on a consistent basis but some are not.

An old adage is appropriate here: practice makes perfect. We do not recommend that an individual researcher take this guidebook and sit down with a large sample of casts or skulls and start scoring away. To initiate research on dental morphology, start slow. Adopt the ASUDAS scoring sheet or develop your own and then score 50 or so casts/skulls for crown and root traits. Go back a month later and score the same casts/skulls for the same trait set and get some idea as to level of intra-observer error. If possible, have someone else score the same sample and determine the level of inter-observer error (e.g., Nichol and Turner 1986). For traits that cannot be scored consistently, either eliminate them from your trait list or determine why they provide problems in recordation.

Beyond evaluating observer error, a good practice is to compare your results with those in the literature. A good starting point would be to compare your frequencies with those reported in Scott and Turner (1997), who provide worldwide frequency ranges for 23 crown and root traits. Although we will judiciously not include the actual references, if you find a cusp 7 frequency of 80% in your sample, you are scoring something other than cusp 7. This trait can get up to 40–50% in some African populations, but outside of Africa it rarely exceeds 10%. Although sampling error in very small samples (e.g., 10) is possible, it is unlikely that a sample would have trait frequencies totally out of line with known patterns of geographic variation.

## References

Aas, I.H.M., and Risnes, S. (1979). The depth of the lingual fossa in permanent incisors of Norwegians. I. Method of measurement, statistical distribution and sex dimorphism. *American Journal of Physical Anthropology* 50, 335–340.

Bailey, S.E., Skinner, M.M., and Hublin, J.-J. (2011). What lies beneath? An evaluation of lower molar trigonid crest patterns based on both dentine and enamel expression. *American Journal of Physical Anthropology* 145, 505–518.

Bang, G., and Hasund, A. (1971). Morphologic characteristics of the Alaskan Eskimo dentition. I. Shovel shape of incisors. *American Journal of Physical Anthropology* 35, 43–48.

(1973). Morphologic characteristics of the Alaskan Eskimo dentition. II. Carabelli's cusp. *American Journal of Physical Anthropology* 37, 35–40.

Barnes, D.S. (1969). Tooth morphology and other aspects of the Teso dentition. *American Journal of Physical Anthropology* 30, 183–194.

Benzécri, J.P. (1992). *Correspondence Analysis Handbook*. New York: Dekker.

Boyd, W.C. (1950). *Genetics and the Races of Man*. Boston: Little, Brown and Company.

Braga, J. (2016). Non-invasive imaging techniques. In *A Companion to Dental Anthropology*, ed. J.D. Irish and G.R. Scott. Chichester, West Sussex: John Wiley & Sons, pp. 514–527.

Braga, J., Thackeray, J.F., Subsol, G., *et al.* (2010). The enamel–dentine junction in the post-canine dentition of *Australopithecus africanus*: individual metameric and antimeric variation. *Journal of Anatomy* 216, 62–79.

Brook, A.H. (1984). A unifying aetiological explanation for anomalies of human tooth number and size. *Archives of Oral Biology* 29, 373–378.

Burnett, S.E. (1998). *Maxillary Premolar Accessory Ridges (Mxpar): Worldwide Occurrence and Utility in Population Differentiation.* MA thesis, Arizona State University, Tempe.

(2016). Crown wear: Identification and categorization. In *A Companion to Dental Anthropology*, ed. J.D. Irish and G.R. Scott, New York: Wiley Blackwell. pp. 415–432.

Burnett, S.E., Irish, J.D., and Fong, M.R. (2013). Wears the problem? Examining the effect of dental wear on studies of crown morphology. In *Anthropological Perspectives on Tooth Morphology: Genetics, Evolution, Variation*, ed. G.R. Scott and J.D. Irish. Cambridge: Cambridge University Press. pp. 535–553.

Cavalli-Sforza, L.L., Menozzi, P., and Piazza, A. (1994). *The History and Geography of Human Genes.* Princeton: Princeton University Press.

Clausen, S.E. (1988). *Applied Correspondence Analysis: An Introduction.* Thousand Oaks: Sage Publications.

Dahlberg, A.A. (1956). Materials for the establishment of standards for classification of tooth characters, attributes, and techniques in morphological studies of the dentition. Zollar Laboratory of Dental Anthropology, University of Chicago (mimeo).

Falconer, D. S. (1960). *Introduction to Quantitative Genetics.* New York: The Ronald Press Company.

Goaz, P.W., and Miller, M.C., III (1966). A preliminary description of the dental morphology of the Peruvian Indian. *Journal of Dental Research* 45, 106–119.

Goldstein, M.S. (1948). Dentition of Indian crania from Texas. *American Journal of Physical Anthropology* 6, 63–84.

Grüneberg, H. (1952). Genetical studies on the skeleton of the mouse. IV. Quasi-continuous variations. *Journal of Genetics* 51, 95–114.

Harris, E.F. (1977). *Anthropologic and Genetic Aspects of the Dental Morphology of Solomon Islanders, Melanesia.* PhD dissertation, Department of Anthropology, Arizona State University, Tempe.

Harris, E.F., and Sjøvold, T. (2004). Calculation of Smith's Mean Measure of Divergence for intergroup comparisons using nonmetric data. *Dental Anthropology* 17, 83–93.

Howells, W.W. (1973). *Cranial Variation in Man: A Study by Multivariate Analysis of Patterns of Difference among Recent Human Populations.* Papers of the Peabody Museum of Archaeology and Ethnology, Harvard University, volume 67. Cambridge, MA: Harvard University.

(1989). *Skull Shapes and the Map: Craniometric Analyses in the Dispersion of Modern Homo.* Papers of the Peabody Museum of Archaeology and Ethnology, Harvard University, volume 79. Cambridge, MA: Harvard University.

Irish, J.D. (1993). *Biological Affinities of Late Pleistocene through Modern African Aboriginal Populations: The Dental Evidence.* PhD dissertation, Department of Anthropology, Arizona State University, Tempe.

(1997). Characteristic high- and low-frequency dental traits in sub-Saharan African populations. *American Journal of Physical Anthropology* 102, 455–467.

(2006). Who were the ancient Egyptians? Dental affinities among Neolithic through postdynastic peoples. *American Journal of Physical Anthropology* 129, 529–543.

(2010). The mean measure of divergence (MMD): its utility in model-free and model-bound analyses relative to the Mahalanobis D2 distance for nonmetric traits. *American Journal of Human Biology* 22, 378–395.

Irish, J.D., Black, W., Sealy, J., and Ackermann, R. (2014). Questions of Khoesan continuity: dental affinities among the indigenous Holocene peoples of South Africa. *American Journal of Physical Anthropology* 155, 33–44.

Keene, H.J. (1965). The relationship between third molar agenesis and the morphologic variability of the molar teeth. *Angle Orthodontist* 35, 289–298.

(1968). The relationship between Carabelli's trait and the size, number and morphology of the maxillary molars. *Archives of Oral Biology* 13, 1023–1025.

Konigsberg, L.W. (1990). Analysis of prehistoric biological variation under a model of isolation by geographic and temporal distance. *Human Biology* 62, 49–70.

Korenhof, C.A.W. (1960). *Morphogenetical Aspects of the Human Upper Molar*. Utrecht: Uitgeversmaatschappij Neerlandia.

Kruskal, J.B., and Wish, M. (1978). *Multidimensional Scaling*. Beverly Hills: Sage Publications.

Livingstone, F.B. (1991). Phylogenies and the forces of evolution. *American Journal of Human Biology* 3, 83–89.

Macchiarelli, R., Bayle, P., Bondioli, L., Mazurier, A., and Zanolli, C. (2013). From outer to inner structural morphology in dental anthropology: integration of the third dimension in the visualization and quantitative analysis of fossil remains. In *Anthropological Perspectives on Dental Morphology: Genetics, Evolution, Variation*, ed. G.R. Scott and J.D. Irish. Cambridge: Cambridge University Press, pp. 250–277.

Mahalanobis, P.C. (1936). On the generalized distance in statistics. *Proceedings of the National Institute of Science, India* 2, 49–55.

Martínez de Pinillos, M., Martinón-Torres, M., Skinner, M.M., *et al.* (2014). Trigonid crests expression in Atapuerca-Sima de los Huesos lower molars: internal and external morphological expression and evolutionary inferences. *Comtes Rendus Palevol* 13, 205–221.

Martinón-Torres, M., Martínez de Pínillos, M., Skinner, M.M., *et al.* (2014). Talonid crests expression at the enamel–dentine junction of hominin lower permanent and deciduous molars. *Comptes Rendus Palevol* 13, 223–234.

Nelson, C.T. (1938). The teeth of the Indians of Pecos Pueblo. *American Journal of Physical Anthropology* 23, 261–293.

Nichol, C.R., and Turner, C.G., II (1986). Intra- and interobserver concordance in scoring dental morphology. *American Journal of Physical Anthropology* 69, 299–315.

Noss, J.F., Scott, G.R., Potter, R.H.Y., Dahlberg, A.A., and Dahlberg, T. (1983). The influence of crown size dimorphism on sex differences in the Carabelli trait and the canine distal accessory ridge in man. *Archives of Oral Biology* 28, 527–530.

Phillips, D. (1995). Correspondence analysis. *Social Research Update* 7, 1–8.
Romesburg, C.H. (1984). *Cluster Analysis for Researchers*. Belmont: Lifetime Learning Publications.
Rosenzweig, K.A., and Zilberman, Y. (1967). Dental morphology of Jews from Yemen and Cochin. *American Journal of Physical Anthropology* 26, 15–22.
(1969). Dentition of Bedouin in Israel. II. Morphology. *American Journal of Physical Anthropology* 31, 199–204.
Scott, G.R. (1973). *Dental Morphology: A Genetic Study of American White Families and Variation in Living Southwest Indians*. PhD dissertation, Department of Anthropology, Arizona State University, Tempe.
(1977a). Classification, sex dimorphism, association, and population variation of the canine distal accessory ridge. *Human Biology* 49, 453–469.
(1977b). Interaction between shoveling of the maxillary and mandibular incisors. *Journal of Dental Research* 56, 1423.
(1978). The relationship between Carabelli's trait and the protostylid. *Journal of Dental Research* 57, 570.
(1979). Association between the hypocone and Carabelli's trait of the maxillary molars. *Journal of Dental Research* 58, 1403–1404.
(1980). Population variation of Carabelli's trait. *Human Biology* 52, 63–78.
Scott, G.R., and Turner, C.G., II (1997). *The Anthropology of Modern Human Teeth: Dental Morphology and its Variation in Recent Human Populations*. Cambridge: Cambridge University Press.
Scott, G.R., Potter, R.H.Y., Noss, J.F., Dahlberg, A.A., and Dahlberg, T. (1983). The dental morphology of Pima Indians. *American Journal of Physical Anthropology* 61, 13–31.
Sjøvold, T. (1977). Non-metrical divergence between skeletal populations: the theoretical foundation and biological importance of C.A.B. Smith's Mean Measure of Divergence. *Ossa* 4 (Suppl. 1), 1–133.
Skinner, M.M., Wood, B.A., Boesch, C., *et al.* (2008). Dental trait expression at the enamel–dentine junction of lower molars in extant and fossil hominoids. *Journal of Human Evolution* 54, 173–186.
Sofaer, J.A., Niswander, J.D., MacLean, C.J., and Workman, P.L. (1972). Population studies on Southwestern Indian tribes. V. Tooth morphology as an indicator of biological distance. *American Journal of Physical Anthropology* 37, 357–366.
Stojanowski, C.M., and Johnson, K.M. (2015). Observer error, dental wear, and the inference of New World Sundadonty. *American Journal of Physical Anthropology* 156, 349–362.
Turner, C.G., II, and Scott, G.R. (1977). Dentition of Easter Islanders. In *Orofacial Growth and Development*, ed. A.A. Dahlberg and T.M. Graber. The Hague: Mouton Publishers, pp. 229–249.
Turner, C.G., II, Nichol, C.R., and Scott, G.R. (1991). Scoring procedures for key morphological traits of the permanent dentition: the Arizona State University dental

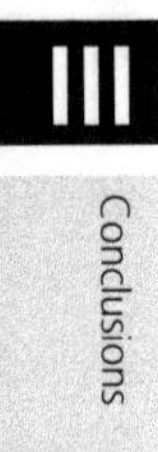



anthropology system. In *Advances in Dental Anthropology*, ed. M.A. Kelley and C.S. Larsen. New York: Wiley-Liss, pp. 13–31.

Zubov, A.A. (1968). *Odontology: A Method of Anthropological Research.* Moscow: Nauka (in Russian).

(1977). Odontoglyphics: the laws of variation of the human molar crown relief. In *Orofacial Growth and Development*, ed. A.A. Dahlberg and T.M. Graber. The Hague: Mouton Publishers, pp. 269–282.

# Appendix: Comparative Data



Over a four-decade period, Christy Turner traveled the world making systematic observations on 30,000 skulls, with special emphasis on crown and root trait morphology. His initial focus was on the Arctic but he expanded his dental "range" to include much of the New World, Asia, Australia, the Pacific, and Europe, with a light touch on Africa and India. Following the untimely death of Dr. Turner in 2013, the first author (GRS) made several visits to Tempe, Arizona, to salvage as much data as possible. In addition to scanning 3000 slides and 30,000 data sheets, he copied hundreds of computer printouts that included the full class frequency distributions for Turner's 29 key dental traits.

While it is not practical at this time to provide the data sheets or many of the summarized tables, the authors appreciate that not all researchers and students have access to comparative material. Moreover, in the literature on dental morphology, some authors use breakpoints to characterize a sample. While this is efficient in terms of space, it is an issue when authors use different breakpoints. For example, Scott and Turner (1997) used grade 3 as a breakpoint for UI1 shoveling because they had both studied numerous Native American samples where shoveling was extremely common and pronounced. Irish (1993), who has focused on dozens of African samples, used grade 2 as a breakpoint because these populations show much less, as well as an alternative form of, shoveling than Native Americans. To help researchers (students and professionals alike) put their data into a broader comparative framework, we are providing tables with full class frequency distributions for 29 key traits in 60 samples representing major geographic regions of the world. In some tables, data on a few traits are missing because they were not included on a particular computer printout. When the individual data sheets are released, researchers can fill in these blanks as needed.

Because of certain idiosyncrasies in how trait frequencies were tabulated on the Turner printouts, we provide a key to describe how each trait should be interpreted. For most traits, the results are straightforward, with 0 = absence and grades 1–*n* equaling varying degrees of trait presence. A few traits, however, require explanation, and those are in the key. For example, even though Turner set up the classification for cusp 6, there is not a category for just cusp 6. However, cusp 6 frequencies are readily available under the rows for lower molar cusp number. For those rows, 0 = 4 cusps,

1 = 5 cusps, and 2 = the presence of cusp 6. Other areas of potential confusion are explained in the key.

## Key to Tables

**Winging UI1** (1 = bilateral winging; 2 = unilateral winging; 3 = straight; 4 = counter-winging); follows Turner *et al.* (1991), as the scale proposed in this volume had not been adopted
**Shoveling UI1** (0 = absence; 1–7 = varying degrees of trait presence)
**Double-shoveling UI1** (0 = absence; 1–6 = varying degrees of trait presence); note: in some tables Turner did not include class frequency distributions but only a total trait frequency under column 6. We moved this so that 1 = presence of double-shoveling
**Interruption grooves UI2** (0 = absence; 1 = trait presence, of any form)
**Tuberculum dentale UI2** (0 = absence; 1–7 = varying degrees of trait presence)
**Bushman canine UC** (0 = absence; 1–3 = varying degrees of trait presence)
**Distal accessory ridge UC** (0 = absence; 1–5 = varying degrees of trait presence)
**Uto-Aztecan UP1** (0 = absence; 1 = presence)
**Hypocone UM2** (0 = absence; 1–3 follows plaque; 3.5 = 4, 4 = 5, 5 = 6)
**Cusp 5 UM1** (0 = absence; 1–5 = varying degrees of trait presence)
**Carabelli's cusp UM1** (0 = absence; 1–7 = varying degrees of trait presence)
**Parastyle UM3** (0 = absence; 1–6 = varying degrees of trait presence); although Turner scored parastyles on UM1, his tables of key traits focused on UM3 expression. The pronounced forms of paramolar tubercles (grades 4, 5, and 6) are found on either UM2 or UM3
**Enamel extension UM1** (0 = absence; 1–3 = varying degrees of trait presence)
**Root number UP1** (1 = one root; 2 = two roots; 3 = three roots)
**Root number UM2** (1 = one root; 2 = two roots; 3 = three roots; 4 = four roots)
**PRM (pegged/reduced/missing) UM3** (0 = absence; 1 = pegged; 2 = reduced; 3 = congenital absence; if there is only one number, it is placed under "3" and combines pegged, reduced, and missing)
**Lingual cusp number LP2** (0 = absence of lingual cusp; 1 = graded 0–1; 2 = grades 2–7; 3 = grades 8–9); note: grade 0 on plaque has lingual cusp, new grade 0 = no lingual cusp)
**Groove pattern LM2** (1 = Y pattern; 2 = X pattern; 3 = + pattern)
**Cusp number LM1** (Grade 0 = 4 cusps; 1 = 5 cusps; 2 = 6 cusps); note: this is the only location in the tables to find four-cusped LM1 and cusp 6 on LM1 (which is present/absent, not ranked)
**Cusp number LM2** (0 = 4 cusps; 1 = 5 cusps; 2 = 6 cusps)
**Deflecting wrinkle LM1** (0 = absent; 1–3 = varying degrees of trait presence)
**Trigonid crest LM1** (= distal trigonid crest) (0 = absent; 1 = present)
**Protostylid LM1** (0 = absence; 1 = buccal pit; 2–7 = varying degrees of trait presence)

**Cusp 7 LM1** (0 = absence; 1 = grade 1; 2 = grade 1A; 3 = grade 2; 4 = grade 3; 5 = grade 4; in a few tables, there is a frequency under 6, which we assume represents a prounounced cusp 7 that exceeds grade 4 in size)
**Tomes' root LP1** (0 = absence; 1–7 = varying degrees of trait presence); note: Turner plaque has grade 0 and five degrees of trait presence, but in the tables there are seven degrees of trait presence (on this scale, Turner used grades 4–7/0–7 as the breakpoint). Before standard plaque was developed, Turner scored Tomes' root as present or absent; in some tables, 0 = absence, 1 = presence
**Root number LC** (1 = one root; 2 = two roots)
**Root number LM1** (1 = one root; 2 = two roots; 3 = three roots, or 3RM1)
**Root number LM2** (1 = one root; 2 = two roots; 3 = three roots)
**Odontome UP1,2/LP1,2** (0 = absence; 1 = presence on any upper or lower premolar)

## Sample Provenance

### Africa

**Egypt:** *n* = 111, Lowie Museum, University of California, Berkeley
**Nubia:** (1) *n* = 45, Nubian Upper Stone Age, 12,000–10,000 BC, Southern Methodist University; (2) *n* = 12, 18,000–14,500 BP, Nubia 67/80, Nubian Upper Stone Age, Southern Methodist University
**West Africa:** *n* = 72, Ashanti, Dahomey, late 1800s, American Museum of Natural History

### Asia

**Ainu:** (1) *n* = 88, Hokkaido, recent and living, University of Tokyo; (2) *n* = 45, Sakhalin, recent and living, Institute of Ethnography, Leningrad, Musée de l'Homme, Paris; (3) *n* = 55, Sakhalin, shellmounds and recent, Kyoto University; (4) *n* = 26, Hokkaido, Ainu, shellmounds and recent, Kyoto University; (5) *n* = 116, Hokkaido, Ainu, middle to late Edo period, AD 1750–1900, Sapporo Medical College
**An-Yang:** *n* = 277, Northern Hunan Province, China, 1100 BC, Academia Sinica, Taipei, Taiwan
**Japan:** (1) *n* = 180, historic to recent, locality unspecified, Musée de l'Homme, Paris; (2) *n* = 131, recent, northern Honshu, Kyoto University
**Jomon:** *n* = 363, Central Japan, Yokohama–Tokyo area, 2500–300 BC, Middle to Late Jomon, University of Tokyo
**Sopka 2:** *n* = 170, Siberia, Krotov Culture, early Bronze Age, 1600–1200 BC, Institute of History, Philology, and Philosophy, Novosibirsk, Siberia
**South China:** *n* = 63, Canton, AD 1900, Smithsonian Institution; other elements N/A
**Urga:** *n* = 190, Ulaanbaatar, Mongolia, Mongolic, AD 1900, San Diego Museum of Man

## Australia

**Lower Murray:** $n = 74$, South Australian Museum, Adelaide, Australia
**North Australia:** $n = 57$, north of Tropic of Capricorn, historic, Smithsonian Institution, Natural History Musuem, London, American Museum of Natural History; other samples N/A
**Queensland:** $n = 96$, Queensland Museum, Australian Museum; other samples N/A
**South Australia:** $n = 133$, age unknown, American Museum of Natural History, Natural History Musuem, London, Duckworth Laboratory
**Swanport:** $n = 95$, Lower Murray River, age unknown, South Australian Museum, Adelaide, Australia

## Circumpolar

**Aleuts (Eastern):** (1) $n = 79$, Pt. Mollar to Unalaska, prehistoric to historic, Paleo- and Neo-Aleut, American Museum of Natural History, Smithsonian Institution, Arizona State University, Field Museum, San Diego Museum of Man; (2) $n = 82$, Kashega to Shiprock, prehistoric to historic, Paleo- and Neo-Aleut, American Museum of Natural History, Smithsonian Institution; (3) $n = 150$; Umnak to Kagamil, Paleo- and Neo-Aleut, Smithsonian Institution, University of Oregon; (4) $n = 7$, unspecified Eastern islands, date unknown, Peabody Museum
**Aleuts (Western):** (1) $n = 51$, Amlia to Kiska, Alaska, prehistoric to early historic, primarily Paleo-Aleut, Smithsonian Institution; (2) $n = 36$, Shemya to Attu, Alaska, prehistoric to early historic, primarily Paleo-Aleut, Smithsonian Institution
**Greenland:** (1) $n = 41$, Southwest Greenland, prehistoric; (2) $n = 17$, West Greenland, prehistoric; (3) $n = 52$, Northeast Greenland, prehistoric; (4) $n = 96$, Southeast Greenland, prehistoric; (5) $n = 4$, East Greenland, prehistoric; Panum Institute, Copenhagen, Denmark
**Kodiak Island:** $n = 225$, Larsen Bay, Kachemak to Koniag, 1000 BC–AD 1500, Smithsonian Institution
**Southampton Island:** $n = 70$, Sadlermiut Eskimo, AD 1902, Archaeological Survey of Canada
**St. Lawrence Island:** $n = 244$; Old Bering Sea to recent times, Smithsonian Institution

## Europe

**Basques (Spain):** $n = 255$, eleventh to eighteenth century AD, Cathedral of Santa Maria, Vitoria, Spain
**Dorestad de Heul, Netherlands:** $n = 72$, early medieval culture, AD 700–800, Institute of Human Biology, Utrecht
**Estonia**: $n = 84$, late Bronze to early Iron Age, 1000–800 BC, Estonian Institute of History, Archaeology, and Art History, Tallinn, Estonia; later sample N/A

**Lapps:** $n$ = 64, Kola Peninsula, Chalmny-Varra, AD 1800–1900, Institute of Ethnography, Leningrad

**Poundbury, England:** $n$ = 131, Romano-British Christian community, AD 150–350, Natural History Museum, London

**Russians:** $n$ = 146, probably Russian Orthodox, AD 1700–1800, Institute of Ethnography, Leningrad

## Melanesia

**Fiji:** (1) $n$ = 73, proto-historic and historic, Bishop Museum, Honolulu, Hawaii, Natural History Museum, London, Musée de l'Homme, Paris; (2) $n$ = 19, prehistoric, Rotuma, Simon Fraser University

**Loyalty Islands:** $n$ = 78, historic, Musée de l'Homme, Paris, Natural History Museum, London

**New Britain:** (1) $n$ = 140, recent, American Museum of Natural History; (2) $n$ = 34, recent, Smithsonian Institution; (3) $n$ = 46, Hoskins Peninsula, West Nakanai, living, University of Washington

**New Hebrides:** (1) $n$ = 12, historic, Bishop Museum, Honolulu, Hawaii; (2) $n$ = 66, Vanuatu, Queensland University

## Mesoamerica

**Cuicuilco:** $n$ = 98, pre-classic, 2500 BC–AD 200, Instituto Nacional Antropología e Historia, Mexico City

**Guasave:** $n$ = 53, N/A, American Museum of Natural History

**Tlatelolco:** $n$ = 261, post-classic/colonial – just before and after the conquest of the Aztecs, AD 1500–1600, Instituto Nacional Antropología e Historia, Mexico City

## Micronesia

**Guam:** (1) $n$ = 195, Micronesian, prehistoric and possibly historic, Bishop Museum, Honolulu, Hawaii; (2) $n$ = 17, 1320 BC to recent, California State University, Los Angeles; (3) $n$ = 10, historic Guam, Musée de l'Homme, Paris

## New Guinea

**New Guinea:** $n$ = 50, Melanesian, historic, Duckworth Laboratory, Los Angeles Country Museum, University of Arkansas

**New Guinea Gulf:** $n$ = 72, New Guinea gulf, historic, Australian Museum, Sydney, Australia

**Torres 1 and 2:** $n$ = 36, Torres Strait, historic, Duckworth Laboratory, Musée de l'Homme, Paris

## North America (Native Populations)

**Alabama:** (1) *n* = 103, Shellmound, archaic component, University of Alabama; (2) *n* = 115, Kroger Island, Mississippian component, University of Alabama; (3) *n* = 50, Three Mile Island, prehistoric, University of Alabama

**Grasshopper:** *n* = 217, Mogollon, AD 1275–1400, University of Arizona

**Greater Northwest Coast:** (1) *n* = 185, Northern Maritime, prehistoric to historic, Smithsonian Institution, American Museum of Natural History, Archaeological Survey of Canada, Simon Frazer; (2) *n* = 133, Central Maritime, prehistoric to historic, American Museum of Natural History; (3) *n* = 325, Gulf of Georgia, prehistoric to historic, American Museum of Natural History, Provincial Museum of British Columbia, University of Washington, Field Museum of Natural History

**Iroquois:** (1) *n* = 98, Roebuck Iroquois, AD 1500+, Archaeological Survey of Canada; (2) *n* = 429, Toronto Iroquois, AD 1585–1615, Kleinberg Ossuary, Protohistoric, University of Toronto Museum

**Maryland:** (1) *n* = 131, Maryland ossuaries, mostly prehistoric, some colonial, Smithsonian Institution; (2) *n* = 71, Nanjemoy/Joule, prehistoric, Late Woodland, Smithsonian Institution

**Northern California:** (1) *n* = 62, Humboldt County, prehistoric, University of California, Berkeley; (2) *n* = 119, Sacramento County, Windmiller period to late prehistoric, University of California, Berkeley; (3) *n* = 61, Alameda Country, prehistoric, University of California, Berkeley

**Point of Pines:** (1) *n* = 12, Point of Pines early, Mogollon, AD 400–1000, University of Arizona, Tucson; (2) *n* = 78, Point of Pines middle, Mogollon, AD 1000–1285, University of Arizona, Tucson; (3) *n* = 92, Point of Pines late, Mogollon and possible Kayenta Anasazi intrusion, AD 1285–1450, University of Arizona, Tucson

## Polynesia

**Marquesas:** (1) *n* = 46, Marquesas No. 1, protohistoric and historic, Bishop Museum, Honolulu, Hawaii; (2) *n* = 194, Marquesas No. 2, prehistoric to historic, American Museum of Natural History

**Mokapu:** *n* = 248, Mokapu, Hawaii, prehistoric, British Museum

**New Zealand:** *n* = 117, mainly historic Polynesian, Natural History Museum, London, Duckworth Laboratory, Musée de l'Homme, Paris

## South America (Native Populations)

**Ayalan:** *n* = 162, western Ecuador, Ayalan, Mailagro Culture, 500 BC–AD 1600, Smithsonian Institution

**Corondo:** *n* = 103, Brazil, Itaipu Phase, 4000 BC–AD 1900, Instituto Arqueologia Brasileira, Rio de Janeiro, Brazil

**Herradura and Teatinos:** (1) $n = 30$, Herradura, Chile, >200 BC, Museo Nacional de Historia Natural, Santiago, Chile; (2) $n = 137$, Punta Teatinos, Chile, pre-ceramic, 3300–2000 BP, Museo Nacional de Historia Natural, Santiago, Chile

**Peru 1 and 2:** (1) $n = 275$, Peruvian coast, Chincha Valley, AD 1 to colonial, San Diego Museum of Man, University of California, Berkeley; (2) $n = 476$, Chicama Valley, AD 600–1580, late Coastal States, Smithsonian Institution

**Preceramic Peru:** $n = 95$, south-central Peruvian coast, 5500–4000 BP, American Museum of Natural History, Centro por Investigaciones de Zona Arida, Lima, Peru

**Sambaqui South:** $n = 119$, Brazil, coastal shellmounds, 4000–3000 BP, Museu Nacional de Brasil, Rio de Janeiro, Brazil

**Santa Elena:** $n = 69$, Santa Elena Peninsula, Ecuador, pre-ceramic, 10,000–6000 BP, Banco Central, Guayaquil, Ecuador

## Southeast Asia

**Borneo:** $n = 39$, historic Borneo, Natural History Museum, London

**Calatagan:** $n = 61$, Calatagan, Philippines, AD 1500–1700, National Museum of the Philippines

**Malay:** $n = 58$, Malaysia, historic and recent, American Museum of Natural History, Natural History Museum, London, Duckworth Laboratory, Field Museum of Natural History, Chicago

**Philippines:** (1) $n = 89$, various tribal associations including negrito, 200 BC–AD 1500, American Museum of Natural History, Field Museum of Natural History, National Museum of the Philippines, Smithsonian Institution; (2) $n = 25$, historic Philippines, Natural History Museum, London, Duckworth Laboratory

**Taiwan:** $n = 66$, prehistoric Taiwan, 4000–1500 BP, National Taiwan University, Taipei, Taiwan

# Samples by Geographic Area

## Africa

### *Egypt*

| | | Grade | | | | | | | |
|---|---|---|---|---|---|---|---|---|---|
| **Trait** | *n* | 0 | 1 | 2 | 3 | 4 | 5 | 6 | 7 |
| Winging UI1 | 9 | | 0.000 | 0.000 | 1.000 | 0.000 | | | |
| Shoveling UI1 | 3 | 0.667 | 0.333 | 0.000 | 0.000 | 0.000 | 0.000 | 0.000 | 0.000 |
| Double-shoveling UI1 | 3 | 1.000 | 0.000 | 0.000 | 0.000 | 0.000 | 0.000 | 0.000 | |
| Interruption grooves UI2 | 3 | 1.000 | 0.000 | | | | | | |
| Tuberculum dentale UI2 | 2 | 0.500 | 0.000 | 0.000 | 0.000 | 0.000 | 0.000 | 0.000 | 0.500 |
| Bushman canine UC | 6 | 0.833 | 0.167 | 0.000 | 0.000 | | | | |
| Distal acc. ridge UC | | | | | | | | | |
| Uto-Aztecan UP1 | | | | | | | | | |
| Hypocone UM2 | 56 | 0.013 | 0.018 | 0.089 | 0.125 | 0.429 | 0.286 | 0.036 | |
| Cusp 5 UM1 | 41 | 0.829 | 0.049 | 0.049 | 0.049 | 0.024 | 0.000 | | |
| Carabelli's cusp UM1 | 26 | 0.038 | 0.462 | 0.038 | 0.077 | 0.039 | 0.115 | 0.115 | 0.115 |
| Parastyle UM3 | 46 | 0.978 | 0.000 | 0.022 | 0.000 | 0.000 | 0.000 | 0.000 | |
| Enamel extension UM1 | 65 | 0.631 | 0.323 | 0.000 | 0.046 | | | | |
| Root number UP1 | 87 | | 0.345 | 0.632 | 0.023 | | | | |
| Root number UM2 | 71 | | 0.056 | 0.141 | 0.803 | | | | |
| PRM UM3 | 74 | 0.838 | 0.000 | 0.014 | 0.149 | | | | |
| Lingual cusp no. LP2 | 8 | 0.000 | 0.125 | 0.750 | 0.125 | | | | |
| Groove pattern LM2 | 24 | | 0.250 | 0.542 | 0.208 | | | | |
| Cusp number LM1 | 12 | 0.000 | 0.917 | 0.083 | | | | | |
| Cusp number LM2 | 19 | 0.684 | 0.316 | 0.000 | | | | | |
| Deflecting wrinkle LM1 | | | | | | | | | |
| Trigonid crest LM1 | | | | | | | | | |
| Protostylid LM1 | | | | | | | | | |
| Cusp 7 LM1 | 20 | 0.750 | 0.000 | 0.050 | 0.050 | 0.100 | 0.050 | | |
| Tomes' root LP1 | 21 | 0.286 | 0.000 | 0.286 | 0.333 | 0.095 | 0.000 | 0.000 | 0.000 |
| Root number LC | | | | | | | | | |
| Root number LM1 | 36 | | 0.000 | 1.000 | 0.000 | | | | |
| Root number LM2 | 33 | | 0.091 | 0.909 | 0.000 | | | | |
| Odontome UP1,2/LP1,2 | 13 | 1.000 | 0.000 | | | | | | |

## Nubia

| Trait | n | Grade 0 | 1 | 2 | 3 | 4 | 5 | 6 | 7 |
|---|---|---|---|---|---|---|---|---|---|
| Winging UI1 | 25 | | 0.320 | 0.200 | 0.480 | 0.000 | | | |
| Shoveling UI1 | 22 | 0.000 | 0.409 | 0.545 | 0.045 | 0.000 | 0.000 | 0.000 | 0.000 |
| Double-shoveling UI1 | 20 | 1.000 | 0.000 | 0.000 | 0.000 | 0.000 | 0.000 | 0.000 | |
| Interruption grooves UI2 | 25 | 0.840 | 0.160 | | | | | | |
| Tuberculum dentale UI2 | 18 | 0.500 | 0.000 | 0.111 | 0.222 | 0.111 | 0.000 | 0.000 | 0.056 |
| Bushman canine UC | 18 | 0.778 | 0.167 | 0.056 | 0.000 | | | | |
| Distal acc. ridge UC | | | | | | | | | |
| Uto-Aztecan UP1 | | | | | | | | | |
| Hypocone UM2 | 27 | 0.000 | 0.037 | 0.037 | 0.111 | 0.333 | 0.444 | 0.037 | |
| Cusp 5 UM1 | 15 | 0.333 | 0.267 | 0.067 | 0.267 | 0.067 | 0.000 | | |
| Carabelli's cusp UM1 | 13 | 0.308 | 0.231 | 0.077 | 0.077 | 0.000 | 0.231 | 0.077 | 0.000 |
| Parastyle UM3 | 34 | 1.000 | 0.000 | 0.000 | 0.000 | 0.000 | 0.000 | 0.000 | |
| Enamel extension UM1 | 35 | 0.400 | 0.600 | 0.000 | 0.000 | | | | |
| Root number UP1 | 28 | | 0.286 | 0.714 | 0.000 | | | | |
| Root number UM2 | 25 | | 0.080 | 0.200 | 0.720 | | | | |
| PRM UM3 | 37 | 0.838 | | | 0.162 | | | | |
| Lingual cusp no. LP2 | 15 | 0.000 | 0.067 | 0.867 | 0.067 | | | | |
| Groove pattern LM2 | 27 | | 0.370 | 0.148 | 0.481 | | | | |
| Cusp number LM1 | 14 | 0.000 | 0.357 | 0.643 | | | | | |
| Cusp number LM2 | 33 | 0.061 | 0.757 | 0.182 | | | | | |
| Deflecting wrinkle LM1 | | | | | | | | | |
| Trigonid crest LM1 | | | | | | | | | |
| Protostylid LM1 | 21 | 0.714 | 0.286 | 0.000 | 0.000 | 0.000 | 0.000 | 0.000 | 0.000 |
| Cusp 7 LM1 | 28 | 0.786 | 0.000 | 0.179 | 0.036 | 0.000 | 0.000 | | |
| Tomes' root LP1 | 17 | 0.353 | 0.000 | 0.118 | 0.059 | 0.412 | 0.000 | 0.059 | 0.000 |
| Root number LC | 14 | | 1.000 | 0.000 | | | | | |
| Root number LM1 | 45 | | 0.000 | 0.844 | 0.156 | | | | |
| Root number LM2 | 35 | | 0.143 | 0.857 | 0.000 | | | | |
| Odontome UP1,2/LP1,2 | 8 | 1.000 | 0.000 | | | | | | |

## West Africa

| Trait | n | Grade 0 | 1 | 2 | 3 | 4 | 5 | 6 | 7 |
|---|---|---|---|---|---|---|---|---|---|
| Winging UI1 | 27 | | 0.074 | 0.037 | 0.889 | 0.000 | | | |
| Shoveling UI1 | 19 | 0.263 | 0.211 | 0.421 | 0.105 | 0.000 | 0.000 | 0.000 | 0.000 |
| Double-shoveling UI1 | 19 | 0.947 | 0.053 | | | | | | |
| Interruption grooves UI2 | 23 | 0.957 | 0.043 | | | | | | |
| Tuberculum dentale UI2 | 24 | 0.583 | 0.000 | 0.208 | 0.000 | 0.125 | 0.042 | 0.000 | 0.042 |
| Bushman canine UC | 37 | 0.676 | 0.216 | 0.108 | 0.000 | | | | |
| Distal acc. ridge UC | | | | | | | | | |
| Uto-Aztecan UP1 | | | | | | | | | |
| Hypocone UM2 | 56 | 0.000 | 0.036 | 0.089 | 0.143 | 0.286 | 0.429 | 0.018 | |
| Cusp 5 UM1 | 42 | 0.381 | 0.310 | 0.190 | 0.095 | 0.024 | 0.000 | | |
| Carabelli's cusp UM1 | 48 | 0.229 | 0.188 | 0.167 | 0.125 | 0.104 | 0.104 | 0.021 | 0.063 |
| Parastyle UM3 | 51 | 0.980 | 0.000 | 0.020 | 0.000 | 0.000 | 0.000 | 0.000 | |
| Enamel extension UM1 | 64 | 0.625 | 0.375 | 0.000 | 0.000 | | | | |
| Root number UP1 | 59 | | 0.356 | 0.644 | 0.000 | | | | |
| Root number UM2 | 57 | | 0.035 | 0.088 | 0.877 | | | | |
| PRM UM3 | 64 | 0.969 | 0.000 | 0.016 | 0.016 | | | | |
| Lingual cusp no. LP2 | 39 | 0.000 | 0.179 | 0.590 | 0.231 | | | | |
| Groove pattern LM2 | 40 | | 0.625 | 0.150 | 0.225 | | | | |
| Cusp number LM1 | 33 | 0.000 | 0.636 | 0.364 | | | | | |
| Cusp number LM2 | 42 | 0.167 | 0..738 | 0.095 | | | | | |
| Deflecting wrinkle LM1 | | | | | | | | | |
| Trigonid crest LM1 | | | | | | | | | |
| Protostylid LM1 | 40 | 0.800 | 0.175 | 0.000 | 0.000 | 0.000 | 0.000 | 0.025 | 0.000 |
| Cusp 7 LM1 | 43 | 0.326 | 0.093 | 0.279 | 0.116 | 0.070 | 0.070 | 0.047 | |
| Tomes' root LP1 | 32 | 0.656 | 0.344 | | | | | | |
| Root number LC | 18 | | 1.000 | 0.000 | | | | | |
| Root number LM1 | 47 | | 0.000 | 1.000 | 0.000 | | | | |
| Root number LM2 | 47 | | 0.043 | 0.957 | 0.000 | | | | |
| Odontome UP1,2/LP1,2 | 48 | 1.000 | 0.000 | | | | | | |

# Asia

## *Ainu*

| Trait | *n* | Grade 0 | 1 | 2 | 3 | 4 | 5 | 6 | 7 |
|---|---|---|---|---|---|---|---|---|---|
| Winging UI1 | 127 | | 0.339 | 0.071 | 0.591 | 0.000 | | | |
| Shoveling UI1 | 127 | 0.016 | 0.260 | 0.370 | 0.236 | 0.071 | 0.031 | 0.016 | 0.000 |
| Double-shoveling UI1 | 119 | 0.824 | 0.118 | 0.025 | 0.000 | 0.008 | 0.025 | 0.000 | |
| Interruption grooves UI2 | 137 | 0.555 | 0.445 | | | | | | |
| Tuberculum dentale UI2 | 148 | 0.764 | 0.000 | 0.081 | 0.014 | 0.000 | 0.014 | 0.007 | 0.122 |
| Bushman canine UC | 169 | 0.988 | 0.006 | 0.006 | 0.000 | | | | |
| Distal acc. ridge UC | 44 | 0.182 | 0.273 | 0.182 | 0.159 | 0.205 | 0.000 | | |
| Uto-Aztecan UP1 | | | | | | | | | |
| Hypocone UM2 | 202 | 0.158 | 0.064 | 0.158 | 0.233 | 0.342 | 0.045 | 0.000 | |
| Cusp 5 UM1 | 163 | 0.914 | 0.025 | 0.031 | 0.000 | 0.012 | 0.018 | | |
| Carabelli's cusp UM1 | 202 | 0.589 | 0.238 | 0.020 | 0.045 | 0.025 | 0.079 | 0.000 | 0.005 |
| Parastyle UM3 | 95 | 0.989 | 0.011 | 0.000 | 0.000 | 0.000 | 0.000 | 0.000 | |
| Enamel extension UM1 | 185 | 0.443 | 0.162 | 0.027 | 0.368 | | | | |
| Root number UP1 | 211 | | 0.844 | 0.147 | 0.009 | | | | |
| Root number UM2 | 175 | | 0.274 | 0.211 | 0.514 | | | | |
| PRM UM3 | 192 | 0.698 | 0.016 | 0.005 | 0.281 | | | | |
| Lingual cusp no. LP2 | 184 | 0.005 | 0.353 | 0.587 | 0.054 | | | | |
| Groove pattern LM2 | 200 | | 0.310 | 0.515 | 0.175 | | | | |
| Cusp number LM1 | 187 | 0.011 | 0.722 | 0.267 | | | | | |
| Cusp number LM2 | 195 | 0.359 | 0.548 | 0.092 | | | | | |
| Deflecting wrinkle LM1 | 45 | 0.467 | 0.111 | 0.289 | 0.133 | | | | |
| Trigonid crest LM1 | | | | | | | | | |
| Protostylid LM1 | 194 | 0.835 | 0.057 | 0.041 | 0.062 | 0.005 | 0.000 | 0.000 | 0.000 |
| Cusp 7 LM1 | 224 | 0.942 | 0.013 | 0.009 | 0.027 | 0.004 | 0.004 | | |
| Tomes' root LP1 | 118 | 0.000 | 0.695 | 0.169 | 0.051 | 0.068 | 0.008 | 0.008 | 0.000 |
| Root number LC | 199 | | 0.985 | 0.015 | | | | | |
| Root number LM1 | 205 | | 0.000 | 0.922 | 0.078 | | | | |
| Root number LM2 | 190 | | 0.326 | 0.668 | 0.005 | | | | |
| Odontome UP1,2/LP1,2 | 124 | 0.976 | 0.024 | | | | | | |

## An-Yang

| Trait | n | Grade 0 | 1 | 2 | 3 | 4 | 5 | 6 | 7 |
|---|---|---|---|---|---|---|---|---|---|
| Winging UI1 | 144 | | 0.174 | 0.042 | 0.778 | 0.007 | | | |
| Shoveling UI1 | 118 | 0.000 | 0.000 | 0.102 | 0.339 | 0.195 | 0.280 | 0.085 | 0.000 |
| Double-shoveling UI1 | 142 | 0.676 | 0.324 | | | | | | |
| Interruption grooves UI2 | 115 | 0.461 | 0.539 | | | | | | |
| Tuberculum dentale UI2 | 146 | 0.808 | 0.000 | 0.116 | 0.014 | 0.007 | 0.000 | 0.007 | 0.048 |
| Bushman canine UC | 132 | 0.962 | 0.015 | 0.023 | 0.000 | | | | |
| Distal acc. ridge UC | 63 | 0.286 | 0.079 | 0.127 | 0.206 | 0.206 | 0.095 | | |
| Uto-Aztecan UP1 | | | | | | | | | |
| Hypocone UM2 | 188 | 0.048 | 0.021 | 0.064 | 0.261 | 0.500 | 0.106 | 0.000 | |
| Cusp 5 UM1 | 23 | .870 | .000 | .087 | .043 | 0.000 | 0.000 | | |
| Carabelli's cusp UM1 | 156 | 0.628 | 0.071 | 0.090 | 0.038 | 0.038 | 0.051 | 0.045 | 0.038 |
| Parastyle UM3 | | | | | | | | | |
| Enamel extension UM1 | 224 | 0.196 | 0.228 | 0.156 | 0.420 | | | | |
| Root number UP1 | 143 | | 0.699 | 0.280 | 0.021 | | | | |
| Root number UM2 | 133 | | 0.113 | 0.083 | 0.782 | 0.023 | | | |
| PRM UM3 | 215 | 0.674 | | | 0.326 | | | | |
| Lingual cusp no. LP2 | 145 | 0.000 | 0.145 | 0.814 | 0.041 | | | | |
| Groove pattern LM2 | 152 | | 0.224 | 0.717 | 0.059 | | | | |
| Cusp number LM1 | 67 | 0.000 | 0.493 | 0.507 | | | | | |
| Cusp number LM2 | 103 | 0.126 | 0.680 | 0.194 | | | | | |
| Deflecting wrinkle LM1 | 8 | 0.125 | 0.000 | 0.250 | 0.625 | | | | |
| Trigonid crest LM1 | | | | | | | | | |
| Protostylid LM1 | 148 | 0.649 | 0.230 | 0.027 | 0.047 | 0.027 | 0.014 | 0.007 | 0.000 |
| Cusp 7 LM1 | 153 | 0.882 | 0.020 | 0.013 | 0.026 | 0.039 | 0.020 | | |
| Tomes' root LP1 | | | | | | | | | |
| Root number LC | 62 | | 1.000 | 0.000 | | | | | |
| Root number LM1 | 172 | | 0.000 | 0.616 | 0.384 | | | | |
| Root number LM2 | 141 | | 0.319 | 0.667 | 0.014 | | | | |
| Odontome UP1,2/LP1,2 | 6 | 1.000 | 0.000 | | | | | | |

## *Japan*

| Trait | *n* | Grade 0 | 1 | 2 | 3 | 4 | 5 | 6 | 7 |
|---|---|---|---|---|---|---|---|---|---|
| Winging UI1 | 25 | | 0.280 | 0.080 | 0.640 | 0.000 | | | |
| Shoveling UI1 | 20 | 0.000 | 0.000 | 0.200 | 0.450 | 0.200 | 0.150 | 0.000 | 0.000 |
| Double-shoveling UI1 | 23 | 0.478 | 0.087 | 0.043 | 0.000 | 0.000 | 0.391 | 0.000 | |
| Interruption grooves UI2 | 38 | 0.737 | 0.263 | | | | | | |
| Tuberculum dentale UI2 | 37 | 0.757 | 0.000 | 0.135 | 0.000 | 0.027 | 0.000 | 0.000 | 0.081 |
| Bushman canine UC | 59 | 1.000 | 0.000 | 0.000 | 0.000 | | | | |
| Distal acc. ridge UC | 38 | 0.316 | 0.184 | 0.184 | 0.211 | 0.105 | 0.000 | | |
| Uto-Aztecan UP1 | 66 | 1.000 | 0.000 | | | | | | |
| Hypocone UM2 | 116 | 0.060 | 0.034 | 0.103 | 0.233 | 0.448 | 0.121 | 0.000 | |
| Cusp 5 UM1 | 98 | 0.745 | 0.071 | 0.102 | 0.071 | 0.010 | 0.000 | | |
| Carabelli's cusp UM1 | 121 | 0.413 | 0.273 | 0.041 | 0.091 | 0.041 | 0.083 | 0.000 | 0.058 |
| Parastyle UM3 | 56 | 0.982 | 0.018 | 0.000 | 0.000 | 0.000 | 0.000 | 0.000 | |
| Enamel extension UM1 | 130 | 0.231 | 0.208 | 0.062 | 0.500 | | | | |
| Root number UP1 | 138 | | 0.725 | 0.268 | 0.007 | | | | |
| Root number UM2 | 126 | | 0.079 | 0.222 | 0.698 | | | | |
| PRM UM3 | 126 | 0.563 | 0.000 | 0.024 | 0.413 | | | | |
| Lingual cusp no. LP2 | 92 | 0.000 | 0.272 | 0.641 | 0.087 | | | | |
| Groove pattern LM2 | 96 | | 0.229 | 0.667 | 0.104 | | | | |
| Cusp number LM1 | 83 | 0.000 | 0.542 | 0.458 | | | | | |
| Cusp number LM2 | 92 | 0.109 | 0.750 | 0.141 | | | | | |
| Deflecting wrinkle LM1 | 64 | 0.469 | 0.047 | 0.219 | 0.266 | | | | |
| Trigonid crest LM1 | 89 | 0.921 | 0.079 | | | | | | |
| Protostylid LM1 | 97 | 0.763 | 0.134 | 0.000 | 0.062 | 0.021 | 0.010 | 0.010 | 0.000 |
| Cusp 7 LM1 | 105 | 0.905 | 0.010 | 0.029 | 0.057 | 0.000 | 0.000 | | |
| Tomes' root LP1 | 41 | 0.268 | 0.293 | 0.171 | 0.146 | 0.073 | 0.024 | 0.000 | 0.024 |
| Root number LC | 98 | | 1.000 | 0.000 | | | | | |
| Root number LM1 | 119 | | 0.000 | 0.731 | 0.269 | | | | |
| Root number LM2 | 120 | | 0.367 | 0.625 | 0.008 | | | | |
| Odontome UP1,2/LP1,2 | 98 | 0.949 | 0.051 | | | | | | |

## Jomon

| Trait | n | Grade 0 | 1 | 2 | 3 | 4 | 5 | 6 | 7 |
|---|---|---|---|---|---|---|---|---|---|
| Winging UI1 | 82 | | 0.183 | 0.024 | 0.793 | 0.000 | | | |
| Shoveling UI1 | 36 | 0.000 | 0.222 | 0.417 | 0.278 | 0.083 | 0.000 | 0.000 | 0.000 |
| Double-shoveling UI1 | 59 | 0.780 | 0.186 | 0.034 | 0.000 | 0.000 | 0.000 | 0.000 | |
| Interruption grooves UI2 | 95 | 0.347 | 0.653 | | | | | | |
| Tuberculum dentale UI2 | 96 | 0.813 | 0.000 | 0.010 | 0.010 | 0.000 | 0.000 | 0.000 | 0.167 |
| Bushman canine UC | 54 | 0.963 | 0.019 | 0.019 | 0.000 | | | | |
| Distal acc. ridge UC | 16 | 0.063 | 0.188 | 0.500 | 0.188 | 0.000 | 0.063 | | |
| Uto-Aztecan UP1 | 112 | 1.000 | 0.000 | | | | | | |
| Hypocone UM2 | 61 | 0.066 | 0.098 | 0.230 | 0.311 | 0.262 | 0.033 | 0.000 | |
| Cusp 5 UM1 | 135 | 0.911 | 0.015 | 0.007 | 0.007 | 0.015 | 0.044 | | |
| Carabelli's cusp UM1 | 58 | 0.724 | 0.172 | 0.069 | 0.017 | 0.000 | 0.000 | 0.017 | 0.000 |
| Parastyle UM3 | 89 | 0.944 | 0.000 | 0.034 | 0.011 | 0.011 | 0.000 | 0.000 | |
| Enamel extension UM1 | 76 | 0.724 | 0.145 | 0.026 | 0.105 | | | | |
| Root number UP1 | 73 | | 0.685 | 0.301 | 0.014 | | | | |
| Root number UM2 | 60 | | 0.100 | 0.200 | 0.683 | 0.017 | | | |
| PRM UM3 | 135 | 0.859 | 0.000 | 0.015 | 0.126 | | | | |
| Lingual cusp no. LP2 | 127 | 0.000 | 0.252 | 0.709 | 0.039 | | | | |
| Groove pattern LM2 | 88 | | 0.352 | 0.364 | 0.284 | | | | |
| Cusp number LM1 | 63 | 0.000 | 0.397 | 0.603 | | | | | |
| Cusp number LM2 | 66 | 0.318 | 0.546 | 0.136 | | | | | |
| Deflecting wrinkle LM1 | 72 | 0.875 | 0.014 | 0.042 | 0.069 | | | | |
| Trigonid crest LM1 | 128 | 0.938 | 0.063 | | | | | | |
| Protostylid LM1 | 76 | 0.697 | 0.211 | 0.066 | 0.026 | 0.000 | 0.000 | 0.000 | 0.000 |
| Cusp 7 LM1 | 82 | 0.976 | 0.000 | 0.000 | 0.012 | 0.012 | 0.000 | | |
| Tomes' root LP1 | 134 | 0.000 | 0.709 | 0.239 | 0.022 | 0.022 | 0.000 | 0.007 | 0.000 |
| Root number LC | 53 | | 1.000 | 0.000 | | | | | |
| Root number LM1 | 100 | | 0.000 | 0.950 | 0.050 | | | | |
| Root number LM2 | 95 | | 0.032 | 0.968 | 0.000 | | | | |
| Odontome UP1,2/LP1,2 | 110 | 1.000 | 0.000 | | | | | | |

## Sopka 2

| Trait | n | Grade | | | | | | | |
|---|---|---|---|---|---|---|---|---|---|
| | | 0 | 1 | 2 | 3 | 4 | 5 | 6 | 7 |
| Winging UI1 | 57 | | 0.158 | 0.018 | 0.825 | 0.000 | | | |
| Shoveling UI1 | 53 | 0.038 | 0.170 | 0.434 | 0.321 | 0.038 | 0.000 | 0.000 | 0.000 |
| Double-shoveling UI1 | 49 | 0.571 | 0.204 | 0.122 | 0.061 | 0.041 | 0.000 | 0.000 | |
| Interruption grooves UI2 | 78 | 0.449 | 0.551 | | | | | | |
| Tuberculum dentale UI2 | 82 | 0.768 | 0.000 | 0.061 | 0.049 | 0.000 | 0.012 | 0.000 | 0.110 |
| Bushman canine UC | 86 | 0.895 | 0.000 | 0.081 | 0.023 | | | | |
| Distal acc. ridge UC | 28 | 0.286 | 0.357 | 0.107 | 0.250 | 0.000 | 0.000 | | |
| Uto-Aztecan UP1 | 94 | 1.000 | 0.000 | | | | | | |
| Hypocone UM2 | 103 | 0.155 | 0.019 | 0.136 | 0.214 | 0.437 | 0.039 | 0.000 | |
| Cusp 5 UM1 | 82 | 0.720 | 0.122 | 0.061 | 0.085 | 0.012 | 0.000 | | |
| Carabelli's cusp UM1 | 81 | 0.407 | 0.259 | 0.173 | 0.099 | 0.025 | 0.037 | 0.000 | 0.000 |
| Parastyle UM3 | 73 | 0.973 | 0.014 | 0.014 | 0.000 | 0.000 | 0.000 | 0.000 | |
| Enamel extension UM1 | 129 | 0.488 | 0.248 | 0.000 | 0.264 | | | | |
| Root number UP1 | 114 | | 0.649 | 0.333 | 0.018 | | | | |
| Root number UM2 | 105 | | 0.314 | 0.238 | 0.448 | | | | |
| PRM UM3 | 120 | 0.767 | 0.025 | 0.042 | 0.167 | | | | |
| Lingual cusp no. LP2 | 105 | 0.000 | 0.524 | 0.457 | 0.019 | | | | |
| Groove pattern LM2 | 139 | | 0.173 | 0.712 | 0.115 | | | | |
| Cusp number LM1 | 90 | 0.000 | 0.733 | 0.267 | | | | | |
| Cusp number LM2 | 114 | 0.447 | 0.518 | 0.035 | | | | | |
| Deflecting wrinkle LM1 | 62 | 0.613 | 0.000 | 0.113 | 0.274 | | | | |
| Trigonid crest LM1 | 101 | 0.891 | 0.109 | | | | | | |
| Protostylid LM1 | 122 | 0.598 | 0.393 | 0.000 | 0.008 | 0.000 | 0.000 | 0.000 | 0.000 |
| Cusp 7 LM1 | 138 | 0.899 | 0.007 | 0.014 | 0.043 | 0.029 | 0.007 | | |
| Tomes' root LP1 | 117 | 0.000 | 0.393 | 0.256 | 0.154 | 0.120 | 0.068 | 0.009 | 0.000 |
| Root number LC | 135 | | 0.963 | 0.037 | | | | | |
| Root number LM1 | 146 | | 0.000 | 0.973 | 0.027 | | | | |
| Root number LM2 | 122 | | 0.574 | 0.426 | 0.000 | | | | |
| Odontome UP1,2/LP1,2 | 78 | 0.987 | 0.013 | | | | | | |

## *South China*

| Trait | n | Grade 0 | 1 | 2 | 3 | 4 | 5 | 6 | 7 |
|---|---|---|---|---|---|---|---|---|---|
| Winging UI1 | 339 | | 0.245 | 0.035 | 0.717 | 0.003 | | | |
| Shoveling UI1 | 350 | 0.000 | 0.063 | 0.289 | 0.411 | 0.171 | 0.054 | 0.011 | 0.000 |
| Double-shoveling UI1 | 340 | 0.415 | 0.306 | 0.194 | 0.056 | 0.023 | 0.006 | 0.000 | |
| Interruption grooves UI2 | 337 | 0.591 | 0.409 | | | | | | |
| Tuberculum dentale UI2 | 352 | 0.824 | 0.000 | 0.063 | 0.014 | 0.000 | 0.003 | 0.000 | 0.097 |
| Bushman canine UC | 375 | 0.965 | 0.019 | 0.013 | 0.003 | | | | |
| Distal acc. ridge UC | | | | | | | | | |
| Uto-Aztecan UP1 | | | | | | | | | |
| Hypocone UM2 | 422 | 0.081 | 0.028 | 0.073 | 0.197 | 0.526 | 0.095 | 0.000 | |
| Cusp 5 UM1 | 373 | 0.914 | 0.043 | 0.019 | 0.011 | 0.008 | 0.005 | | |
| Carabelli's cusp UM1 | 427 | 0.384 | 0.265 | 0.035 | 0.103 | 0.026 | 0.119 | 0.030 | 0.037 |
| Parastyle UM3 | 233 | 0.966 | 0.009 | 0.009 | 0.009 | 0.004 | 0.004 | 0.000 | |
| Enamel extension UM1 | 237 | 0.266 | 0.173 | 0.042 | 0.519 | | | | |
| Root number UP1 | 273 | | 0.641 | 0.341 | 0.018 | | | | |
| Root number UM2 | 240 | | 0.075 | 0.183 | 0.737 | 0.004 | | | |
| PRM UM3 | 405 | 0.684 | 0.002 | 0.012 | 0.301 | | | | |
| Lingual cusp no. LP2 | 422 | 0.000 | 0.320 | 0.621 | 0.059 | | | | |
| Groove pattern LM2 | 334 | | 0.237 | 0.674 | 0.090 | | | | |
| Cusp number LM1 | 346 | 0.000 | 0.650 | 0.350 | | | | | |
| Cusp number LM2 | 397 | 0.232 | 0.615 | 0.154 | | | | | |
| Deflecting wrinkle LM1 | | | | | | | | | |
| Trigonid crest LM1 | | | | | | | | | |
| Protostylid LM1 | 389 | 0.787 | 0.087 | 0.000 | 0.057 | 0.031 | 0.028 | 0.008 | 0.003 |
| Cusp 7 LM1 | 410 | 0.912 | 0.015 | 0.012 | 0.041 | 0.015 | 0.005 | | |
| Tomes' root LP1 | 172 | 0.000 | 0.477 | 0.198 | 0.128 | 0.157 | 0.041 | 0.000 | 0.000 |
| Root number LC | 230 | | 1.000 | 0.000 | | | | | |
| Root number LM1 | 245 | | 0.000 | 0.837 | 0.163 | | | | |
| Root number LM2 | 229 | | 0.354 | 0.646 | 0.000 | | | | |
| Odontome UP1,2/LP1,2 | 442 | 0.943 | 0.057 | | | | | | |

## *Urga*

| Trait | *n* | Grade | | | | | | | |
|---|---|---|---|---|---|---|---|---|---|
| | | 0 | 1 | 2 | 3 | 4 | 5 | 6 | 7 |
| Winging UI1 | 68 | | 0.309 | 0.000 | 0.676 | 0.015 | | | |
| Shoveling UI1 | 42 | 0.000 | 0.000 | 0.167 | 0.595 | 0.095 | 0.119 | 0.024 | 0.000 |
| Double-shoveling UI1 | 45 | 0.667 | 0.333 | | | | | | |
| Interruption grooves UI2 | 53 | 0.623 | 0.377 | | | | | | |
| Tuberculum dentale UI2 | 48 | 0.813 | 0.000 | 0.125 | 0.000 | 0.000 | 0.042 | 0.000 | 0.021 |
| Bushman canine UC | 67 | 1.000 | 0.000 | 0.000 | 0.000 | | | | |
| Distal acc. ridge UC | 37 | 0.162 | 0.081 | 0.270 | 0.216 | 0.216 | 0.054 | | |
| Uto-Aztecan UP1 | | | | | | | | | |
| Hypocone UM2 | 90 | 0.011 | 0.067 | 0.100 | 0.267 | 0.500 | 0.056 | 0.000 | |
| Cusp 5 UM1 | 87 | 0.713 | 0.092 | 0.080 | 0.103 | 0.011 | 0.000 | | |
| Carabelli's cusp UM1 | 90 | 0.444 | 0.278 | 0.089 | 0.033 | 0.044 | 0.089 | 0.000 | 0.022 |
| Parastyle UM3 | 63 | 0.857 | 0.063 | 0.032 | 0.032 | 0.016 | 0.000 | 0.000 | |
| Enamel extension UM1 | 118 | 0.415 | 0.186 | 0.127 | 0.271 | | | | |
| Root number UP1 | 89 | | 0.798 | 0.202 | 0.000 | | | | |
| Root number UM2 | 77 | | 0.208 | 0.208 | 0.584 | | | | |
| PRM UM3 | 108 | 0.528 | 0.037 | 0.028 | 0.407 | | | | |
| Lingual cusp no. LP2 | 65 | 0.000 | 0.108 | 0.846 | 0.046 | | | | |
| Groove pattern LM2 | 83 | | 0.265 | 0.699 | 0.036 | | | | |
| Cusp number LM1 | 71 | 0.000 | 0.732 | 0.268 | | | | | |
| Cusp number LM2 | 62 | 0.145 | 0.790 | 0.065 | | | | | |
| Deflecting wrinkle LM1 | 23 | 0.696 | 0.000 | 0.000 | 0.304 | | | | |
| Trigonid crest LM1 | 69 | 1.000 | 0.000 | | | | | | |
| Protostylid LM1 | 82 | 0.805 | 0.134 | 0.037 | 0.000 | 0.024 | 0.000 | 0.000 | 0.000 |
| Cusp 7 LM1 | 81 | 0.877 | 0.000 | 0.037 | 0.025 | 0.037 | 0.025 | | |
| Tomes' root LP1 | 46 | 0.957 | 0.000 | 0.000 | 0.000 | 0.000 | 0.000 | 0.000 | 0.043 |
| Root number LC | 38 | | 1.000 | 0.000 | | | | | |
| Root number LM1 | 88 | | 0.000 | 0.614 | 0.386 | | | | |
| Root number LM2 | 86 | | 0.419 | 0.570 | 0.012 | | | | |
| Odontome UP1,2/LP1,2 | 133 | 0.985 | 0.015 | | | | | | |

# Australia

## *Lower Murray*

| Trait | *n* | Grade 0 | 1 | 2 | 3 | 4 | 5 | 6 | 7 |
|---|---|---|---|---|---|---|---|---|---|
| Winging UI1 | 115 | | 0.043 | 0.009 | 0.948 | 0.000 | | | |
| Shoveling UI1 | 74 | 0.000 | 0.149 | 0.689 | 0.122 | 0.041 | 0.000 | 0.000 | 0.000 |
| Double-shoveling UI1 | 69 | 0.986 | 0.014 | 0.000 | 0.000 | 0.000 | 0.000 | 0.000 | |
| Interruption grooves UI2 | 86 | 0.884 | 0.116 | | | | | | |
| Tuberculum dentale UI2 | 96 | 0.729 | 0.000 | 0.094 | 0.104 | 0.021 | 0.000 | 0.000 | 0.052 |
| Bushman canine UC | 89 | 1.000 | 0.000 | 0.000 | 0.000 | | | | |
| Distal acc. ridge UC | | | | | | | | | |
| Uto-Aztecan UP1 | | | | | | | | | |
| Hypocone UM2 | 129 | 0.008 | 0.016 | 0.070 | 0.109 | 0.519 | 0.279 | 0.000 | |
| Cusp 5 UM1 | 88 | 0.398 | 0.080 | 0.148 | 0.159 | 0.148 | 0.068 | | |
| Carabelli's cusp UM1 | 70 | 0.329 | 0.171 | 0.114 | 0.086 | 0.029 | 0.214 | 0.043 | 0.014 |
| Parastyle UM3 | 116 | 0.966 | 0.009 | 0.009 | 0.017 | 0.000 | 0.000 | 0.000 | |
| Enamel extension UM1 | 171 | 0.713 | 0.181 | 0.035 | 0.070 | | | | |
| Root number UP1 | 101 | | 0.564 | 0.416 | 0.020 | | | | |
| Root number UM2 | 124 | | 0.040 | 0.153 | 0.806 | | | | |
| PRM UM3 | 169 | 0.935 | 0.018 | 0.018 | 0.030 | | | | |
| Lingual cusp no. LP2 | 61 | 0.000 | 0.246 | 0.705 | 0.049 | | | | |
| Groove pattern LM2 | 102 | | 0.235 | 0.696 | 0.069 | | | | |
| Cusp number LM1 | 58 | 0.000 | 0.362 | 0.638 | | | | | |
| Cusp number LM2 | 89 | 0.101 | 0.697 | 0.202 | | | | | |
| Deflecting wrinkle LM1 | | | | | | | | | |
| Trigonid crest LM1 | | | | | | | | | |
| Protostylid LM1 | 64 | 0.906 | 0.078 | 0.000 | 0.000 | 0.000 | 0.016 | 0.000 | 0.000 |
| Cusp 7 LM1 | 106 | 0.962 | 0.000 | 0.009 | 0.019 | 0.000 | 0.009 | | |
| Tomes' root LP1 | 71 | 0.000 | 0.620 | 0.028 | 0.028 | 0.268 | 0.028 | 0.028 | 0.000 |
| Root number LC | 34 | | 1.000 | 0.000 | | | | | |
| Root number LM1 | 137 | | 0.007 | 0.927 | 0.066 | | | | |
| Root number LM2 | 92 | | 0.076 | 0.924 | 0.000 | | | | |
| Odontome UP1,2/LP1,2 | 79 | 0.962 | 0.038 | | | | | | |

## North Australia

| Trait | n | Grade 0 | 1 | 2 | 3 | 4 | 5 | 6 | 7 |
|---|---|---|---|---|---|---|---|---|---|
| Winging UI1 | 100 | | 0.150 | 0.010 | 0.840 | 0.000 | | | |
| Shoveling UI1 | 36 | 0.028 | 0.167 | 0.667 | 0.111 | 0.028 | 0.000 | 0.000 | 0.000 |
| Double-shoveling UI1 | 33 | 0.879 | 0.121 | 0.000 | 0.000 | 0.000 | 0.000 | 0.000 | |
| Interruption grooves UI2 | 48 | 0.771 | 0.229 | | | | | | |
| Tuberculum dentale UI2 | 52 | 0.615 | 0.000 | 0.173 | 0.077 | 0.019 | 0.000 | 0.000 | 0.115 |
| Bushman canine UC | 56 | 0.964 | 0.036 | 0.000 | 0.000 | | | | |
| Distal acc. ridge UC | | | | | | | | | |
| Uto-Aztecan UP1 | | | | | | | | | |
| Hypocone UM2 | 121 | 0.008 | 0.008 | 0.050 | 0.099 | 0.504 | 0.331 | 0.000 | |
| Cusp 5 UM1 | 81 | 0.395 | 0.173 | 0.123 | 0.160 | 0.111 | 0.037 | | |
| Carabelli's cusp UM1 | 54 | 0.222 | 0.296 | 0.093 | 0.111 | 0.019 | 0.167 | 0.037 | 0.056 |
| Parastyle UM3 | 117 | 0.932 | 0.017 | 0.017 | 0.017 | 0.017 | 0.000 | 0.000 | |
| Enamel extension UM1 | 148 | 0.588 | 0.318 | 0.020 | 0.074 | | | | |
| Root number UP1 | 121 | | 0.537 | 0.463 | 0.000 | | | | |
| Root number UM2 | 119 | | 0.109 | 0.151 | 0.739 | | | | |
| PRM UM3 | 153 | 0.954 | 0.013 | 0.007 | 0.026 | | | | |
| Lingual cusp no. LP2 | 51 | 0.000 | 0.157 | 0.667 | 0.176 | | | | |
| Groove pattern LM2 | 83 | | 0.325 | 0.506 | 0.169 | | | | |
| Cusp number LM1 | 31 | 0.000 | 0.194 | 0.806 | | | | | |
| Cusp number LM2 | 68 | 0.088 | 0.691 | 0.221 | | | | | |
| Deflecting wrinkle LM1 | | | | | | | | | |
| Trigonid crest LM1 | | | | | | | | | |
| Protostylid LM1 | 42 | 0.929 | 0.071 | 0.000 | 0.000 | 0.000 | 0.000 | 0.000 | 0.000 |
| Cusp 7 LM1 | 54 | 0.926 | 0.000 | 0.037 | 0.000 | 0.037 | 0.000 | | |
| Tomes' root LP1 | 88 | 0.170 | 0.511 | 0.034 | 0.045 | 0.159 | 0.011 | 0.023 | 0.045 |
| Root number LC | 80 | | 1.000 | 0.000 | | | | | |
| Root number LM1 | 112 | | 0.000 | 0.964 | 0.036 | | | | |
| Root number LM2 | 104 | | 0.096 | 0.904 | 0.000 | | | | |
| Odontome UP1,2/LP1,2 | 52 | 1.000 | 0.000 | | | | | | |

## *Queensland*

| Trait | *n* | Grade 0 | 1 | 2 | 3 | 4 | 5 | 6 | 7 |
|---|---|---|---|---|---|---|---|---|---|
| Winging UI1 | 80 | | 0.037 | 0.000 | 0.962 | 0.000 | | | |
| Shoveling UI1 | 22 | 0.318 | 0.364 | 0.273 | 0.045 | 0.000 | 0.000 | 0.000 | 0.000 |
| Double-shoveling UI1 | 20 | 1.000 | 0.000 | 0.000 | 0.000 | 0.000 | 0.000 | 0.000 | |
| Interruption grooves UI2 | 22 | 0.909 | 0.091 | | | | | | |
| Tuberculum dentale UI2 | 29 | 0.724 | 0.000 | 0.103 | 0.103 | 0.034 | 0.000 | 0.000 | 0.034 |
| Bushman canine UC | 40 | 0.975 | 0.000 | 0.025 | 0.000 | | | | |
| Distal acc. ridge UC | 14 | 0.143 | 0.000 | 0.286 | 0.429 | 0.143 | 0.000 | | |
| Uto-Aztecan UP1 | 71 | 1.000 | 0.000 | | | | | | |
| Hypocone UM2 | 115 | 0.000 | 0.017 | 0.035 | 0.130 | 0.417 | 0.400 | 0.000 | |
| Cusp 5 UM1 | 71 | 0.338 | 0.183 | 0.085 | 0.239 | 0.155 | 0.000 | | |
| Carabelli's cusp UM1 | 70 | 0.371 | 0.300 | 0.029 | 0.129 | 0.014 | 0.100 | 0.029 | 0.029 |
| Parastyle UM3 | 97 | 0.959 | 0.000 | 0.021 | 0.010 | 0.000 | 0.010 | 0.000 | |
| Enamel extension UM1 | 141 | 0.702 | 0.241 | 0.000 | 0.057 | | | | |
| Root number UP1 | 145 | | 0.607 | 0.386 | 0.007 | | | | |
| Root number UM2 | 147 | | 0.020 | 0.156 | 0.823 | | | | |
| PRM UM3 | 155 | 0.994 | 0.000 | 0.000 | 0.006 | | | | |
| Lingual cusp no. LP2 | 45 | 0.000 | 0.133 | 0.711 | 0.156 | | | | |
| Groove pattern LM2 | 87 | | 0.322 | 0.540 | 0.138 | | | | |
| Cusp number LM1 | 47 | 0.000 | 0.234 | 0.766 | | | | | |
| Cusp number LM2 | 80 | 0.137 | 0.613 | 0.250 | | | | | |
| Deflecting wrinkle LM1 | 22 | 0.727 | 0.000 | 0.227 | 0.045 | | | | |
| Trigonid crest LM1 | 45 | 0.933 | 0.067 | | | | | | |
| Protostylid LM1 | | | | | | | | | |
| Cusp 7 LM1 | 65 | 0.909 | 0.015 | 0.015 | 0.000 | 0.046 | 0.015 | | |
| Tomes' root LP1 | 107 | 0.000 | 0.570 | 0.093 | 0.028 | 0.262 | 0.019 | 0.028 | |
| Root number LC | 112 | | 1.000 | 0.000 | | | | | |
| Root number LM1 | 118 | | 0.000 | 0.924 | 0.076 | | | | |
| Root number LM2 | 110 | | 0.018 | 0.982 | 0.000 | | | | |
| Odontome UP1,2/LP1,2 | 62 | 0.903 | 0.097 | | | | | | |

| Trait | n | Grade 0 | 1 | 2 | 3 | 4 | 5 | 6 | 7 |
|---|---|---|---|---|---|---|---|---|---|
| Winging UI1 | 84 | | 0.095 | 0.012 | 0.893 | 0.000 | | | |
| Shoveling UI1 | 59 | 0.017 | 0.169 | 0.559 | 0.203 | 0.034 | 0.017 | 0.000 | 0.000 |
| Double-shoveling UI1 | 56 | 0.964 | 0.036 | 0.000 | 0.000 | 0.000 | 0.000 | 0.000 | |
| Interruption grooves UI2 | 78 | 0.872 | 0.128 | | | | | | |
| Tuberculum dentale UI2 | 80 | 0.550 | 0.000 | 0.100 | 0.250 | 0.025 | 0.000 | 0.000 | 0.075 |
| Bushman canine UC | 72 | 1.000 | 0.000 | 0.000 | 0.000 | | | | |
| Distal acc. ridge UC | 26 | 0.308 | 0.000 | 0.346 | 0.192 | 0.154 | 0.000 | | |
| Uto-Aztecan UP1 | 80 | 0.987 | 0.012 | | | | | | |
| Hypocone UM2 | 91 | 0.033 | 0.000 | 0.011 | 0.209 | 0.429 | 0.319 | 0.000 | |
| Cusp 5 UM1 | 61 | 0.410 | 0.180 | 0.262 | 0.115 | 0.033 | 0.000 | | |
| Carabelli's cusp UM1 | 51 | 0.490 | 0.118 | 0.098 | 0.078 | 0.020 | 0.157 | 0.020 | 0.020 |
| Parastyle UM3 | 72 | 0.958 | 0.014 | 0.014 | 0.000 | 0.014 | 0.000 | 0.000 | |
| Enamel extension UM1 | 94 | 0.734 | 0.170 | 0.021 | 0.074 | | | | |
| Root number UP1 | 71 | | 0.577 | 0.423 | 0.000 | | | | |
| Root number UM2 | 64 | | 0.000 | 0.141 | 0.859 | | | | |
| PRM UM3 | 90 | 0.922 | 0.033 | 0.011 | 0.033 | | | | |
| Lingual cusp no. LP2 | 63 | 0.000 | 0.270 | 0.587 | 0.143 | | | | |
| Groove pattern LM2 | 72 | | 0.208 | 0.639 | 0.153 | | | | |
| Cusp number LM1 | 40 | 0.000 | 0.400 | 0.600 | | | | | |
| Cusp number LM2 | 65 | 0.185 | 0.700 | 0.215 | | | | | |
| Deflecting wrinkle LM1 | 28 | 0.679 | 0.000 | 0.107 | 0.214 | | | | |
| Trigonid crest LM1 | 43 | 0.930 | 0.070 | | | | | | |
| Protostylid LM1 | | | | | | | | | |
| Cusp 7 LM1 | 64 | 0.938 | 0.016 | 0.000 | 0.016 | 0.016 | 0.016 | | |
| Tomes' root LP1 | 46 | 0.000 | 0.630 | 0.087 | 0.043 | 0.152 | 0.087 | 0.000 | 0.000 |
| Root number LC | 45 | | 1.000 | 0.000 | | | | | |
| Root number LM1 | 83 | | 0.000 | 0.940 | 0.060 | | | | |
| Root number LM2 | 70 | | 0.043 | 0.943 | 0.014 | | | | |
| Odontome UP1,2/LP1,2 | 59 | 0.965 | 0.034 | | | | | | |

## *Swanport*

| Trait | *n* | Grade 0 | 1 | 2 | 3 | 4 | 5 | 6 | 7 |
|---|---|---|---|---|---|---|---|---|---|
| Winging UI1 | 67 | | 0.045 | 0.015 | 0.940 | 0.000 | | | |
| Shoveling UI1 | 21 | 0.000 | 0.143 | 0.714 | 0.095 | 0.048 | 0.000 | 0.000 | 0.000 |
| Double-shoveling UI1 | 19 | 0.947 | 0.053 | 0.000 | 0.000 | 0.000 | 0.000 | | |
| Interruption grooves UI2 | 38 | 0.868 | 0.132 | | | | | | |
| Tuberculum dentale UI2 | 42 | 0.786 | 0.000 | 0.048 | 0.119 | 0.024 | 0.000 | 0.000 | 0.024 |
| Bushman canine UC | 36 | 1.000 | 0.000 | 0.000 | 0.000 | | | | |
| Distal acc. ridge UC | 13 | 0.231 | 0.000 | 0.308 | 0.462 | 0.000 | 0.000 | | |
| Uto-Aztecan UP1 | 48 | 1.000 | 0.000 | | | | | | |
| Hypocone UM2 | 68 | 0.000 | 0.000 | 0.074 | 0.044 | 0.588 | 0.294 | 0.000 | |
| Cusp 5 UM1 | 49 | 0.347 | 0.143 | 0.204 | 0.245 | 0.061 | 0.000 | | |
| Carabelli's cusp UM1 | 38 | 0.368 | 0.132 | 0.053 | 0.079 | 0.026 | 0.289 | 0.053 | 0.000 |
| Parastyle UM3 | 56 | 0.982 | 0.000 | 0.000 | 0.018 | 0.000 | 0.000 | 0.000 | |
| Enamel extension UM1 | 96 | 0.646 | 0.208 | 0.063 | 0.083 | | | | |
| Root number UP1 | 68 | | 0.651 | 0.349 | 0.000 | | | | |
| Root number UM2 | 64 | | 0.031 | 0.172 | 0.797 | | | | |
| PRM UM3 | 92 | 0.946 | 0.011 | 0.011 | 0.033 | | | | |
| Lingual cusp no. LP2 | 17 | 0.000 | 0.294 | 0.706 | 0.000 | | | | |
| Groove pattern LM2 | 45 | | 0.200 | 0.711 | 0.089 | | | | |
| Cusp number LM1 | 18 | 0.000 | 0.389 | 0.611 | | | | | |
| Cusp number LM2 | 37 | 0.135 | 0.703 | 0.162 | | | | | |
| Deflecting wrinkle LM1 | 19 | 0.632 | 0.000 | 0.211 | 0.158 | | | | |
| Trigonid crest LM1 | 25 | 1.000 | 0.000 | | | | | | |
| Protostylid LM1 | | | | | | | | | |
| Cusp 7 LM1 | 38 | 0.974 | 0.000 | 0.000 | 0.000 | 0.000 | 0.026 | | |
| Tomes' root LP1 | 32 | 0.000 | 0.594 | 0.031 | 0.031 | 0.313 | 0.000 | 0.031 | 0.000 |
| Root number LC | 28 | | 1.000 | 0.000 | | | | | |
| Root number LM1 | 54 | | 0.000 | 0.889 | 0.111 | | | | |
| Root number LM2 | 44 | | 0.045 | 0.955 | 0.000 | | | | |
| Odontome UP1,2/LP1,2 | 37 | 0.973 | 0.027 | | | | | | |

# Circumpolar

## *Aleuts (Eastern)*

| Trait | n | Grade 0 | 1 | 2 | 3 | 4 | 5 | 6 | 7 |
|---|---|---|---|---|---|---|---|---|---|
| Winging UI1 | 52 | | 0.365 | 0.000 | 0.635 | 0.000 | | | |
| Shoveling UI1 | 28 | 0.000 | 0.036 | 0.250 | 0.571 | 0.143 | 0.000 | 0.000 | 0.000 |
| Double-shoveling UI1 | 27 | 0.407 | 0.593 | | | | | | |
| Interruption grooves UI2 | 52 | 0.327 | 0.673 | | | | | | |
| Tuberculum dentale UI2 | 45 | 0.844 | 0.000 | 0.067 | 0.000 | 0.000 | 0.022 | 0.022 | 0.044 |
| Bushman canine UC | 56 | 1.000 | 0.000 | 0.000 | 0.000 | | | | |
| Distal acc. ridge UC | | | | | | | | | |
| Uto-Aztecan UP1 | | | | | | | | | |
| Hypocone UM2 | 74 | 0.284 | 0.122 | 0.257 | 0.243 | 0.095 | 0.000 | 0.000 | |
| Cusp 5 UM1 | 64 | 0.828 | 0.078 | 0.047 | 0.047 | 0.000 | 0.000 | | |
| Carabelli's cusp UM1 | 70 | 0.557 | 0.400 | 0.014 | 0.029 | 0.000 | 0.000 | 0.000 | 0.000 |
| Parastyle UM3 | 56 | 0.911 | 0.018 | 0.071 | 0.000 | 0.000 | 0.000 | 0.000 | |
| Enamel extension UM1 | 144 | 0.264 | 0.257 | 0.083 | 0.396 | | | | |
| Root number UP1 | 157 | | 0.930 | 0.070 | | | | | |
| Root number UM2 | 108 | | 0.315 | 0.204 | 0.481 | | | | |
| PRM UM3 | 126 | 0.746 | 0.032 | 0.071 | 0.151 | | | | |
| Lingual cusp no. LP2 | 47 | 0.000 | 0.596 | 0.383 | 0.021 | | | | |
| Groove pattern LM2 | 101 | | 0.287 | 0.505 | 0.208 | | | | |
| Cusp number LM1 | 68 | 0.000 | 0.500 | 0.500 | | | | | |
| Cusp number LM2 | 77 | 0.104 | 0.727 | 0.169 | | | | | |
| Deflecting wrinkle LM1 | 35 | 0.343 | 0.114 | 0.229 | 0.314 | | | | |
| Trigonid crest LM1 | 63 | 0.968 | 0.032 | | | | | | |
| Protostylid LM1 | 76 | 0.776 | 0.224 | 0.000 | 0.000 | 0.000 | 0.000 | 0.000 | 0.000 |
| Cusp 7 LM1 | 86 | 0.919 | 0.012 | 0.058 | 0.000 | 0.012 | 0.000 | | |
| Tomes' root LP1 | 149 | 0.966 | 0.007 | 0.013 | 0.000 | 0.007 | 0.000 | 0.000 | 0.007 |
| Root number LC | | | | | | | | | |
| Root number LM1 | 203 | | 0.000 | 0.601 | 0.399 | | | | |
| Root number LM2 | 175 | | 0.309 | 0.680 | 0.011 | | | | |
| Odontome UP1,2/LP1,2 | 86 | 0.953 | 0.047 | | | | | | |

## Aleuts (Western)

| Trait | n | Grade 0 | 1 | 2 | 3 | 4 | 5 | 6 | 7 |
|---|---|---|---|---|---|---|---|---|---|
| Winging UI1 | 14 | | 0.429 | 0.000 | 0.571 | 0.000 | | | |
| Shoveling UI1 | 11 | 0.000 | 0.000 | 0.273 | 0.636 | 0.091 | 0.000 | 0.000 | 0.000 |
| Double-shoveling UI1 | 11 | 0.727 | 0.273 | | | | | | |
| Interruption grooves UI2 | 18 | 0.556 | 0.444 | | | | | | |
| Tuberculum dentale UI2 | 17 | 0.706 | 0.000 | 0.118 | 0.118 | 0.000 | 0.059 | 0.000 | 0.000 |
| Bushman canine UC | 16 | 1.000 | 0.000 | 0.000 | 0.000 | | | | |
| Distal acc. ridge UC | | | | | | | | | |
| Uto-Aztecan UP1 | | | | | | | | | |
| Hypocone UM2 | 28 | 0.321 | 0.107 | 0.179 | 0.286 | 0.107 | 0.000 | 0.000 | |
| Cusp 5 UM1 | 33 | 0.939 | 0.030 | 0.000 | 0.000 | 0.000 | 0.030 | | |
| Carabelli's cusp UM1 | 30 | 0.767 | 0.133 | 0.033 | 0.000 | 0.000 | 0.033 | 0.033 | 0.000 |
| Parastyle UM3 | 13 | 1.000 | 0.000 | 0.000 | 0.000 | 0.000 | 0.000 | 0.000 | |
| Enamel extension UM1 | 49 | 0.449 | 0.265 | 0.143 | 0.143 | | | | |
| Root number UP1 | 47 | | 0.894 | 0.106 | 0.000 | | | | |
| Root number UM2 | 30 | | 0.333 | 0.200 | 0.467 | | | | |
| PRM UM3 | 35 | 0.714 | 0.000 | 0.057 | 0.229 | | | | |
| Lingual cusp no. LP2 | 13 | 0.000 | 0.692 | 0.308 | 0.000 | | | | |
| Groove pattern LM2 | 33 | | 0.303 | 0.515 | 0.182 | | | | |
| Cusp number LM1 | 25 | 0.000 | 0.680 | 0.320 | | | | | |
| Cusp number LM2 | 25 | 0.120 | 0.560 | 0.320 | | | | | |
| Deflecting wrinkle LM1 | 12 | 0.333 | 0.000 | 0.000 | 0.667 | | | | |
| Trigonid crest LM1 | 20 | 1.000 | 0.000 | | | | | | |
| Protostylid LM1 | 32 | 0.688 | 0.281 | 0.000 | 0.031 | 0.000 | 0.000 | 0.000 | 0.000 |
| Cusp 7 LM1 | 36 | 0.889 | 0.028 | 0.083 | 0.000 | 0.000 | 0.000 | | |
| Tomes' root LP1 | 27 | 1.000 | 0.000 | 0.000 | 0.000 | 0.000 | 0.000 | 0.000 | 0.000 |
| Root number LC | | | | | | | | | |
| Root number LM1 | 48 | | 0.000 | 0.479 | 0.521 | | | | |
| Root number LM2 | 40 | | 0.275 | 0.725 | 0.000 | | | | |
| Odontome UP1,2/LP1,2 | 26 | 1.000 | 0.000 | | | | | | |

| Trait | n | Grade 0 | 1 | 2 | 3 | 4 | 5 | 6 | 7 |
|---|---|---|---|---|---|---|---|---|---|
| Winging UI1 | 45 | | 0.089 | 0.022 | 0.889 | 0.000 | | | |
| Shoveling UI1 | 32 | 0.000 | 0.094 | 0.344 | 0.344 | 0.094 | 0.063 | 0.063 | 0.000 |
| Double-shoveling UI1 | 29 | 0.655 | 0.207 | 0.034 | 0.069 | 0.000 | 0.034 | 0.000 | |
| Interruption grooves UI2 | 44 | 0.432 | 0.568 | | | | | | |
| Tuberculum dentale UI2 | 50 | 0.740 | 0.000 | 0.020 | 0.040 | 0.000 | 0.000 | 0.100 | 0.100 |
| Bushman canine UC | 99 | 1.000 | 0.000 | | | | | | |
| Distal acc. ridge UC | 42. | 0.238 | 0.095 | 0.238 | 0.357 | 0.071 | 0.000 | | |
| Uto-Aztecan UP1 | 68 | 1.000 | 0.000 | | | | | | |
| Hypocone UM2 | 201 | 0.299 | 0.035 | 0.219 | 0.214 | 0.219 | 0.010 | 0.005 | |
| Cusp 5 UM1 | 131 | 0.870 | 0.053 | 0.061 | 0.015 | 0.000 | 0.000 | | |
| Carabelli's cusp UM1 | 158 | 0.576 | 0.285 | 0.076 | 0.038 | 0.006 | 0.019 | 0.000 | 0.000 |
| Parastyle UM3 | 153 | 0.922 | 0.007 | 0.059 | 0.013 | 0.000 | 0.000 | 0.000 | |
| Enamel extension UM1 | 295 | 0.224 | 0.271 | 0.044 | 0.461 | | | | |
| Root number UP1 | 314 | | 0.955 | 0.045 | 0.000 | | | | |
| Root number UM2 | 285 | | 0.372 | 0.246 | 0.383 | | | | |
| PRM UM3 | 322 | 0.845 | 0.012 | 0.034 | 0.109 | | | | |
| Lingual cusp no. LP2 | 95 | 0.000 | 0.747 | 0.211 | 0.042 | | | | |
| Groove pattern LM2 | 179 | | 0.263 | 0.536 | 0.201 | | | | |
| Cusp number LM1 | 106 | 0.000 | 0.396 | 0.604 | | | | | |
| Cusp number LM2 | 170 | 0.029 | 0.759 | 0.212 | | | | | |
| Deflecting wrinkle LM1 | 76 | 0.553 | 0.026 | 0.237 | 0.184 | | | | |
| Trigonid crest LM1 | 119 | 0.647 | 0.353 | | | | | | |
| Protostylid LM1 | 148 | 0.797 | 0.196 | 0.000 | 0.000 | 0.007 | 0.000 | 0.000 | 0.000 |
| Cusp 7 LM1 | 193 | 0.959 | 0.000 | 0.026 | 0.005 | 0.005 | 0.005 | | |
| Tomes' root LP1 | 121 | 0.058 | 0.595 | 0.215 | 0.074 | 0.025 | 0.025 | 0.008 | 0.000 |
| Root number LC | 239 | | 0.992 | 0.008 | | | | | |
| Root number LM1 | 279 | | 0.000 | 0.731 | 0.269 | | | | |
| Root number LM2 | 255 | | 0.298 | 0.694 | 0.008 | | | | |
| Odontome UP1,2/LP1,2 | 90 | 0.911 | 0.089 | | | | | | |

## *Kodiak Island*

| Trait | *n* | Grade 0 | 1 | 2 | 3 | 4 | 5 | 6 | 7 |
|---|---|---|---|---|---|---|---|---|---|
| Winging UI1 | 76 | | 0.329 | 0.066 | 0.579 | 0.026 | | | |
| Shoveling UI1 | 74 | 0.000 | 0.027 | 0.135 | 0.689 | 0.122 | 0.014 | 0.014 | 0.000 |
| Double-shoveling UI1 | 62 | 0.339 | 0.661 | | | | | | |
| Interruption grooves UI2 | 72 | 0.361 | 0.639 | | | | | | |
| Tuberculum dentale UI2 | 75 | 0.600 | 0.000 | 0.093 | 0.027 | 0.027 | 0.000 | 0.000 | 0.253 |
| Bushman canine UC | 96 | 1.000 | 0.000 | 0.000 | 0.000 | | | | |
| Distal acc. ridge UC | 41. | 0.195 | 0.024 | 0.244 | 0.415 | 0.073 | 0.049 | | |
| Uto-Aztecan UP1 | 107 | 0.991 | 0.009 | | | | | | |
| Hypocone UM2 | 109 | 0.092 | 0.055 | 0.229 | 0.239 | 0.367 | 0.018 | 0.000 | |
| Cusp 5 UM1 | 97 | 0.907 | 0.041 | 0.031 | 0.021 | 0.000 | 0.000 | | |
| Carabelli's cusp UM1 | 95 | 0.347 | 0.411 | 0.105 | 0.084 | 0.032 | 0.000 | 0.021 | 0.000 |
| Parastyle UM3 | 93 | 0.989 | 0.011 | 0.000 | 0.000 | 0.000 | 0.000 | 0.000 | |
| Enamel extension UM1 | 150 | 0.240 | 0.273 | 0.053 | 0.433 | | | | |
| Root number UP1 | 113 | | 0.947 | 0.044 | 0.009 | | | | |
| Root number UM2 | 71 | | 0.535 | 0.141 | 0.324 | | | | |
| PRM UM3 | 119 | 0.748 | 0.109 | 0.034 | 0.109 | | | | |
| Lingual cusp no. LP2 | 87 | 0.000 | 0.471 | 0.460 | 0.069 | | | | |
| Groove pattern LM2 | 115 | | 0.287 | 0.557 | 0.157 | | | | |
| Cusp number LM1 | 84 | 0.000 | 0.488 | 0.512 | | | | | |
| Cusp number LM2 | 109 | 0.028 | 0.743 | 0.229 | | | | | |
| Deflecting wrinkle LM1 | 54 | 0.389 | 0.056 | 0.278 | 0.278 | | | | |
| Trigonid crest LM1 | 93 | 0.946 | 0.054 | | | | | | |
| Protostylid LM1 | 119 | 0.748 | 0.176 | 0.042 | 0.000 | 0.034 | 0.000 | 0.000 | 0.000 |
| Cusp 7 LM1 | 117 | 0.863 | 0.051 | 0.051 | 0.026 | 0.000 | 0.009 | | |
| Tomes' root LP1 | 76 | 0.947 | 0.053 | | | | | | |
| Root number LC | 81 | | 1.000 | 0.000 | | | | | |
| Root number LM1 | 142 | | 0.000 | 0.761 | 0.239 | | | | |
| Root number LM2 | 115 | | 0.313 | 0.687 | 0.000 | | | | |
| Odontome UP1,2/LP1,2 | 130 | 0.954 | 0.046 | | | | | | |

## Southampton Island

| Trait | n | Grade 0 | 1 | 2 | 3 | 4 | 5 | 6 | 7 |
|---|---|---|---|---|---|---|---|---|---|
| Winging UI1 | 11 | | 0.182 | 0.091 | 0.727 | 0.000 | | | |
| Shoveling UI1 | 12 | 0.000 | 0.083 | 0.250 | 0.417 | 0.250 | 0.000 | 0.000 | 0.000 |
| Double-shoveling UI1 | 11 | 0.818 | 0.182 | | | | | | |
| Interruption grooves UI2 | 13 | 0.308 | 0.692 | | | | | | |
| Tuberculum dentale UI2 | 15 | 0.000 | 0.000 | 0.067 | 0.133 | 0.000 | 0.200 | 0.000 | 0.600 |
| Bushman canine UC | 21 | 1.000 | 0.000 | 0.000 | 0.000 | | | | |
| Distal acc. ridge UC | 11. | 0.273 | 0.000 | 0.091 | 0.455 | 0.091 | 0.091 | | |
| Uto-Aztecan UP1 | 19 | 1.000 | 0.000 | | | | | | |
| Hypocone UM2 | 23 | 0.087 | 0.043 | 0.174 | 0.522 | 0.130 | 0.043 | 0.000 | |
| Cusp 5 UM1 | 27 | 0.481 | 0.074 | 0.296 | 0.111 | 0.000 | 0.037 | | |
| Carabelli's cusp UM1 | 35 | 0.457 | 0.429 | 0.029 | 0.029 | 0.057 | 0.000 | 0.000 | 0.000 |
| Parastyle UM3 | 21 | 0.952 | 0.000 | 0.048 | 0.000 | 0.000 | 0.000 | 0.000 | |
| Enamel extension UM1 | 42 | 0.381 | 0.310 | 0.048 | 0.262 | | | | |
| Root number UP1 | 45 | | 0.978 | 0.022 | 0.000 | | | | |
| Root number UM2 | 38 | | 0.658 | 0.184 | 0.158 | | | | |
| PRM UM3 | 44 | 0.705 | 0.023 | 0.000 | 0.273 | | | | |
| Lingual cusp no. LP2 | 21 | 0.000 | 0.714 | 0.238 | 0.048 | | | | |
| Groove pattern LM2 | 28 | | 0.357 | 0.429 | 0.214 | | | | |
| Cusp number LM1 | 32 | 0.000 | 0.531 | 0.469 | | | | | |
| Cusp number LM2 | 29 | 0.000 | 0.621 | 0.379 | | | | | |
| Deflecting wrinkle LM1 | 24 | 0.167 | 0.083 | 0.417 | 0.333 | | | | |
| Trigonid crest LM1 | 38 | 0.974 | 0.026 | | | | | | |
| Protostylid LM1 | 41 | 0.902 | 0.098 | 0.000 | 0.000 | 0.000 | 0.000 | 0.000 | 0.000 |
| Cusp 7 LM1 | 42 | 0.952 | 0.000 | 0.048 | 0.000 | 0.000 | 0.000 | | |
| Tomes' root LP1 | | | | | | | | | |
| Root number LC | 54 | | 1.000 | 0.000 | | | | | |
| Root number LM1 | 50 | | 0.000 | 0.840 | 0.160 | | | | |
| Root number LM2 | 41 | | 0.463 | 0.537 | 0.000 | | | | |
| Odontome UP1,2/LP1,2 | 28 | 0.929 | 0.071 | | | | | | |

| Trait | n | Grade 0 | 1 | 2 | 3 | 4 | 5 | 6 | 7 |
|---|---|---|---|---|---|---|---|---|---|
| Winging UI1 | 61 | | 0.262 | 0.000 | 0.721 | 0.016 | | | |
| Shoveling UI1 | 62 | 0.000 | 0.016 | 0.290 | 0.565 | 0.081 | 0.016 | 0.032 | 0.000 |
| Double-shoveling UI1 | 56 | 0.232 | 0.375 | 0.304 | 0.089 | 0.000 | 0.000 | 0.000 | |
| Interruption grooves UI2 | 61 | 0.508 | 0.492 | | | | | | |
| Tuberculum dentale UI2 | 66 | 0.606 | 0.000 | 0.091 | 0.015 | 0.015 | 0.045 | 0.000 | 0.227 |
| Bushman canine UC | 97 | 1.000 | 0.000 | 0.000 | 0.000 | | | | |
| Distal acc. ridge UC | 37. | 0.378 | 0.081 | 0.108 | 0.270 | 0.162 | 0.000 | | |
| Uto-Aztecan UP1 | 108 | 1.000 | 0.000 | | | | | | |
| Hypocone UM2 | 135 | 0.296 | 0.000 | 0.141 | 0.252 | 0.289 | 0.015 | 0.007 | |
| Cusp 5 UM1 | 94 | 0.840 | 0.032 | 0.085 | 0.043 | 0.000 | 0.000 | | |
| Carabelli's cusp UM1 | 104 | 0.317 | 0.481 | 0.096 | 0.087 | 0.010 | 0.000 | 0.010 | 0.000 |
| Parastyle UM3 | 106 | 0.925 | 0.009 | 0.028 | 0.000 | 0.019 | 0.019 | 0.000 | |
| Enamel extension UM1 | 187 | 0.182 | 0.385 | 0.027 | 0.406 | | | | |
| Root number UP1 | 180 | | 0.944 | 0.050 | 0.006 | | | | |
| Root number UM2 | 153 | | 0.320 | 0.275 | 0.405 | | | | |
| PRM UM3 | 197 | 0.843 | 0.000 | 0.015 | 0.142 | | | | |
| Lingual cusp no. LP2 | 57 | 0.000 | 0.719 | 0.263 | 0.018 | | | | |
| Groove pattern LM2 | 89 | | 0.225 | 0.573 | 0.202 | | | | |
| Cusp number LM1 | 73 | 0.000 | 0.507 | 0.493 | | | | | |
| Cusp number LM2 | 84 | 0.036 | 0.738 | 0.226 | | | | | |
| Deflecting wrinkle LM1 | 52 | 0.442 | 0.038 | 0.269 | 0.250 | | | | |
| Trigonid crest LM1 | 82 | 0.756 | 0.244 | | | | | | |
| Protostylid LM1 | 90 | 0.867 | 0.122 | 0.000 | 0.110 | 0.000 | 0.000 | 0.000 | 0.000 |
| Cusp 7 LM1 | 97 | 0.887 | 0.000 | 0.103 | 0.010 | 0.000 | 0.000 | | |
| Tomes' root LP1 | 30 | 0.000 | 0.567 | 0.167 | 0.233 | 0.033 | 0.000 | 0.000 | 0.000 |
| Root number LC | 90 | | 1.000 | 0.000 | | | | | |
| Root number LM1 | 117 | | 0.000 | 0.632 | 0.368 | | | | |
| Root number LM2 | 104 | | 0.308 | 0.683 | 0.010 | | | | |
| Odontome UP1,2/LP1,2 | 74 | 0.905 | 0.095 | | | | | | |

# Europe

## *Basques (Spain)*

| Trait | *n* | Grade 0 | 1 | 2 | 3 | 4 | 5 | 6 | 7 |
|---|---|---|---|---|---|---|---|---|---|
| Winging UI1 | 136 | | 0.096 | 0.000 | 0.904 | 0.000 | | | |
| Shoveling UI1 | 135 | 0.674 | 0.200 | 0.081 | 0.044 | 0.000 | 0.000 | 0.000 | 0.000 |
| Double-shoveling UI1 | 133 | 0.970 | 0.015 | 0.008 | 0.008 | 0.000 | 0.000 | 0.000 | |
| Interruption grooves UI2 | 139 | 0.705 | 0.295 | | | | | | |
| Tuberculum dentale UI2 | 129 | 0.541 | 0.238 | 0.131 | 0.046 | 0.030 | 0.015 | | |
| Bushman canine UC | 92 | 0.978 | 0.022 | 0.000 | 0.000 | | | | |
| Distal acc. ridge UC | 82 | 0.561 | 0.037 | 0.110 | 0.256 | 0.037 | 0.000 | | |
| Uto-Aztecan UP1 | | | | | | | | | |
| Hypocone UM2 | 200 | 0.265 | 0.055 | 0.030 | 0.425 | 0.215 | 0.001 | 0.000 | |
| Cusp 5 UM1 | 138 | 0.775 | 0.094 | 0.116 | 0.007 | 0.007 | 0.000 | | |
| Carabelli's cusp UM1 | 144 | 0.285 | 0.056 | 0.097 | 0.160 | 0.194 | 0.125 | 0.063 | 0.021 |
| Parastyle UM3 | | | | | | | | | |
| Enamel extension UM1 | 156 | 0.968 | 0.013 | 0.019 | | | | | |
| Root number UP1 | 169 | | 0.484 | 0.516 | 0.000 | | | | |
| Root number UM2 | 169 | | 0.172 | 0.219 | 0.609 | | | | |
| PRM UM3 | 86 | 0.884 | | | 0.116 | | | | |
| Lingual cusp no. LP2 | 181 | 0.000 | 0.481 | 0.519 | | | | | |
| Groove pattern LM2 | 149 | | 0.148 | 0.812 | 0.040 | | | | |
| Cusp number LM1 | 198 | .076 | .845 | .079 | | | | | |
| Cusp number LM2 | 189 | 0.868 | 0.132 | .000 | | | | | |
| Deflecting wrinkle LM1 | 89 | 0.798 | 0.202 | | | | | | |
| Trigonid crest LM1 | | | | | | | | | |
| Protostylid LM1 | 147 | 1.000 | 0.000 | 0.000 | 0.000 | 0.000 | 0.000 | 0.000 | 0.000 |
| Cusp 7 LM1 | 185 | 0.930 | 0.000 | 0.027 | 0.027 | 0.005 | 0.011 | | |
| Tomes' root LP1 | 234 | 0.607 | 0.040 | 0.138 | 0.076 | 0.098 | 0.040 | 0.000 | 0.000 |
| Root number LC | 295 | | 0.915 | 0.088 | 0.004 | | | | |
| Root number LM1 | 152 | | 0.000 | 0.987 | 0.013 | | | | |
| Root number LM2 | 198 | | 0.237 | 0.763 | 0.000 | | | | |
| Odontome UP1,2/LP1,2 | 226 | 1.000 | 0.000 | | | | | | |

## *Dorestad de Heul, Netherlands*

| | | Grade | | | | | | | |
|---|---|---|---|---|---|---|---|---|---|
| **Trait** | ***n*** | **0** | **1** | **2** | **3** | **4** | **5** | **6** | **7** |
| Winging UI1 | 17 | | 0.176 | 0.000 | 0.824 | 0.000 | | | |
| Shoveling UI1 | 16 | 0.250 | 0.500 | 0.250 | 0.000 | 0.000 | 0.000 | 0.000 | 0.000 |
| Double-shoveling UI1 | 15 | 0.800 | 0.200 | 0.000 | 0.000 | 0.000 | 0.000 | 0.000 | |
| Interruption grooves UI2 | 24 | 0.583 | 0.417 | | | | | | |
| Tuberculum dentale UI2 | 22 | 0.500 | 0.000 | 0.091 | 0.091 | 0.045 | 0.000 | 0.000 | 0.272 |
| Bushman canine UC | 29 | 0.966 | 0.034 | 0.000 | 0.000 | | | | |
| Distal acc. ridge UC | 14 | 0.286 | 0.071 | 0.214 | 0.429 | 0.000 | 0.000 | | |
| Uto-Aztecan UP1 | 31 | 1.000 | 0.000 | | | | | | |
| Hypocone UM2 | 31 | 0.323 | 0.000 | 0.000 | 0.097 | 0.484 | 0.097 | 0.000 | |
| Cusp 5 UM1 | 19 | 0.947 | 0.000 | 0.000 | 0.053 | 0.000 | 0.000 | | |
| Carabelli's cusp UM1 | 21 | 0.143 | 0.286 | 0.095 | 0.143 | 0.000 | 0.190 | 0.048 | 0.095 |
| Parastyle UM3 | 24 | 1.000 | 0.000 | 0.000 | 0.000 | 0.000 | 0.000 | 0.000 | |
| Enamel extension UM1 | 36 | 0.556 | 0.389 | 0.000 | 0.056 | | | | |
| Root number UP1 | 33 | | 0.697 | 0.303 | 0.000 | | | | |
| Root number UM2 | 33 | | 0.182 | 0.364 | 0.455 | | | | |
| PRM UM3 | 33 | 0.788 | | | 0.212 | | | | |
| Lingual cusp no. LP2 | 46 | 0.000 | 0.435 | 0.543 | 0.022 | | | | |
| Groove pattern LM2 | 47 | | 0.170 | 0.553 | 0.277 | | | | |
| Cusp number LM1 | 32 | 0.063 | 0.906 | 0.031 | | | | | |
| Cusp number LM2* | 47 | 0.915 | 0.085 | 0.000 | | | | | |
| Deflecting wrinkle LM1 | 15 | 1.000 | 0.000 | 0.000 | 0.000 | | | | |
| Trigonid crest LM1 | 33 | 0.936 | 0.064 | | | | | | |
| Protostylid LM1 | 33 | 0.939 | 0.061 | 0.000 | 0.000 | 0.000 | 0.000 | 0.000 | 0.000 |
| Cusp 7 LM1 | 41 | 0.927 | 0.000 | 0.024 | 0.024 | 0.024 | 0.000 | | |
| Tomes' root LP1 | 53 | 0.906 | 0.094 | | | | | | |
| Root number LC | 55 | | 0.906 | 0.094 | | | | | |
| Root number LM1 | 55 | | 0.000 | 1.000 | 0.000 | | | | |
| Root number LM2 | 50 | | 0.300 | 0.700 | 0.000 | | | | |
| Odontome UP1,2/LP1,2 | 29 | 0.966 | 0.034 | | | | | | |

## Estonia



| Trait | n | 0 | 1 | 2 | 3 | 4 | 5 | 6 | 7 |
|---|---|---|---|---|---|---|---|---|---|
| | | Grade | | | | | | | |
| Winging UI1 | 99 | | 0.040 | 0.000 | 0.949 | 0.010 | | | |
| Shoveling UI1 | 116 | 0.216 | 0.543 | 0.172 | 0.069 | 0.000 | 0.000 | 0.000 | 0.000 |
| Double-shoveling UI1 | 113 | 0.805 | 0.159 | 0.009 | 0.009 | 0.018 | 0.000 | 0.000 | |
| Interruption grooves UI2 | 118 | 0.602 | 0.398 | | | | | | |
| Tuberculum dentale UI2 | 125 | 0.792 | 0.000 | 0.048 | 0.032 | 0.016 | 0.000 | 0.000 | 0.112 |
| Bushman canine UC | 126 | 0.976 | 0.000 | 0.008 | 0.016 | | | | |
| Distal acc. ridge UC | | | | | | | | | |
| Uto-Aztecan UP1 | | | | | | | | | |
| Hypocone UM2 | 151 | 0.159 | 0.040 | 0.119 | 0.146 | 0.417 | 0.119 | 0.000 | |
| Cusp 5 UM1 | 131 | 0.824 | 0.160 | 0.008 | 0.008 | 0.000 | 0.000 | | |
| Carabelli's cusp UM1 | 131 | 0.252 | 0.252 | 0.076 | 0.168 | 0.015 | 0.145 | 0.038 | 0.053 |
| Parastyle UM3 | 105 | 0.962 | 0.019 | 0.010 | 0.000 | 0.000 | 0.010 | 0.000 | |
| Enamel extension UM1 | 184 | 0.793 | 0.141 | 0.038 | 0.027 | | | | |
| Root number UP1 | 169 | | 0.580 | 0.414 | 0.006 | | | | |
| Root number UM2 | 135 | | 0.222 | 0.178 | 0.600 | | | | |
| PRM UM3 | 159 | 0.780 | 0.019 | 0.019 | 0.182 | | | | |
| Lingual cusp no. LP2 | 120 | 0.000 | 0.375 | 0.592 | 0.033 | | | | |
| Groove pattern LM2 | 153 | | 0.294 | 0.575 | 0.131 | | | | |
| Cusp number LM1 | 131 | 0.053 | 0.878 | 0.069 | | | | | |
| Cusp number LM2 | 144 | 0.785 | 0.188 | 0.028 | | | | | |
| Deflecting wrinkle LM1 | | | | | | | | | |
| Trigonid crest LM1 | | | | | | | | | |
| Protostylid LM1 | 114 | 0.904 | 0.096 | 0.000 | 0.000 | 0.000 | 0.000 | 0.000 | 0.000 |
| Cusp 7 LM1 | 155 | 0.968 | 0.006 | 0.013 | 0.000 | 0.006 | 0.006 | | |
| Tomes' root LP1 | 173 | 0.000 | 0.526 | 0.249 | 0.121 | 0.092 | 0.000 | 0.012 | 0.000 |
| Root number LC | 174 | | 0.926 | 0.074 | | | | | |
| Root number LM1 | 186 | | 0.000 | 0.989 | 0.011 | | | | |
| Root number LM2 | 153 | | 0.216 | 0.771 | 0.013 | | | | |
| Odontome UP1,2/LP1,2 | 129 | 0.984 | 0.016 | | | | | | |

## *Lapps*

| Trait | *n* | Grade 0 | 1 | 2 | 3 | 4 | 5 | 6 | 7 |
|---|---|---|---|---|---|---|---|---|---|
| Winging UI1 | 11 | | 0.545 | 0.000 | 0.455 | 0.000 | | | |
| Shoveling UI1 | 6 | 0.000 | 0.667 | 0.167 | 0.167 | 0.000 | 0.000 | 0.000 | 0.000 |
| Double-shoveling UI1 | 6 | 0.333 | 0.500 | 0.167 | 0.000 | 0.000 | 0.000 | 0.000 | |
| Interruption grooves UI2 | 20 | 0.550 | 0.450 | | | | | | |
| Tuberculum dentale UI2 | 21 | 0.667 | 0.000 | 0.048 | 0.048 | 0.000 | 0.048 | 0.000 | 0.191 |
| Bushman canine UC | 24 | 0.958 | 0.000 | 0.042 | 0.000 | | | | |
| Distal acc. ridge UC | 8 | 0.500 | 0.250 | 0.125 | 0.125 | 0.000 | 0.000 | | |
| Uto-Aztecan UP1 | 35 | 1.000 | 0.000 | | | | | | |
| Hypocone UM2 | 37 | 0.270 | 0.000 | 0.189 | 0.189 | 0.324 | 0.027 | 0.000 | |
| Cusp 5 UM1 | 31 | 0.806 | 0.065 | 0.129 | 0.000 | 0.000 | 0.000 | | |
| Carabelli's cusp UM1 | 34 | 0.353 | 0.441 | 0.059 | 0.059 | 0.059 | 0.029 | 0.000 | 0.000 |
| Parastyle UM3 | 27 | 0.963 | 0.000 | 0.000 | 0.037 | 0.000 | 0.000 | 0.000 | |
| Enamel extension UM1 | 55 | 0.527 | 0.309 | 0.000 | 0.164 | | | | |
| Root number UP1 | 61 | | 0.721 | 0.262 | 0.016 | | | | |
| Root number UM2 | 49 | | 0.408 | 0.122 | 0.449 | 0.020 | | | |
| PRM UM3 | 54 | 0.833 | 0.019 | 0.019 | 0.130 | | | | |
| Lingual cusp no. LP2 | 45 | 0.000 | 0.333 | 0.622 | 0.044 | | | | |
| Groove pattern LM2 | 45 | | 0.222 | 0.422 | 0.356 | | | | |
| Cusp number LM1 | 30 | 0.033 | 0.767 | 0.200 | | | | | |
| Cusp number LM2* | 39 | 0.615 | 0.332 | 0.051 | | | | | |
| Deflecting wrinkle LM1 | 24 | 0.708 | 0.083 | 0.208 | | | | | |
| Trigonid crest LM1 | 38 | 0.921 | 0.079 | | | | | | |
| Protostylid LM1 | 38 | 0.632 | 0.342 | 0.000 | 0.000 | 0.000 | 0.026 | 0.000 | |
| Cusp 7 LM1 | 44 | 0.977 | 0.000 | 0.023 | 0.000 | 0.000 | | | |
| Tomes' root LP1 | 45 | 0.000 | 0.756 | 0.156 | 0.044 | 0.022 | 0.022 | 0.000 | |
| Root number LC | 62 | | 0.968 | 0.032 | | | | | |
| Root number LM1 | 60 | | 0.000 | 1.000 | 0.000 | | | | |
| Root number LM2 | 60 | | 0.367 | 0.633 | 0.000 | | | | |
| Odontome UP1,2/LP1,2 | 27 | 1.000 | 0.000 | | | | | | |

## Poundbury, England

| Trait | n | Grade 0 | 1 | 2 | 3 | 4 | 5 | 6 | 7 |
|---|---|---|---|---|---|---|---|---|---|
| Winging UI1 | 85 | | 0.024 | 0.000 | 0.929 | 0.047 | | | |
| Shoveling UI1 | 107 | 0.196 | 0.673 | 0.103 | 0.028 | 0.000 | 0.000 | 0.000 | 0.000 |
| Double-shoveling UI1 | 109 | 0.807 | 0.138 | 0.055 | 0.000 | 0.000 | 0.000 | 0.000 | |
| Interruption grooves UI2 | 103 | 0.563 | 0.437 | | | | | | |
| Tuberculum dentale UI2 | 102 | 0.745 | 0.000 | 0.176 | 0.020 | 0.000 | 0.000 | 0.000 | 0.059 |
| Bushman canine UC | 84 | 0.952 | 0.000 | 0.048 | 0.000 | | | | |
| Distal acc. ridge UC | 70 | 0.429 | 0.200 | 0.229 | 0.143 | 0.000 | 0.000 | | |
| Uto-Aztecan UP1 | 100 | 1.000 | 0.000 | | | | | | |
| Hypocone UM2 | 113 | 0.230 | 0.044 | 0.080 | 0.221 | 0.354 | 0.071 | 0.000 | |
| Cusp 5 UM1 | 115 | 0.878 | 0.061 | 0.061 | 0.000 | 0.000 | 0.000 | | |
| Carabelli's cusp UM1 | 115 | 0.122 | 0.270 | 0.113 | 0.070 | 0.061 | 0.217 | 0.078 | 0.070 |
| Parastyle UM3 | 63 | 0.921 | 0.048 | 0.016 | 0.016 | 0.000 | 0.000 | 0.000 | |
| Enamel extension UM1 | 119 | 0.613 | 0.378 | 0.000 | 0.008 | | | | |
| Root number UP1 | 71 | | 0.690 | 0.296 | 0.014 | | | | |
| Root number UM2 | 72 | | 0.139 | 0.264 | 0.597 | | | | |
| PRM UM3 | 79 | 0.886 | | | 0.114 | | | | |
| Lingual cusp no. LP2 | 59 | 0.000 | 0.407 | 0.542 | 0.051 | | | | |
| Groove pattern LM2 | 77 | | 0.312 | 0.481 | 0.208 | | | | |
| Cusp number LM1 | 76 | 0.092 | 0.816 | 0.092 | | | | | |
| Cusp number LM2 | 78 | 0.731 | 0.244 | 0.026 | | | | | |
| Deflecting wrinkle LM1 | 63 | 0.778 | 0.063 | 0.079 | 0.079 | | | | |
| Trigonid crest LM1 | 72 | 0.903 | 0.097 | | | | | | |
| Protostylid LM1 | 75 | 0.800 | 0.147 | 0.000 | 0.013 | 0.000 | 0.027 | 0.013 | 0.000 |
| Cusp 7 LM1 | 80 | 0.962 | 0.000 | 0.025 | 0.012 | 0.000 | 0.000 | | |
| Tomes' root LP1 | 34 | 0.000 | 0.647 | 0.088 | 0.147 | 0.118 | 0.000 | 0.000 | 0.000 |
| Root number LC | 38 | | 0.947 | 0.053 | | | | | |
| Root number LM1 | 76 | | 0.000 | 0.987 | 0.013 | | | | |
| Root number LM2 | 60 | | 0.233 | 0.767 | 0.000 | | | | |
| Odontome UP1,2/LP1,2 | 108 | 1.000 | 0.000 | | | | | | |

## Russians

| Trait | n | Grade 0 | 1 | 2 | 3 | 4 | 5 | 6 | 7 |
|---|---|---|---|---|---|---|---|---|---|
| Winging UI1 | 51 | | 0.118 | 0.000 | 0.843 | 0.039 | | | |
| Shoveling UI1 | 42 | 0.238 | 0.643 | 0.095 | 0.024 | 0.000 | 0.000 | 0.000 | 0.000 |
| Double-shoveling UI1 | 42 | 0.833 | 0.143 | 0.024 | 0.000 | 0.000 | 0.000 | 0.000 | |
| Interruption grooves UI2 | 47 | 0.681 | 0.319 | | | | | | |
| Tuberculum dentale UI2 | 48 | 0.833 | 0.000 | 0.083 | 0.042 | 0.021 | 0.000 | 0.000 | 0.021 |
| Bushman canine UC | 58 | 1.000 | 0.000 | 0.000 | 0.000 | | | | |
| Distal acc. ridge UC | 22 | 0.364 | 0.045 | 0.318 | 0.273 | 0.000 | 0.000 | | |
| Uto-Aztecan UP1 | 75 | 1.000 | 0.000 | | | | | | |
| Hypocone UM2 | 85 | 0.282 | 0.047 | 0.094 | 0.106 | 0.353 | 0.118 | 0.000 | |
| Cusp 5 UM1 | 37 | 0.892 | 0.081 | 0.027 | 0.000 | 0.000 | 0.000 | | |
| Carabelli's cusp UM1 | 39 | 0.179 | 0.359 | 0.103 | 0.179 | 0.051 | 0.051 | 0.026 | 0.051 |
| Parastyle UM3 | 52 | 0.981 | 0.019 | 0.000 | 0.000 | 0.000 | 0.000 | 0.000 | |
| Enamel extension UM1 | 94 | 0.787 | 0.191 | 0.011 | 0.011 | | | | |
| Root number UP1 | 123 | | 0.447 | 0.537 | 0.016 | | | | |
| Root number UM2 | 100 | | 0.150 | 0.150 | 0.700 | | | | |
| PRM UM3 | 116 | 0.672 | 0.009 | 0.009 | 0.310 | | | | |
| Lingual cusp no. LP2 | 40 | 0.000 | 0.300 | 0.700 | 0.000 | | | | |
| Groove pattern LM2 | 83 | | 0.217 | 0.518 | 0.265 | | | | |
| Cusp number LM1 | 33 | 0.121 | 0.788 | 0.091 | | | | | |
| Cusp number LM2 | 66 | 0.924 | 0.076 | 0.000 | | | | | |
| Deflecting wrinkle LM1 | 29 | 0.893 | 0.000 | 0.036 | 0.071 | | | | |
| Trigonid crest LM1 | 33 | 1.000 | 0.000 | | | | | | |
| Protostylid LM1 | 34 | 0.882 | 0.118 | 0.000 | 0.000 | 0.000 | 0.000 | 0.000 | 0.000 |
| Cusp 7 LM1 | 55 | 0.982 | 0.000 | 0.000 | 0.018 | 0.000 | 0.000 | | |
| Tomes' root LP1 | 108 | 0.000 | 0.481 | 0.194 | 0.213 | 0.102 | 0.000 | 0.009 | 0.000 |
| Root number LC | 112 | | 0.946 | 0.054 | | | | | |
| Root number LM1 | 98 | | 0.000 | 1.000 | 0.000 | | | | |
| Root number LM2 | 97 | | 0.134 | 0.866 | 0.000 | | | | |
| Odontome UP1,2/LP1,2 | 50 | 1.000 | 0.000 | | | | | | |

# Melanesia

## *Fiji*

| Trait | *n* | Grade 0 | 1 | 2 | 3 | 4 | 5 | 6 | 7 |
|---|---|---|---|---|---|---|---|---|---|
| Winging UI1 | 23 | | 0.174 | 0.000 | 0.826 | 0.000 | | | |
| Shoveling UI1 | 16 | 0.125 | 0.188 | 0.438 | 0.250 | 0.000 | 0.000 | 0.000 | 0.000 |
| Double-shoveling UI1 | 19 | 0.842 | 0.158 | 0.000 | 0.000 | 0.000 | 0.000 | 0.000 | |
| Interruption grooves UI2 | 29 | 0.621 | 0.379 | | | | | | |
| Tuberculum dentale UI2 | 29 | 0.655 | 0.000 | 0.069 | 0.103 | 0.000 | 0.000 | 0.000 | 0.172 |
| Bushman canine UC | 31 | 1.000 | 0.000 | 0.000 | 0.000 | | | | |
| Distal acc. ridge UC | 18 | 0.167 | 0.056 | 0.389 | 0.222 | 0.167 | 0.000 | | |
| Uto-Aztecan UP1 | 37 | 1.000 | 0.000 | | | | | | |
| Hypocone UM2 | 59 | 0.068 | 0.068 | 0.051 | 0.102 | 0.542 | 0.169 | 0.000 | |
| Cusp 5 UM1 | 58 | 0.552 | 0.052 | 0.155 | 0.121 | 0.034 | 0.086 | | |
| Carabelli's cusp UM1 | 61 | 0.525 | 0.230 | 0.066 | 0.066 | 0.033 | 0.066 | 0.016 | 0.000 |
| Parastyle UM3 | 37 | 0.919 | 0.000 | 0.027 | 0.054 | 0.000 | 0.000 | 0.000 | |
| Enamel extension UM1 | 64 | 0.516 | 0.313 | 0.031 | 0.141 | | | | |
| Root number UP1 | 83 | | 0.566 | 0.422 | 0.012 | | | | |
| Root number UM2 | 65 | | 0.031 | 0.262 | 0.708 | | | | |
| PRM UM3 | 65 | 0.938 | | | 0.062 | | | | |
| Lingual cusp no. LP2 | 36 | 0.000 | 0.083 | 0.778 | 0.139 | | | | |
| Groove pattern LM2 | 49 | | 0.184 | 0.571 | 0.245 | | | | |
| Cusp number LM1 | 45 | 0.044 | 0.467 | 0.489 | | | | | |
| Cusp number LM2 | 45 | 0.311 | 0.555 | 0.133 | | | | | |
| Deflecting wrinkle LM1 | 40 | 0.600 | 0.075 | 0.175 | 0.150 | | | | |
| Trigonid crest LM1 | 49 | 0.918 | 0.082 | | | | | | |
| Protostylid LM1 | 26 | 0.808 | 0.154 | 0.000 | 0.000 | 0.038 | 0.000 | 0.000 | 0.000 |
| Cusp 7 LM1 | 49 | 0.816 | 0.020 | 0.020 | 0.102 | 0.020 | 0.000 | 0.020 | |
| Tomes' root LP1 | 30 | 0.167 | 0.367 | 0.167 | 0.133 | 0.100 | 0.033 | 0.033 | 0.000 |
| Root number LC | 51 | | 1.000 | 0.000 | | | | | |
| Root number LM1 | 60 | | 0.000 | 0.967 | 0.033 | | | | |
| Root number LM2 | 59 | | 0.203 | 0.797 | 0.000 | | | | |
| Odontome UP1,2/LP1,2 | 50 | 1.000 | 0.000 | | | | | | |

## *Loyalty Islands*

| | | Grade | | | | | | | |
|---|---|---|---|---|---|---|---|---|---|
| **Trait** | *n* | **0** | **1** | **2** | **3** | **4** | **5** | **6** | **7** |
| Winging UI1 | 20 | | 0.250 | 0.000 | 0.750 | 0.000 | | | |
| Shoveling UI1 | 17 | 0.000 | 0.294 | 0.529 | 0.176 | 0.000 | 0.000 | 0.000 | 0.000 |
| Double-shoveling UI1 | 17 | 0.706 | 0.235 | 0.059 | 0.000 | 0.000 | 0.000 | 0.000 | |
| Interruption grooves UI2 | 26 | 0.692 | 0.308 | | | | | | |
| Tuberculum dentale UI2 | 28 | 0.571 | 0.000 | 0.250 | 0.036 | 0.000 | 0.000 | 0.000 | 0.143 |
| Bushman canine UC | 32 | 0.969 | 0.031 | 0.000 | 0.000 | | | | |
| Distal acc. ridge UC | 17 | 0.412 | 0.059 | 0.059 | 0.471 | 0.000 | 0.000 | | |
| Uto-Aztecan UP1 | 40 | 1.000 | 0.000 | | | | | | |
| Hypocone UM2 | 55 | 0.018 | 0.018 | 0.036 | 0.127 | 0.582 | 0.218 | 0.000 | |
| Cusp 5 UM1 | 50 | 0.580 | 0.060 | 0.080 | 0.240 | 0.020 | 0.020 | | |
| Carabelli's cusp UM1 | 51 | 0.510 | 0.216 | 0.059 | 0.039 | 0.000 | 0.039 | 0.020 | 0.118 |
| Parastyle UM3 | 43 | 0.977 | 0.000 | 0.000 | 0.023 | 0.000 | 0.000 | 0.000 | |
| Enamel extension UM1 | 60 | 0.433 | 0.450 | 0.000 | 0.117 | | | | |
| Root number UP1 | 56 | | 0.607 | 0.375 | 0.018 | | | | |
| Root number UM2 | 57 | | 0.035 | 0.228 | 0.737 | | | | |
| PRM UM3 | 62 | 0.935 | | | 0.065 | | | | |
| Lingual cusp no. LP2 | 31 | 0.000 | 0.226 | 0.613 | 0.161 | | | | |
| Groove pattern LM2 | 39 | | 0.308 | 0.333 | 0.359 | | | | |
| Cusp number LM1 | 35 | 0.000 | 0.543 | 0.457 | | | | | |
| Cusp number LM2 | 37 | 0.378 | 0.514 | 0.108 | | | | | |
| Deflecting wrinkle LM1 | 26 | 0.731 | 0.000 | 0.115 | 0.154 | | | | |
| Trigonid crest LM1 | 29 | 0.931 | 0.069 | | | | | | |
| Protostylid LM1 | 39 | 0.949 | 0.051 | 0.000 | 0.000 | 0.000 | 0.000 | 0.000 | 0.000 |
| Cusp 7 LM1 | 38 | 0.816 | 0.000 | 0.000 | 0.053 | 0.079 | 0.053 | | |
| Tomes' root LP1 | 18 | 0.000 | 0.611 | 0.000 | 0.000 | 0.278 | 0.056 | 0.056 | 0.000 |
| Root number LC | 31 | | 1.000 | 0.000 | | | | | |
| Root number LM1 | 43 | | 0.000 | 0.930 | 0.070 | | | | |
| Root number LM2 | 44 | | 0.205 | 0.795 | 0.000 | | | | |
| Odontome UP1,2/LP1,2 | 40 | 0.925 | 0.075 | | | | | | |

| Trait | n | 0 | 1 | 2 | 3 | 4 | 5 | 6 | 7 |
|---|---|---|---|---|---|---|---|---|---|
| | | Grade | | | | | | | |
| Winging UI1 | 155 | | 0.232 | 0.000 | 0.761 | 0.006 | | | |
| Shoveling UI1 | 124 | 0.218 | 0.516 | 0.177 | 0.073 | 0.016 | 0.000 | 0.000 | 0.000 |
| Double-shoveling UI1 | 123 | 0.967 | 0.033 | | | | | | |
| Interruption grooves UI2 | 146 | 0.808 | 0.192 | | | | | | |
| Tuberculum dentale UI2 | 144 | 0.632 | 0.000 | 0.194 | 0.076 | 0.014 | 0.021 | 0.000 | 0.063 |
| Bushman canine UC | 154 | 0.981 | 0.006 | 0.013 | 0.000 | | | | |
| Distal acc. ridge UC | 66 | 0.333 | 0.288 | 0.167 | 0.167 | 0.045 | 0.000 | | |
| Uto-Aztecan UP1 | 30 | 1.000 | 0.000 | | | | | | |
| Hypocone UM2 | 214 | 0.033 | 0.037 | 0.056 | 0.145 | 0.514 | 0.210 | 0.005 | |
| Cusp 5 UM1 | 183 | 0.317 | 0.093 | 0.153 | 0.257 | 0.180 | 0.000 | | |
| Carabelli's cusp UM1 | 207 | 0.314 | 0.275 | 0.034 | 0.087 | 0.092 | 0.087 | 0.010 | 0.102 |
| Parastyle UM3 | 151 | 0.934 | 0.000 | 0.046 | 0.013 | 0.007 | 0.000 | 0.000 | |
| Enamel extension UM1 | 195 | 0.785 | 0.179 | 0.010 | 0.026 | | | | |
| Root number UP1 | 175 | | 0.509 | 0.486 | 0.006 | | | | |
| Root number UM2 | 184 | | 0.054 | 0.120 | 0.810 | 0.016 | | | |
| PRM UM3 | 207 | 0.870 | 0.024 | 0.029 | 0.077 | | | | |
| Lingual cusp no. LP2 | 172 | 0.000 | 0.093 | 0.605 | 0.302 | | | | |
| Groove pattern LM2 | 211 | | 0.431 | 0.289 | 0.280 | | | | |
| Cusp number LM1 | 170 | 0.018 | 0.453 | 0.529 | | | | | |
| Cusp number LM2 | 193 | 0.497 | 0.425 | 0.078 | | | | | |
| Deflecting wrinkle LM1 | 124 | 0.444 | 0.097 | 0.258 | 0.202 | | | | |
| Trigonid crest LM1 | 138 | 0.993 | 0.007 | | | | | | |
| Protostylid LM1 | 73 | 0.781 | 0.137 | 0.027 | 0.000 | 0.041 | 0.000 | 0.014 | 0.000 |
| Cusp 7 LM1 | 221 | 0.869 | 0.014 | 0.027 | 0.032 | 0.018 | 0.041 | | |
| Tomes' root LP1 | 126 | 0.667 | 0.175 | 0.016 | 0.008 | 0.048 | 0.000 | 0.008 | 0.079 |
| Root number LC | 130 | | 1.000 | 0.000 | | | | | |
| Root number LM1 | 193 | | 0.000 | 0.964 | 0.036 | | | | |
| Root number LM2 | 184 | | 0.043 | 0.951 | 0.005 | | | | |
| Odontome UP1,2/LP1,2 | 168 | 0.982 | 0.018 | | | | | | |

## New Hebrides

| Trait | n | Grade 0 | 1 | 2 | 3 | 4 | 5 | 6 | 7 |
|---|---|---|---|---|---|---|---|---|---|
| Winging UI1 | 31 | | 0.065 | 0.000 | 0.935 | 0.000 | | | |
| Shoveling UI1 | 7 | 0.143 | 0.571 | 0.286 | 0.000 | 0.000 | 0.000 | 0.000 | 0.000 |
| Double-shoveling UI1 | 7 | 1.000 | 0.000 | | | | | | |
| Interruption grooves UI2 | 13 | 0.846 | 0.154 | | | | | | |
| Tuberculum dentale UI2 | 14 | 0.786 | 0.000 | 0.071 | 0.143 | 0.000 | 0.000 | 0.000 | 0.000 |
| Bushman canine UC | 16 | 0.938 | 0.063 | 0.000 | 0.000 | | | | |
| Distal acc. ridge UC | | | | | | | | | |
| Uto-Aztecan UP1 | | | | | | | | | |
| Hypocone UM2 | 50 | 0.080 | 0.040 | 0.000 | 0.120 | 0.460 | 0.300 | 0.000 | |
| Cusp 5 UM1 | 40 | 0.600 | 0.125 | 0.075 | 0.175 | 0.025 | 0.000 | | |
| Carabelli's cusp UM1 | 50 | 0.320 | 0.240 | 0.100 | 0.040 | 0.020 | 0.100 | 0.060 | 0.120 |
| Parastyle UM3 | 36 | 0.972 | 0.000 | 0.000 | 0.028 | 0.000 | 0.000 | 0.000 | |
| Enamel extension UM1 | 55 | 0.691 | 0.273 | 0.000 | 0.036 | | | | |
| Root number UP1 | 68 | | 0.588 | 0.412 | 0.000 | | | | |
| Root number UM2 | 67 | | 0.149 | 0.209 | 0.642 | | | | |
| PRM UM3 | 67 | 0.940 | 0.015 | 0.015 | 0.030 | | | | |
| Lingual cusp no. LP2 | 10 | 0.000 | 0.200 | 0.700 | 0.100 | | | | |
| Groove pattern LM2 | 18 | | 0.333 | 0.500 | 0.167 | | | | |
| Cusp number LM1 | 17 | 0.059 | 0.588 | 0.353 | | | | | |
| Cusp number LM2 | 17 | 0.529 | 0.353 | 0.118 | | | | | |
| Deflecting wrinkle LM1 | | | | | | | | | |
| Trigonid crest LM1 | | | | | | | | | |
| Protostylid LM1 | 20 | 0.950 | 0.000 | 0.000 | 0.000 | 0.000 | 0.050 | 0.000 | 0.000 |
| Cusp 7 LM1 | 20 | 0.850 | 0.000 | 0.000 | 0.000 | 0.100 | 0.050 | | |
| Tomes' root LP1 | 13 | 0.000 | 0.692 | 0.077 | 0.077 | 0.154 | 0.000 | 0.000 | 0.000 |
| Root number LC | 19 | | 1.000 | 0.000 | | | | | |
| Root number LM1 | 22 | | 0.000 | 1.000 | 0.000 | | | | |
| Root number LM2 | 23 | | 0.087 | 0.913 | 0.000 | | | | |
| Odontome UP1,2/LP1,2 | 34 | 0.941 | 0.059 | | | | | | |

## Mesoamerica

### *Cuicuilco*

| | | Grade | | | | | | | |
|---|---|---|---|---|---|---|---|---|---|
| **Trait** | *n* | **0** | **1** | **2** | **3** | **4** | **5** | **6** | **7** |
| Winging UI1 | 24 | | 0.583 | 0.083 | 0.333 | 0.000 | | | |
| Shoveling UI1 | 29 | 0.000 | 0.000 | 0.069 | 0.517 | 0.241 | 0.069 | 0.103 | 0.000 |
| Double-shoveling UI1 | 25 | 0.120 | 0.520 | 0.200 | 0.160 | 0.000 | 0.000 | 0.000 | |
| Interruption grooves UI2 | 28 | 0.571 | 0.429 | | | | | | |
| Tuberculum dentale UI2 | 27 | 0.778 | 0.000 | 0.000 | 0.000 | 0.000 | 0.000 | 0.000 | 0.222 |
| Bushman canine UC | 29 | 0.966 | 0.000 | 0.034 | 0.000 | | | | |
| Distal acc. ridge UC | 13 | 0.231 | 0.077 | 0.231 | 0.385 | 0.077 | 0.000 | | |
| Uto-Aztecan UP1 | 50 | 0.980 | 0.020 | | | | | | |
| Hypocone UM2 | 39 | 0.077 | 0.000 | 0.205 | 0.410 | 0.256 | 0.051 | 0.000 | |
| Cusp 5 UM1 | 29 | 1.000 | 0.000 | 0.000 | 0.000 | 0.000 | 0.000 | | |
| Carabelli's cusp UM1 | 27 | 0.185 | 0.370 | 0.333 | 0.111 | 0.000 | 0.000 | 0.000 | 0.000 |
| Parastyle UM3 | 36 | 0.972 | 0.000 | 0.000 | 0.000 | 0.028 | 0.000 | 0.000 | |
| Enamel extension UM1 | 56 | 0.429 | 0.232 | 0.054 | 0.286 | | | | |
| Root number UP1 | 45 | | 0.889 | 0.111 | 0.000 | | | | |
| Root number UM2 | 42 | | 0.190 | 0.190 | 0.619 | | | | |
| PRM UM3 | 53 | 0.849 | 0.038 | 0.019 | 0.094 | | | | |
| Lingual cusp no. LP2 | 62 | 0.000 | 0.790 | 0.194 | 0.016 | | | | |
| Groove pattern LM2 | 69 | | 0.217 | 0.710 | 0.072 | | | | |
| Cusp number LM1 | 38 | 0.000 | 0.474 | 0.526 | | | | | |
| Cusp number LM2 | 59 | 0.017 | 0.864 | 0.119 | | | | | |
| Deflecting wrinkle LM1 | 28 | 0.357 | 0.036 | 0.357 | 0.250 | | | | |
| Trigonid crest LM1 | 59 | 1.000 | 0.000 | | | | | | |
| Protostylid LM1 | 64 | 0.719 | 0.266 | 0.016 | 0.000 | 0.000 | 0.000 | 0.000 | 0.000 |
| Cusp 7 LM1 | 72 | 0.917 | 0.014 | 0.014 | 0.028 | 0.014 | 0.014 | | |
| Tomes' root LP1 | 46 | 0.000 | 0.370 | 0.261 | 0.109 | 0.196 | 0.065 | 0.000 | 0.000 |
| Root number LC | 50 | | 1.000 | 0.000 | | | | | |
| Root number LM1 | 72 | | 0.000 | 0.917 | 0.083 | | | | |
| Root number LM2 | 66 | | 0.258 | 0.727 | 0.015 | | | | |
| Odontome UP1,2/LP1,2 | 37 | 0.946 | 0.054 | | | | | | |

## *Guasave*

| Trait | *n* | Grade 0 | 1 | 2 | 3 | 4 | 5 | 6 | 7 |
|---|---|---|---|---|---|---|---|---|---|
| Winging UI1 | 16 | | 0.375 | 0.000 | 0.625 | 0.000 | | | |
| Shoveling UI1 | 14 | 0.000 | 0.000 | 0.071 | 0.571 | 0.071 | 0.143 | 0.143 | 0.000 |
| Double-shoveling UI1 | 10 | 0.200 | 0.200 | 0.500 | 0.100 | 0.000 | 0.000 | 0.000 | |
| Interruption grooves UI2 | 14 | 0.357 | 0.643 | | | | | | |
| Tuberculum dentale UI2 | 18 | 0.667 | 0.000 | 0.167 | 0.000 | 0.000 | 0.000 | 0.000 | 0.167 |
| Bushman canine UC | 18 | 1.000 | 0.000 | 0.000 | 0.000 | | | | |
| Distal acc. ridge UC | 4 | 0.000 | 0.000 | 0.250 | 0.500 | 0.250 | 0.000 | | |
| Uto-Aztecan UP1 | 28 | 0.964 | 0.036 | | | | | | |
| Hypocone UM2 | 24 | 0.000 | 0.042 | 0.042 | 0.083 | 0.750 | 0.042 | 0.042 | |
| Cusp 5 UM1 | 14 | 0.786 | 0.000 | 0.071 | 0.143 | 0.000 | 0.000 | | |
| Carabelli's cusp UM1 | 15 | 0.067 | 0.133 | 0.333 | 0.267 | 0.000 | 0.133 | 0.000 | 0.067 |
| Parastyle UM3 | 14 | 1.000 | 0.000 | 0.000 | 0.000 | 0.000 | 0.000 | 0.000 | |
| Enamel extension UM1 | 21 | 0.714 | 0.190 | 0.048 | 0.048 | | | | |
| Root number UP1 | 16 | | 0.813 | 0.188 | 0.000 | | | | |
| Root number UM2 | 14 | | 0.286 | 0.214 | 0.500 | | | | |
| PRM UM3 | 24 | 0.958 | | | 0.042 | | | | |
| Lingual cusp no. LP2 | 19 | 0.105 | 0.684 | 0.211 | 0.000 | | | | |
| Groove pattern LM2 | 19 | | 0.421 | 0.579 | 0.000 | | | | |
| Cusp number LM1 | 11 | 0.000 | 0.545 | 0.455 | | | | | |
| Cusp number LM2 | 18 | 0.111 | 0.889 | 0.000 | | | | | |
| Deflecting wrinkle LM1 | 9 | 0.444 | 0.000 | 0.111 | 0.444 | | | | |
| Trigonid crest LM1 | 18 | 1.000 | 0.000 | | | | | | |
| Protostylid LM1 | 15 | 0.733 | 0.200 | 0.000 | 0.067 | 0.000 | 0.000 | 0.000 | 0.000 |
| Cusp 7 LM1 | 21 | 0.905 | 0.000 | 0.000 | 0.000 | 0.048 | 0.048 | | |
| Tomes' root LP1 | 14 | 0.000 | 0.643 | 0.214 | 0.000 | 0.071 | 0.071 | 0.000 | 0.000 |
| Root number LC | 18 | | 1.000 | 0.000 | | | | | |
| Root number LM1 | 23 | | 0.000 | 1.000 | 0.000 | | | | |
| Root number LM2 | 16 | | 0.313 | 0.688 | 0.000 | | | | |
| Odontome UP1,2/LP1,2 | 22 | 1.000 | 0.000 | | | | | | |

| Trait | *n* | Grade 0 | 1 | 2 | 3 | 4 | 5 | 6 | 7 |
|---|---|---|---|---|---|---|---|---|---|
| Winging UI1 | 31 | | 0.645 | 0.000 | 0.355 | 0.000 | | | |
| Shoveling UI1 | 39 | 0.000 | 0.000 | 0.000 | 0.333 | 0.256 | 0.410 | 0.000 | 0.000 |
| Double-shoveling UI1 | 36 | 0.056 | 0.194 | 0.500 | 0.056 | 0.167 | 0.028 | 0.000 | |
| Interruption grooves UI2 | 47 | 0.553 | 0.447 | | | | | | |
| Tuberculum dentale UI2 | 47 | 0.617 | 0.000 | 0.064 | 0.064 | 0.021 | 0.000 | 0.000 | 0.234 |
| Bushman canine UC | 63 | 1.000 | 0.000 | 0.000 | 0.000 | | | | |
| Distal acc. ridge UC | 34 | 0.265 | 0.029 | 0.206 | 0.265 | 0.206 | 0.029 | | |
| Uto-Aztecan UP1 | 85 | 1.000 | 0.000 | | | | | | |
| Hypocone UM2 | 102 | 0.088 | 0.059 | 0.157 | 0.265 | 0.382 | 0.049 | 0.000 | |
| Cusp 5 UM1 | 91 | 0.769 | 0.088 | 0.077 | 0.055 | 0.011 | 0.000 | | |
| Carabelli's cusp UM1 | 99 | 0.212 | 0.333 | 0.192 | 0.141 | 0.051 | 0.061 | 0.010 | 0.000 |
| Parastyle UM3 | 54 | 0.981 | 0.019 | 0.000 | 0.000 | 0.000 | 0.000 | 0.000 | |
| Enamel extension UM1 | 122 | 0.377 | 0.115 | 0.057 | 0.451 | | | | |
| Root number UP1 | 122 | | 0.910 | 0.090 | 0.000 | | | | |
| Root number UM2 | 88 | | 0.318 | 0.102 | 0.580 | | | | |
| PRM UM3 | 114 | 0.807 | 0.026 | 0.026 | 0.140 | | | | |
| Lingual cusp no. LP2 | 90 | 0.000 | 0.667 | 0.311 | 0.022 | | | | |
| Groove pattern LM2 | 154 | | 0.214 | 0.617 | 0.169 | | | | |
| Cusp number LM1 | 132 | 0.000 | 0.348 | 0.652 | | | | | |
| Cusp number LM2 | 151 | 0.053 | 0.722 | 0.225 | | | | | |
| Deflecting wrinkle LM1 | 96 | 0.333 | 0.031 | 0.375 | 0.260 | | | | |
| Trigonid crest LM1 | 170 | 0.976 | 0.024 | | | | | | |
| Protostylid LM1 | 172 | 0.773 | 0.186 | 0.006 | 0.017 | 0.012 | 0.006 | 0.000 | 0.000 |
| Cusp 7 LM1 | 173 | 0.832 | 0.046 | 0.075 | 0.023 | 0.012 | 0.012 | | |
| Tomes' root LP1 | 86 | 0.000 | 0.198 | 0.384 | 0.174 | 0.186 | 0.023 | 0.035 | 0.000 |
| Root number LC | 154 | | 0.968 | 0.032 | | | | | |
| Root number LM1 | 173 | | 0.006 | 0.942 | 0.052 | | | | |
| Root number LM2 | 163 | | 0.255 | 0.745 | 0.000 | | | | |
| Odontome UP1,2/LP1,2 | 140 | 0.986 | 0.014 | | | | | | |

# Micronesia

## *Guam*

| Trait | *n* | Grade 0 | 1 | 2 | 3 | 4 | 5 | 6 | 7 |
|---|---|---|---|---|---|---|---|---|---|
| Winging UI1 | 51 | | 0.490 | 0.059 | 0.451 | 0.000 | | | |
| Shoveling UI1 | 52 | 0.019 | 0.365 | 0.327 | 0.231 | 0.058 | 0.000 | 0.000 | 0.000 |
| Double-shoveling UI1 | 61 | 0.885 | 0.082 | 0.016 | 0.000 | 0.000 | 0.016 | 0.000 | |
| Interruption grooves UI2 | 66 | 0.742 | 0.258 | | | | | | |
| Tuberculum dentale UI2 | 72 | 0.708 | 0.000 | 0.014 | 0.069 | 0.069 | 0.014 | 0.000 | 0.124 |
| Bushman canine UC | 75 | 0.960 | 0.013 | 0.027 | 0.000 | | | | |
| Distal acc. ridge UC | 32 | 0.063 | 0.125 | 0.375 | 0.438 | 0.000 | 0.000 | | |
| Uto-Aztecan UP1 | 85 | 1.000 | 0.000 | | | | | | |
| Hypocone UM2 | 114 | 0.070 | 0.053 | 0.088 | 0.184 | 0.526 | 0.079 | 0.000 | |
| Cusp 5 UM1 | 102 | 0.716 | 0.127 | 0.118 | 0.039 | 0.000 | 0.000 | | |
| Carabelli's cusp UM1 | 93 | 0.290 | 0.290 | 0.032 | 0.097 | 0.022 | 0.140 | 0.086 | 0.043 |
| Parastyle UM3 | 41 | 0.976 | 0.000 | 0.024 | 0.000 | 0.000 | 0.000 | 0.000 | |
| Enamel extension UM1 | 107 | 0.701 | 0.280 | 0.019 | 0.000 | | | | |
| Root number UP1 | 111 | | 0.459 | 0.541 | 0.000 | | | | |
| Root number UM2 | 104 | | 0.000 | 0.250 | 0.740 | 0.010 | | | |
| PRM UM3 | 105 | 0.495 | 0.019 | 0.010 | 0.476 | | | | |
| Lingual cusp no. LP2 | 79 | 0.000 | 0.152 | 0.722 | 0.127 | | | | |
| Groove pattern LM2 | 95 | | 0.379 | 0.421 | 0.200 | | | | |
| Cusp number LM1 | 89 | 0.000 | 0.584 | 0.416 | | | | | |
| Cusp number LM2 | 95 | 0.168 | 0.621 | 0.211 | | | | | |
| Deflecting wrinkle LM1 | 94 | 0.372 | 0.064 | 0.309 | 0.255 | | | | |
| Trigonid crest LM1 | 112 | 0.973 | 0.027 | | | | | | |
| Protostylid LM1 | 84 | 0.750 | 0.083 | 0.000 | 0.107 | 0.048 | 0.012 | 0.000 | 0.000 |
| Cusp 7 LM1 | 100 | 0.920 | 0.010 | 0.020 | 0.030 | 0.020 | 0.000 | | |
| Tomes' root LP1 | 74 | 0.000 | 0.378 | 0.297 | 0.176 | 0.122 | 0.014 | 0.014 | 0.000 |
| Root number LC | 120 | | 1.000 | 0.000 | | | | | |
| Root number LM1 | 114 | | 0.000 | 0.982 | 0.018 | | | | |
| Root number LM2 | 128 | | 0.133 | 0.867 | 0.000 | | | | |
| Odontome UP1,2/LP1,2 | 103 | 1.000 | 0.000 | | | | | | |

# New Guinea

## *New Guinea*

| Trait | n | Grade 0 | 1 | 2 | 3 | 4 | 5 | 6 | 7 |
|---|---|---|---|---|---|---|---|---|---|
| Winging UI1 | 170 | | 0.076 | 0.000 | 0.924 | 0.000 | | | |
| Shoveling UI1 | 30 | 0.200 | 0.500 | 0.300 | 0.000 | 0.000 | 0.000 | 0.000 | 0.000 |
| Double-shoveling UI1 | 32 | 0.969 | 0.031 | | | | | | |
| Interruption grooves UI2 | 56 | 0.839 | 0.161 | | | | | | |
| Tuberculum dentale UI2 | 56 | 0.768 | 0.000 | 0.071 | 0.054 | 0.018 | 0.000 | 0.000 | 0.090 |
| Bushman canine UC | 54 | 0.981 | 0.000 | 0.019 | 0.000 | | | | |
| Distal acc. ridge UC | | | | | | | | | |
| Uto-Aztecan UP1 | | | | | | | | | |
| Hypocone UM2 | 191 | 0.047 | 0.000 | 0.052 | 0.099 | 0.476 | 0.325 | 0.000 | |
| Cusp 5 UM1 | 151 | 0.543 | 0.099 | 0.159 | 0.139 | 0.046 | 0.013 | | |
| Carabelli's cusp UM1 | 197 | 0.345 | 0.264 | 0.046 | 0.127 | 0.030 | 0.096 | 0.020 | 0.071 |
| Parastyle UM3 | 128 | 0.953 | 0.000 | 0.016 | 0.016 | 0.016 | 0.000 | 0.000 | |
| Enamel extension UM1 | 240 | 0.650 | 0.300 | 0.008 | 0.042 | | | | |
| Root number UP1 | 278 | | 0.698 | 0.302 | 0.000 | | | | |
| Root number UM2 | 260 | | 0.192 | 0.254 | 0.554 | | | | |
| PRM UM3 | 286 | 0.934 | 0.010 | 0.007 | 0.049 | | | | |
| Lingual cusp no. LP2 | 47 | 0.000 | 0.298 | 0.617 | 0.085 | | | | |
| Groove pattern LM2 | 102 | | 0.422 | 0.186 | 0.392 | | | | |
| Cusp number LM1 | 66 | 0.045 | 0.803 | 0.152 | | | | | |
| Cusp number LM2 | 93 | 0.591 | 0.376 | 0.033 | | | | | |
| Deflecting wrinkle LM1 | | | | | | | | | |
| Trigonid crest LM1 | | | | | | | | | |
| Protostylid LM1 | 97 | 0.028 | 0.041 | 0.000 | 0.000 | 0.010 | 0.010 | 0.010 | 0.000 |
| Cusp 7 LM1 | 100 | 0.920 | 0.010 | 0.010 | 0.010 | 0.030 | 0.020 | | |
| Tomes' root LP1 | 126 | 0.000 | 0.825 | 0.048 | 0.056 | 0.056 | 0.000 | 0.016 | 0.000 |
| Root number LC | 148 | | 0.986 | 0.014 | | | | | |
| Root number LM1 | 157 | | 0.000 | 1.000 | 0.000 | | | | |
| Root number LM2 | 142 | | 0.162 | 0.831 | 0.007 | | | | |
| Odontome UP1,2/LP1,2 | 119 | 1.000 | 0.000 | | | | | | |

## *New Guinea Gulf*

| Trait | *n* | Grade 0 | 1 | 2 | 3 | 4 | 5 | 6 | 7 |
|---|---|---|---|---|---|---|---|---|---|
| Winging UI1 | 56 | | 0.054 | 0.000 | 0.946 | 0.000 | | | |
| Shoveling UI1 | 7 | 0.429 | 0.571 | 0.000 | 0.000 | 0.000 | 0.000 | 0.000 | 0.000 |
| Double-shoveling UI1 | 8 | 1.000 | 0.000 | | | | | | |
| Interruption grooves UI2 | 7 | 0.857 | 0.143 | | | | | | |
| Tuberculum dentale UI2 | 7 | 0.857 | 0.000 | 0.000 | 0.000 | 0.000 | 0.000 | 0.000 | 0.143 |
| Bushman canine UC | 9 | 1.000 | 0.000 | 0.000 | 0.000 | | | | |
| Distal acc. ridge UC | 5 | 0.400 | 0.000 | 0.600 | 0.000 | 0.000 | 0.000 | | |
| Uto-Aztecan UP1 | 15 | 1.000 | 0.000 | | | | | | |
| Hypocone UM2 | 46 | 0.087 | 0.000 | 0.022 | 0.130 | 0.457 | 0.304 | 0.000 | |
| Cusp 5 UM1 | 40 | 0.375 | 0.125 | 0.125 | 0.250 | 0.125 | 0.000 | | |
| Carabelli's cusp UM1 | 45 | 0.467 | 0.333 | 0.000 | 0.067 | 0.022 | 0.067 | 0.044 | 0.000 |
| Parastyle UM3 | 22 | 0.909 | 0.000 | 0.045 | 0.045 | 0.000 | 0.000 | 0.000 | |
| Enamel extension UM1 | 55 | 0.655 | 0.291 | 0.018 | 0.036 | | | | |
| Root number UP1 | 61 | | 0.705 | 0.295 | 0.000 | | | | |
| Root number UM2 | 59 | | 0.102 | 0.305 | 0.593 | | | | |
| PRM UM3 | 64 | 0.906 | 0.000 | 0.016 | 0.078 | | | | |
| Lingual cusp no. LP2 | 10 | 0.000 | 0.400 | 0.600 | 0.000 | | | | |
| Groove pattern LM2 | 32 | | 0.563 | 0.125 | 0.313 | | | | |
| Cusp number LM1 | 19 | 0.000 | 0.842 | 0.158 | | | | | |
| Cusp number LM2 | 26 | 0.692 | 0.308 | 0.000 | | | | | |
| Deflecting wrinkle LM1 | 12 | 0.917 | 0.000 | 0.083 | 0.000 | | | | |
| Trigonid crest LM1 | 25 | 1.000 | 0.000 | | | | | | |
| Protostylid LM1 | | | | | | | | | |
| Cusp 7 LM1 | 35 | 1.000 | 0.000 | 0.000 | 0.000 | 0.000 | 0.000 | | |
| Tomes' root LP1 | 50 | 0.000 | 0.900 | 0.000 | 0.020 | 0.040 | 0.000 | 0.040 | 0.000 |
| Root number LC | 54 | | 0.981 | 0.019 | | | | | |
| Root number LM1 | 54 | | 0.000 | 1.000 | 0.000 | | | | |
| Root number LM2 | 51 | | 0.137 | 0.843 | 0.020 | | | | |
| Odontome UP1,2/LP1,2 | 23 | 1.000 | 0.000 | | | | | | |

## *Torres 1 and 2*

| | | Grade | | | | | | | |
|---|---|---|---|---|---|---|---|---|---|
| **Trait** | *n* | **0** | **1** | **2** | **3** | **4** | **5** | **6** | **7** |
| Winging UI1 | 18 | | 0.056 | 0.000 | 0.944 | 0.000 | | | |
| Shoveling UI1 | 10 | 0.100 | 0.500 | 0.400 | 0.000 | 0.000 | 0.000 | 0.000 | 0.000 |
| Double-shoveling UI1 | 11 | 1.000 | 0.000 | | | | | | |
| Interruption grooves UI2 | 9 | 0.889 | 0.111 | | | | | | |
| Tuberculum dentale UI2 | 9 | 0.667 | 0.000 | 0.111 | 0.111 | 0.000 | 0.000 | 0.000 | 0.111 |
| Bushman canine UC | 9 | 1.000 | 0.000 | 0.000 | 0.000 | | | | |
| Distal acc. ridge UC | 5 | 0.600 | 0.000 | 0.200 | 0.200 | 0.000 | 0.000 | | |
| Uto-Aztecan UP1 | 23 | 1.000 | 0.000 | | | | | | |
| Hypocone UM2 | 27 | 0.074 | 0.000 | 0.074 | 0.037 | 0.407 | 0.407 | 0.000 | |
| Cusp 5 UM1 | 28 | 0.357 | 0.036 | 0.179 | 0.321 | 0.071 | 0.036 | | |
| Carabelli's cusp UM1 | 30 | 0.333 | 0.200 | 0.067 | 0.167 | 0.033 | 0.133 | 0.000 | 0.067 |
| Parastyle UM3 | 22 | 1.000 | 0.000 | 0.000 | 0.000 | 0.000 | 0.000 | 0.000 | |
| Enamel extension UM1 | 37 | 0.541 | 0.405 | 0.000 | 0.054 | | | | |
| Root number UP1 | 39 | | 0.769 | 0.231 | 0.000 | | | | |
| Root number UM2 | 36 | | 0.222 | 0.222 | 0.556 | | | | |
| PRM UM3 | 42 | 0.952 | 0.024 | 0.000 | 0.024 | | | | |
| Lingual cusp no. LP2 | 9 | 0.000 | 0.333 | 0.667 | 0.000 | | | | |
| Groove pattern LM2 | 13 | | 0.308 | 0.308 | 0.385 | | | | |
| Cusp number LM1 | 8 | 0.125 | 0.625 | 0.250 | | | | | |
| Cusp number LM2 | 14 | 0.571 | 0.357 | 0.071 | | | | | |
| Deflecting wrinkle LM1 | 9 | 0.667 | 0.222 | 0.111 | 0.000 | | | | |
| Trigonid crest LM1 | 11 | 1.000 | 0.000 | | | | | | |
| Protostylid LM1 | | | | | | | | | |
| Cusp 7 LM1 | 13 | 0.923 | 0.000 | 0.000 | 0.000 | 0.077 | 0.000 | | |
| Tomes' root LP1 | 9 | 0.000 | 0.778 | 0.000 | 0.222 | 0.000 | 0.000 | 0.000 | 0.000 |
| Root number LC | 18 | | 1.000 | 0.000 | | | | | |
| Root number LM1 | 19 | | 0.000 | 1.000 | 0.000 | | | | |
| Root number LM2 | 16 | | 0.250 | 0.750 | 0.000 | | | | |
| Odontome UP1,2/LP1,2 | 20 | 1.000 | 0.000 | | | | | | |

## North America (Native Populations)

### *Alabama*

| | | Grade | | | | | | | |
|---|---|---|---|---|---|---|---|---|---|
| **Trait** | *n* | **0** | **1** | **2** | **3** | **4** | **5** | **6** | **7** |
| Winging UI1 | 121 | | 0.496 | 0.017 | 0.488 | 0.000 | | | |
| Shoveling UI1 | 159 | 0.000 | 0.000 | 0.145 | 0.270 | 0.270 | 0.208 | 0.050 | 0.057 |
| Double-shoveling UI1 | 135 | 0.104 | 0.896 | | | | | | |
| Interruption grooves UI2 | 155 | 0.413 | 0.587 | | | | | | |
| Tuberculum dentale UI2 | 153 | 0.582 | 0.000 | 0.078 | 0.033 | 0.000 | 0.007 | 0.000 | 0.300 |
| Bushman canine UC | 137 | 0.985 | 0.007 | 0.007 | 0.000 | 0.000 | | | |
| Distal acc. ridge UC | 76 | 0.026 | 0.118 | 0.237 | 0.329 | 0.263 | 0.026 | | |
| Uto-Aztecan UP1 | 159 | 0.981 | 0.019 | | | | | | |
| Hypocone UM2 | 143 | 0.035 | 0.028 | 0.133 | 0.420 | 0.315 | 0.070 | 0.000 | |
| Cusp 5 UM1 | 109 | 0.761 | 0.092 | 0.092 | 0.000 | 0.018 | 0.018 | | |
| Carabelli's cusp UM1 | 128 | 0.305 | 0.305 | 0.156 | 0.102 | 0.039 | 0.055 | 0.023 | 0.016 |
| Parastyle UM3 | 113 | 0.956 | 0.018 | 0.018 | 0.000 | 0.000 | 0.009 | 0.000 | |
| Enamel extension UM1 | 203 | 0.409 | 0.251 | 0.069 | 0.271 | | | | |
| Root number UP1 | 108 | | 0.778 | 0.194 | 0.028 | | | | |
| Root number UM2 | 100 | | 0.070 | 0.180 | 0.750 | | | | |
| PRM UM3 | 164 | 0.817 | | | 0.183 | | | | |
| Lingual cusp no. LP2 | 113 | 0.000 | 0.522 | 0.381 | 0.097 | | | | |
| Groove pattern LM2 | 170 | | 0.265 | 0.659 | 0.076 | | | | |
| Cusp number LM1 | 127 | 0.000 | 0.323 | 0.677 | | | | | |
| Cusp number LM2 | 178 | 0.034 | 0.966 | 0.000 | | | | | |
| Deflecting wrinkle LM1 | 100 | 0.240 | 0.020 | 0.270 | 0.470 | | | | |
| Trigonid crest LM1 | 129 | 0.977 | 0.023 | | | | | | |
| Protostylid LM1 | 175 | 0.617 | 0.257 | 0.029 | 0.034 | 0.046 | 0.017 | 0.000 | 0.000 |
| Cusp 7 LM1 | 180 | 0.939 | 0.011 | 0.022 | 0.022 | 0.000 | 0.006 | | |
| Tomes' root LP1 | 59 | 0.831 | 0.169 | | | | | | |
| Root number LC | 99 | | 1.000 | 0.000 | | | | | |
| Root number LM1 | 178 | | 0.000 | 0.949 | 0.051 | | | | |
| Root number LM2 | 119 | | 0.294 | 0.706 | 0.000 | | | | |
| Odontome UP1,2/LP1,2 | 124 | 0.919 | 0.081 | | | | | | |

## Grasshopper

| Trait | n | Grade 0 | 1 | 2 | 3 | 4 | 5 | 6 | 7 |
|---|---|---|---|---|---|---|---|---|---|
| Winging UI1 | 99 | | 0.384 | 0.091 | 0.515 | 0.010 | | | |
| Shoveling UI1 | 133 | 0.000 | 0.000 | 0.090 | 0.459 | 0.158 | 0.218 | 0.075 | 0.000 |
| Double-shoveling UI1 | 127 | 0.283 | 0.717 | | | | | | |
| Interruption grooves UI2 | 128 | 0.492 | 0.508 | | | | | | |
| Tuberculum dentale UI2 | 128 | 0.625 | 0.000 | 0.070 | 0.016 | 0.000 | 0.008 | 0.000 | 0.281 |
| Bushman canine UC | 108 | 0.972 | 0.028 | 0.000 | 0.000 | | | | |
| Distal acc. ridge UC | 63 | 0.238 | 0.206 | 0.222 | 0.190 | 0.143 | 0.000 | | |
| Uto-Aztecan UP1 | 124 | 0.960 | 0.040 | | | | | | |
| Hypocone UM2 | 134 | 0.246 | 0.067 | 0.142 | 0.231 | 0.276 | 0.037 | 0.000 | |
| Cusp 5 UM1 | 104 | 0.981 | 0.019 | 0.000 | 0.000 | 0.000 | 0.000 | | |
| Carabelli's cusp UM1 | 167 | 0.431 | 0.263 | 0.186 | 0.078 | 0.006 | 0.012 | 0.012 | 0.012 |
| Parastyle UM3 | 81 | 0.938 | 0.025 | 0.000 | 0.000 | 0.012 | 0.025 | 0.000 | |
| Enamel extension UM1 | 157 | 0.274 | 0.236 | 0.045 | 0.446 | | | | |
| Root number UP1 | 81 | | 0.840 | 0.160 | 0.000 | | | | |
| Root number UM2 | 55 | | 0.364 | 0.200 | 0.436 | | | | |
| PRM UM3 | 104 | 0.769 | 0.077 | 0.019 | 0.135 | | | | |
| Lingual cusp no. LP2 | 84 | 0.000 | 0.679 | 0.274 | 0.048 | | | | |
| Groove pattern LM2 | 119 | | 0.336 | 0.487 | 0.176 | | | | |
| Cusp number LM1 | 148 | 0.000 | 0.527 | 0.473 | | | | | |
| Cusp number LM2 | 104 | 0.173 | 0.750 | 0.077 | | | | | |
| Deflecting wrinkle LM1 | 121 | 0.182 | 0.099 | 0.372 | 0.347 | | | | |
| Trigonid crest LM1 | 148 | 0.973 | 0.027 | | | | | | |
| Protostylid LM1 | 150 | 0.387 | 0.587 | 0.007 | 0.000 | 0.013 | 0.007 | 0.000 | 0.000 |
| Cusp 7 LM1 | 181 | 0.912 | 0.017 | 0.039 | 0.028 | 0.006 | 0.000 | | |
| Tomes' root LP1 | 54 | 0.833 | 0.167 | | | | | | |
| Root number LC | 58 | | 1.000 | 0.000 | | | | | |
| Root number LM1 | 135 | | 0.000 | 0.933 | 0.067 | | | | |
| Root number LM2 | 78 | | 0.231 | 0.769 | 0.000 | | | | |
| Odontome UP1,2/LP1,2 | 108 | 0.981 | 0.019 | | | | | | |

## Greater Northwest Coast

| | | Grade | | | | | | | |
|---|---|---|---|---|---|---|---|---|---|
| **Trait** | *n* | **0** | **1** | **2** | **3** | **4** | **5** | **6** | **7** |
| Winging UI1 | 226 | | 0.358 | 0.031 | 0.593 | 0.018 | | | |
| Shoveling UI1 | 172 | 0.000 | 0.012 | 0.157 | 0.622 | 0.140 | 0.041 | 0.029 | 0.000 |
| Double-shoveling UI1 | 158 | 0.418 | 0.025 | 0.013 | 0.006 | 0.000 | 0.538 | 0.000 | |
| Interruption grooves UI2 | 223 | 0.350 | 0.650 | | | | | | |
| Tuberculum dentale UI2 | 206 | 0.617 | 0.005 | 0.097 | 0.039 | 0.029 | 0.044 | 0.000 | 0.171 |
| Bushman canine UC | 263 | 0.996 | 0.004 | 0.000 | 0.000 | | | | |
| Distal acc. ridge UC | 110 | 0.155 | 0.036 | 0.209 | 0.373 | 0.200 | 0.027 | | |
| Uto-Aztecan UP1 | 171 | 0.994 | 0.006 | | | | | | |
| Hypocone UM2 | 459 | 0.081 | 0.061 | 0.153 | 0.312 | 0.349 | 0.046 | 0.000 | |
| Cusp 5 UM1 | 378 | 0.786 | 0.048 | 0.074 | 0.069 | 0.008 | 0.017 | | |
| Carabelli's cusp UM1 | 388 | 0.389 | 0.363 | 0.082 | 0.090 | 0.021 | 0.026 | 0.026 | 0.003 |
| Parastyle UM3 | 361 | 0.972 | 0.006 | 0.019 | 0.000 | 0.003 | 0.000 | 0.000 | |
| Enamel extension UM1 | 699 | 0.303 | 0.187 | 0.084 | 0.425 | | | | |
| Root number UP1 | 693 | | 0.932 | 0.063 | 0.004 | | | | |
| Root number UM2 | 523 | | 0.428 | 0.155 | 0.415 | 0.002 | | | |
| PRM UM3 | 577 | 0.818 | 0.038 | 0.016 | 0.128 | | | | |
| Lingual cusp no. LP2 | 278 | 0.004 | 0.507 | 0.442 | 0.047 | | | | |
| Groove pattern LM2 | 498 | | 0.211 | 0.671 | 0.118 | | | | |
| Cusp number LM1 | 332 | 0.000 | 0.497 | 0.503 | | | | | |
| Cusp number LM2 | 477 | 0.044 | 0.688 | 0.268 | | | | | |
| Deflecting wrinkle LM1 | 192 | 0.422 | 0.036 | 0.177 | 0.365 | | | | |
| Trigonid crest LM1 | 294 | 0.922 | 0.078 | | | | | | |
| Protostylid LM1 | 456 | 0.664 | 0.248 | 0.053 | 0.015 | 0.018 | 0.000 | 0.002 | 0.000 |
| Cusp 7 LM1 | 473 | 0.913 | 0.019 | 0.038 | 0.013 | 0.011 | 0.006 | | |
| Tomes' root LP1 | 494 | 0.826 | 0.022 | 0.045 | 0.014 | 0.002 | 0.010 | 0.002 | 0.079 |
| Root number LC | 500 | | 1.000 | 0.000 | | | | | |
| Root number LM1 | 741 | | 0.001 | 0.834 | 0.165 | | | | |
| Root number LM2 | 659 | | 0.387 | 0.613 | 0.000 | | | | |
| Odontome UP1,2/LP1,2 | 371 | 0.933 | 0.065 | | | | | | |

## Iroquois

| Trait | n | Grade | | | | | | | |
|---|---|---|---|---|---|---|---|---|---|
| | | 0 | 1 | 2 | 3 | 4 | 5 | 6 | 7 |
| Winging UI1 | 48 | | 0.313 | 0.104 | 0.583 | 0.000 | | | |
| Shoveling UI1 | 53 | 0.000 | 0.075 | 0.264 | 0.377 | 0.189 | 0.019 | 0.057 | 0.019 |
| Double-shoveling UI1 | 38 | 0.632 | 0.368 | | | | | | |
| Interruption grooves UI2 | 55 | 0.527 | 0.473 | | | | | | |
| Tuberculum dentale UI2 | 52 | 0.654 | 0.019 | 0.115 | 0.058 | 0.000 | 0.038 | 0.000 | 0.115 |
| Bushman canine UC | 67 | 0.955 | 0.045 | 0.000 | 0.000 | | | | |
| Distal acc. ridge UC | 26 | 0.115 | 0.115 | 0.154 | 0.423 | 0.192 | 0.000 | | |
| Uto-Aztecan UP1 | 116 | 1.000 | 0.000 | | | | | | |
| Hypocone UM2 | 176 | 0.040 | 0.034 | 0.176 | 0.278 | 0.369 | 0.102 | 0.000 | |
| Cusp 5 UM1 | 225 | 0.849 | 0.084 | 0.040 | 0.000 | 0.013 | 0.004 | | |
| Carabelli's cusp UM1 | 231 | 0.342 | 0.333 | 0.182 | 0.043 | 0.048 | 0.026 | 0.022 | 0.004 |
| Parastyle UM3 | 63 | 0.952 | 0.032 | 0.016 | 0.000 | 0.000 | 0.000 | 0.000 | |
| Enamel extension UM1 | 231 | 0.511 | 0.229 | 0.074 | 0.186 | | | | |
| Root number UP1 | 207 | | 0.754 | 0.237 | 0.010 | | | | |
| Root number UM2 | 148 | | 0.189 | 0.176 | 0.628 | 0.007 | | | |
| PRM UM3 | 194 | 0.840 | | | 0.160 | | | | |
| Lingual cusp no. LP2 | 79 | 0.000 | 0.494 | 0.443 | 0.063 | | | | |
| Groove pattern LM2 | 167 | | 0.311 | 0.557 | 0.132 | | | | |
| Cusp number LM1 | 152 | 0.000 | 0.434 | 0.566 | | | | | |
| Cusp number LM2 | 163 | 0.092 | 0.908 | 0.000 | | | | | |
| Deflecting wrinkle LM1 | 131 | 0.275 | 0.069 | 0.176 | 0.481 | | | | |
| Trigonid crest LM1 | 165 | 0.958 | 0.042 | | | | | | |
| Protostylid LM1 | 186 | 0.624 | 0.306 | 0.022 | 0.032 | 0.016 | 0.000 | 0.000 | 0.000 |
| Cusp 7 LM1 | 194 | 0.830 | 0.021 | 0.088 | 0.026 | 0.026 | 0.010 | | |
| Tomes' root LP1 | 208 | 0.784 | 0.216 | | | | | | |
| Root number LC | 204 | | 1.000 | 0.000 | | | | | |
| Root number LM1 | 230 | | 0.000 | 0.926 | 0.074 | | | | |
| Root number LM2 | 172 | | 0.326 | 0.674 | 0.000 | | | | |
| Odontome UP1,2/LP1,2 | 152 | 0.993 | 0.007 | | | | | | |

## *Maryland*

| Trait | *n* | Grade | | | | | | | |
|---|---|---|---|---|---|---|---|---|---|
| | | 0 | 1 | 2 | 3 | 4 | 5 | 6 | 7 |
| Winging UI1 | 33 | | 0.727 | 0.061 | 0.212 | 0.000 | | | |
| Shoveling UI1 | 58 | 0.000 | 0.000 | 0.086 | 0.552 | 0.310 | 0.017 | 0.034 | 0.000 |
| Double-shoveling UI1 | 41 | 0.317 | 0.683 | | | | | | |
| Interruption grooves UI2 | 53 | 0.396 | 0.604 | | | | | | |
| Tuberculum dentale UI2 | 50 | 0.560 | 0.000 | 0.040 | 0.060 | 0.000 | 0.040 | 0.000 | 0.300 |
| Bushman canine UC | 47 | 1.000 | 0.000 | 0.000 | 0.000 | | | | |
| Distal acc. ridge UC | 23 | 0.087 | 0.130 | 0.261 | 0.304 | 0.217 | 0.000 | | |
| Uto-Aztecan UP1 | 54 | 1.000 | 0.000 | | | | | | |
| Hypocone UM2 | 73 | 0.014 | 0.055 | 0.178 | 0.288 | 0.315 | 0.151 | 0.000 | |
| Cusp 5 UM1 | 84 | 0.762 | 0.143 | 0.060 | 0.000 | 0.036 | 0.000 | | |
| Carabelli's cusp UM1 | 98 | 0.480 | 0.235 | 0.112 | 0.061 | 0.031 | 0.061 | 0.020 | 0.000 |
| Parastyle UM3 | 38 | 0.895 | 0.026 | 0.026 | 0.000 | 0.053 | 0.000 | 0.000 | |
| Enamel extension UM1 | 106 | 0.387 | 0.255 | 0.094 | 0.264 | | | | |
| Root number UP1 | 94 | | 0.819 | 0.181 | 0.000 | | | | |
| Root number UM2 | 63 | | 0.175 | 0.143 | 0.683 | | | | |
| PRM UM3 | 60 | 0.867 | | | 0.133 | | | | |
| Lingual cusp no. LP2 | 54 | 0.000 | 0.426 | 0.500 | 0.074 | | | | |
| Groove pattern LM2 | 96 | | 0.375 | 0.500 | 0.125 | | | | |
| Cusp number LM1 | 101 | 0.000 | 0.624 | 0.376 | | | | | |
| Cusp number LM2 | 102 | 0.088 | 0.912 | | | | | | |
| Deflecting wrinkle LM1 | 69 | 0.333 | 0.072 | 0.188 | 0.406 | | | | |
| Trigonid crest LM1 | 93 | 0.968 | 0.032 | | | | | | |
| Protostylid LM1 | 116 | 0.543 | 0.362 | 0.052 | 0.009 | 0.017 | 0.017 | 0.000 | 0.000 |
| Cusp 7 LM1 | 120 | 0.908 | 0.017 | 0.008 | 0.042 | 0.017 | 0.008 | | |
| Tomes' root LP1 | 119 | 0.840 | 0.160 | | | | | | |
| Root number LC | 107 | | 1.000 | 0.000 | | | | | |
| Root number LM1 | 150 | | 0.000 | 0.940 | 0.060 | | | | |
| Root number LM2 | 118 | | 0.322 | 0.678 | 0.000 | | | | |
| Odontome UP1,2/LP1,2 | 96 | 1.000 | 0.000 | | | | | | |

## *Northern California*

| Trait | n | Grade 0 | 1 | 2 | 3 | 4 | 5 | 6 | 7 |
|---|---|---|---|---|---|---|---|---|---|
| Winging UI1 | 66 | | 0.424 | 0.000 | 0.576 | 0.000 | | | |
| Shoveling UI1 | 62 | 0.000 | 0.000 | 0.016 | 0.403 | 0.177 | 0.274 | 0.129 | 0.000 |
| Double-shoveling UI1 | 57 | 0.123 | 0.877 | | | | | | |
| Interruption grooves UI2 | 70 | 0.486 | 0.514 | | | | | | |
| Tuberculum dentale UI2 | 70 | 0.571 | 0.000 | 0.100 | 0.057 | 0.014 | 0.029 | 0.000 | 0.228 |
| Bushman canine UC | 78 | 0.974 | 0.026 | 0.000 | 0.000 | | | | |
| Distal acc. ridge UC | 31 | 0.194 | 0.032 | 0.194 | 0.387 | 0.194 | 0.000 | | |
| Uto-Aztecan UP1 | 91 | 0.989 | 0.011 | | | | | | |
| Hypocone UM2 | 108 | 0.056 | 0.009 | 0.139 | 0.407 | 0.343 | 0.046 | 0.000 | |
| Cusp 5 UM1 | 47 | 0.979 | 0.021 | 0.000 | 0.000 | 0.000 | 0.000 | | |
| Carabelli's cusp UM1 | 49 | 0.245 | 0.367 | 0.184 | 0.102 | 0.020 | 0.061 | 0.020 | 0.000 |
| Parastyle UM3 | 102 | 0.971 | 0.010 | 0.010 | 0.010 | 0.000 | 0.000 | 0.000 | |
| Enamel extension UM1 | 148 | 0.331 | 0.216 | 0.068 | 0.385 | | | | |
| Root number UP1 | 139 | | 0.827 | 0.158 | 0.014 | | | | |
| Root number UM2 | 121 | | 0.174 | 0.165 | 0.661 | | | | |
| PRM UM3 | 156 | 0.821 | 0.019 | 0.058 | 0.103 | | | | |
| Lingual cusp no. LP2 | 87 | 0.000 | 0.575 | 0.391 | 0.034 | | | | |
| Groove pattern LM2 | 125 | | 0.280 | 0.584 | 0.136 | | | | |
| Cusp number LM1 | 66 | 0.000 | 0.348 | 0.652 | | | | | |
| Cusp number LM2 | 131 | 0.069 | 0.764 | 0.168 | | | | | |
| Deflecting wrinkle LM1 | 35 | 0.457 | 0.029 | 0.171 | 0.343 | | | | |
| Trigonid crest LM1 | 57 | 0.930 | 0.070 | | | | | | |
| Protostylid LM1 | 101 | 0.634 | 0.317 | 0.000 | 0.020 | 0.020 | 0.010 | 0.000 | 0.000 |
| Cusp 7 LM1 | 106 | 0.972 | 0.009 | 0.009 | 0.000 | 0.000 | 0.009 | | |
| Tomes' root LP1 | 119 | 0.210 | 0.000 | 0.395 | 0.160 | 0.160 | 0.059 | 0.017 | 0.000 |
| Root number LC | 139 | | 1.000 | 0.000 | | | | | |
| Root number LM1 | 176 | | 0.000 | 0.932 | 0.068 | | | | |
| Root number LM2 | 157 | | 0.268 | 0.732 | 0.000 | | | | |
| Odontome UP1,2/LP1,2 | 50 | 0.920 | 0.080 | | | | | | |

## Point of Pines

| Trait | *n* | Grade 0 | 1 | 2 | 3 | 4 | 5 | 6 | 7 |
|---|---|---|---|---|---|---|---|---|---|
| Winging UI1 | 54 | | 0.574 | 0.019 | 0.407 | 0.000 | | | |
| Shoveling UI1 | 76 | 0.000 | 0.013 | 0.092 | 0.316 | 0.329 | 0.224 | 0.013 | 0.013 |
| Double-shoveling UI1 | 46 | 0.326 | 0.674 | | | | | | |
| Interruption grooves UI2 | 73 | 0.531 | 0.469 | | | | | | |
| Tuberculum dentale UI2 | | | | | | | | | |
| Bushman canine UC | 76 | 1.000 | 0.000 | 0.000 | 0.000 | 0.000 | | | |
| Distal acc. ridge UC | 32 | 0.125 | 0.125 | 0.250 | 0.281 | 0.219 | 0.000 | | |
| Uto-Aztecan UP1 | 82 | 0.988 | 0.012 | | | | | | |
| Hypocone UM2 | 90 | 0.144 | 0.067 | 0.089 | 0.344 | 0.311 | 0.044 | 0.000 | |
| Cusp 5 UM1 | 68 | 0.824 | 0.059 | 0.059 | 0.000 | 0.059 | 0.000 | | |
| Carabelli's cusp UM1 | 70 | 0.329 | 0.443 | 0.157 | 0.029 | 0.014 | 0.029 | 0.000 | 0.000 |
| Parastyle UM3 | 69 | 0.957 | 0.014 | 0.000 | 0.000 | 0.014 | 0.014 | 0.000 | |
| Enamel extension UM1 | 112 | 0.286 | 0.179 | 0.089 | 0.446 | | | | |
| Root number UP1 | 95 | | 0.853 | 0.147 | 0.000 | | | | |
| Root number UM2 | 78 | | 0.205 | 0.141 | 0.654 | | | | |
| PRM UM3 | 102 | 0.784 | | | 0.216 | | | | |
| Lingual cusp no. LP2 | 80 | 0.000 | 0.550 | 0.375 | 0.075 | | | | |
| Groove pattern LM2 | 113 | | 0.257 | 0.681 | 0.062 | | | | |
| Cusp number LM1 | 76 | 0.000 | 0.553 | 0.447 | | | | | |
| Cusp number LM2 | 104 | 0.144 | 0.856 | 0.000 | | | | | |
| Deflecting wrinkle LM1 | 61 | 0.164 | 0.066 | 0.410 | 0.361 | | | | |
| Trigonid crest LM1 | 87 | 0.943 | 0.057 | | | | | | |
| Protostylid LM1 | 121 | 0.455 | 0.496 | 0.033 | 0.017 | 0.000 | 0.000 | 0.000 | 0.000 |
| Cusp 7 LM1 | 127 | 0.913 | 0.016 | 0.039 | 0.031 | 0.000 | 0.000 | | |
| Tomes' root LP1 | 79 | 0.899 | 0.101 | | | | | | |
| Root number LC | 94 | | 1.000 | 0.000 | | | | | |
| Root number LM1 | 128 | | 0.000 | 0.906 | 0.094 | | | | |
| Root number LM2 | 104 | | 0.269 | 0.721 | 0.010 | | | | |
| Odontome UP1,2/LP1,2 | 53 | 0.962 | 0.038 | | | | | | |

# Polynesia

## *Marquesas*

| Trait | n | Grade | | | | | | | |
|---|---|---|---|---|---|---|---|---|---|
| | | 0 | 1 | 2 | 3 | 4 | 5 | 6 | 7 |
| Winging UI1 | 50 | | 0.180 | 0.000 | 0.820 | 0.000 | | | |
| Shoveling UI1 | 62 | 0.145 | 0.371 | 0.290 | 0.194 | 0.000 | 0.000 | 0.000 | 0.000 |
| Double-shoveling UI1 | 64 | 0.859 | 0.063 | 0.031 | 0.000 | 0.000 | 0.047 | 0.000 | |
| Interruption grooves UI2 | 68 | 0.765 | 0.235 | | | | | | |
| Tuberculum dentale UI2 | 68 | 0.676 | 0.000 | 0.162 | 0.074 | 0.029 | 0.015 | 0.000 | 0.044 |
| Bushman canine UC | 77 | 0.987 | 0.013 | 0.000 | 0.000 | | | | |
| Distal acc. ridge UC | 23 | 0.217 | 0.217 | 0.130 | 0.174 | 0.217 | 0.043 | | |
| Uto-Aztecan UP1 | | | | | | | | | |
| Hypocone UM2 | 122 | 0.016 | 0.057 | 0.016 | 0.148 | 0.582 | 0.180 | 0.000 | |
| Cusp 5 UM1 | 99 | 0.737 | 0.101 | 0.030 | 0.071 | 0.030 | 0.030 | | |
| Carabelli's cusp UM1 | 155 | 0.510 | 0.123 | 0.077 | 0.052 | 0.026 | 0.090 | 0.026 | 0.096 |
| Parastyle UM3 | 75 | 0.960 | 0.013 | 0.027 | 0.000 | 0.000 | 0.000 | 0.000 | |
| Enamel extension UM1 | 146 | 0.548 | 0.274 | 0.021 | 0.158 | | | | |
| Root number UP1 | 142 | | 0.599 | 0.394 | 0.007 | | | | |
| Root number UM2 | 129 | | 0.147 | 0.256 | 0.597 | | | | |
| PRM UM3 | 121 | 0.669 | 0.008 | 0.033 | 0.289 | | | | |
| Lingual cusp no. LP2 | 90 | 0.000 | 0.111 | 0.800 | 0.089 | | | | |
| Groove pattern LM2 | 116 | | 0.276 | 0.560 | 0.164 | | | | |
| Cusp number LM1 | 109 | 0.009 | 0.450 | 0.541 | | | | | |
| Cusp number LM2 | 107 | 0.393 | 0.467 | 0.140 | | | | | |
| Deflecting wrinkle LM1 | 58 | 0.448 | 0.069 | 0.103 | 0.379 | | | | |
| Trigonid crest LM1 | 72 | 0.986 | 0.014 | | | | | | |
| Protostylid LM1 | 122 | 0.861 | 0.066 | 0.008 | 0.041 | 0.000 | 0.016 | 0.000 | 0.008 |
| Cusp 7 LM1 | 125 | 0.928 | 0.008 | 0.048 | 0.008 | 0.008 | 0.000 | | |
| Tomes' root LP1 | 104 | 0.817 | 0.087 | 0.038 | 0.000 | 0.038 | 0.019 | 0.000 | 0.000 |
| Root number LC | 108 | | 1.000 | 0.000 | | | | | |
| Root number LM1 | 141 | | 0.000 | 0.979 | 0.021 | | | | |
| Root number LM2 | 127 | | 0.244 | 0.724 | 0.024 | 0.008 | | | |
| Odontome UP1,2/LP1,2 | 164 | 0.970 | 0.030 | | | | | | |

## Mokapu

| Trait | n | Grade 0 | 1 | 2 | 3 | 4 | 5 | 6 | 7 |
|---|---|---|---|---|---|---|---|---|---|
| Winging UI1 | 166 | | 0.229 | 0.048 | 0.723 | 0.000 | | | |
| Shoveling UI1 | 166 | 0.024 | 0.325 | 0.416 | 0.187 | 0.048 | 0.000 | 0.000 | 0.000 |
| Double-shoveling UI1 | 171 | 0.737 | 0.170 | 0.035 | 0.047 | 0.012 | 0.000 | 0.000 | |
| Interruption grooves UI2 | 180 | 0.617 | 0.383 | | | | | | |
| Tuberculum dentale UI2 | 184 | 0.875 | 0.000 | 0.049 | 0.027 | 0.005 | 0.000 | 0.000 | 0.044 |
| Bushman canine UC | 186 | 0.973 | 0.005 | 0.022 | 0.000 | | | | |
| Distal acc. ridge UC | 110 | 0.464 | 0.118 | 0.273 | 0.145 | 0.000 | 0.000 | | |
| Uto-Aztecan UP1 | 200 | 1.000 | 0.000 | | | | | | |
| Hypocone UM2 | 195 | 0.046 | 0.021 | 0.077 | 0.097 | 0.579 | 0.179 | 0.000 | |
| Cusp 5 UM1 | 187 | 0.610 | 0.176 | 0.086 | 0.091 | 0.037 | 0.000 | | |
| Carabelli's cusp UM1 | 204 | 0.485 | 0.118 | 0.039 | 0.039 | 0.000 | 0.147 | 0.083 | 0.088 |
| Parastyle UM3 | 115 | 0.948 | 0.000 | 0.035 | 0.009 | 0.009 | 0.000 | 0.000 | |
| Enamel extension UM1 | 205 | 0.459 | 0.337 | 0.024 | 0.180 | | | | |
| Root number UP1 | 171 | | 0.608 | 0.380 | 0.012 | | | | |
| Root number UM2 | 166 | | 0.157 | 0.386 | 0.458 | | | | |
| PRM UM3 | 200 | 0.655 | 0.000 | 0.035 | 0.310 | | | | |
| Lingual cusp no. LP2 | 181 | 0.000 | 0.171 | 0.773 | 0.055 | | | | |
| Groove pattern LM2 | 186 | | 0.290 | 0.559 | 0.151 | | | | |
| Cusp number LM1 | 166 | 0.006 | 0.482 | 0.512 | | | | | |
| Cusp number LM2 | 176 | 0.352 | 0.574 | 0.074 | | | | | |
| Deflecting wrinkle LM1 | 142 | 0.739 | 0.014 | 0.169 | 0.077 | | | | |
| Trigonid crest LM1 | 179 | 0.950 | 0.050 | | | | | | |
| Protostylid LM1 | 184 | 0.837 | 0.060 | 0.000 | 0.049 | 0.016 | 0.022 | 0.016 | 0.000 |
| Cusp 7 LM1 | 190 | 0.921 | 0.005 | 0.042 | 0.032 | 0.000 | 0.000 | | |
| Tomes' root LP1 | 111 | 0.000 | 0.405 | 0.279 | 0.108 | 0.153 | 0.018 | 0.036 | 0.000 |
| Root number LC | 106 | | 0.981 | 0.019 | | | | | |
| Root number LM1 | 184 | | 0.000 | 0.880 | 0.120 | | | | |
| Root number LM2 | 170 | | 0.300 | 0.688 | 0.012 | | | | |
| Odontome UP1,2/LP1,2 | 227 | 0.969 | 0.031 | | | | | | |

| Trait | n | Grade 0 | 1 | 2 | 3 | 4 | 5 | 6 | 7 |
|---|---|---|---|---|---|---|---|---|---|
| Winging UI1 | 26 | | 0.154 | 0.038 | 0.808 | 0.000 | | | |
| Shoveling UI1 | 17 | 0.118 | 0.353 | 0.412 | 0.118 | 0.000 | 0.000 | 0.000 | 0.000 |
| Double-shoveling UI1 | 14 | 0.786 | 0.214 | 0.000 | 0.000 | 0.000 | 0.000 | 0.000 | |
| Interruption grooves UI2 | 21 | 0.609 | 0.391 | | | | | | |
| Tuberculum dentale UI2 | 26 | 0.654 | 0.000 | 0.192 | 0.038 | 0.000 | 0.000 | 0.000 | 0.115 |
| Bushman canine UC | 38 | 0.921 | 0.053 | 0.026 | 0.000 | | | | |
| Distal acc. ridge UC | | | | | | | | | |
| Uto-Aztecan UP1 | | | | | | | | | |
| Hypocone UM2 | 119 | 0.025 | 0.008 | 0.034 | 0.185 | 0.513 | 0.235 | 0.000 | |
| Cusp 5 UM1 | 53 | 0.849 | 0.019 | 0.075 | 0.000 | 0.057 | 0.000 | | |
| Carabelli's cusp UM1 | 57 | 0.544 | 0.193 | 0.035 | 0.035 | 0.018 | 0.088 | 0.035 | 0.053 |
| Parastyle UM3 | 55 | 0.964 | 0.000 | 0.018 | 0.018 | 0.000 | 0.000 | 0.000 | |
| Enamel extension UM1 | 152 | 0.237 | 0.533 | 0.059 | 0.171 | | | | |
| Root number UP1 | 169 | | 0.757 | 0.231 | 0.012 | | | | |
| Root number UM2 | 152 | | 0.237 | 0.303 | 0.461 | | | | |
| PRM UM3 | 177 | 0.718 | 0.023 | 0.034 | 0.226 | | | | |
| Lingual cusp no. LP2 | 28 | 0.000 | 0.214 | 0.786 | 0.000 | | | | |
| Groove pattern LM2 | 54 | | 0.315 | 0.426 | 0.259 | | | | |
| Cusp number LM1 | 27 | 0.000 | 0.259 | 0.741 | | | | | |
| Cusp number LM2 | 50 | 0.220 | 0.580 | 0.200 | | | | | |
| Deflecting wrinkle LM1 | | | | | | | | | |
| Trigonid crest LM1 | | | | | | | | | |
| Protostylid LM1 | 37 | 0.946 | 0.054 | 0.000 | 0.000 | 0.000 | 0.000 | 0.000 | 0.000 |
| Cusp 7 LM1 | 37 | 0.946 | 0.000 | 0.000 | 0.027 | 0.027 | 0.000 | | |
| Tomes' root LP1 | 30 | 0.000 | 0.533 | 0.200 | 0.100 | 0.133 | 0.033 | 0.000 | 0.000 |
| Root number LC | 73 | | 1.000 | 0.000 | | | | | |
| Root number LM1 | 83 | | 0.000 | 0.976 | 0.024 | | | | |
| Root number LM2 | 82 | | 0.427 | 0.573 | 0.000 | | | | |
| Odontome UP1,2/LP1,2 | 24 | 1.000 | 0.000 | | | | | | |

# South America (Native Populations)

## *Ayalan*

| Trait | *n* | Grade 0 | 1 | 2 | 3 | 4 | 5 | 6 | 7 |
|---|---|---|---|---|---|---|---|---|---|
| Winging UI1 | 26 | | 0.654 | 0.000 | 0.346 | 0.000 | | | |
| Shoveling UI1 | 51 | 0.000 | 0.000 | 0.098 | 0.294 | 0.196 | 0.294 | 0.118 | 0.000 |
| Double-shoveling UI1 | 41 | 0.171 | 0.195 | 0.366 | 0.171 | 0.073 | 0.024 | 0.000 | |
| Interruption grooves UI2 | 42 | 0.381 | 0.619 | | | | | | |
| Tuberculum dentale UI2 | 41 | 0.463 | 0.000 | 0.098 | 0.049 | 0.000 | 0.024 | 0.000 | 0.366 |
| Bushman canine UC | 55 | 1.000 | 0.000 | 0.000 | 0.000 | | | | |
| Distal acc. ridge UC | 38 | 0.026 | 0.079 | 0.211 | 0.342 | 0.316 | 0.026 | | |
| Uto-Aztecan UP1 | 74 | 0.986 | 0.014 | | | | | | |
| Hypocone UM2 | 81 | 0.123 | 0.025 | 0.111 | 0.099 | 0.519 | 0.123 | 0.000 | |
| Cusp 5 UM1 | 70 | 0.943 | 0.014 | 0.029 | 0.014 | 0.000 | 0.000 | | |
| Carabelli's cusp UM1 | 82 | 0.293 | 0.256 | 0.244 | 0.134 | 0.037 | 0.037 | 0.000 | 0.000 |
| Parastyle UM3 | 44 | 0.932 | 0.068 | 0.000 | 0.000 | 0.000 | 0.000 | 0.000 | |
| Enamel extension UM1 | 106 | 0.349 | 0.208 | 0.085 | 0.358 | | | | |
| Root number UP1 | 97 | | 0.887 | 0.103 | 0.010 | | | | |
| Root number UM2 | 82 | | 0.329 | 0.207 | 0.451 | 0.012 | | | |
| PRM UM3 | 83 | 0.675 | 0.060 | 0.048 | 0.217 | | | | |
| Lingual cusp no. LP2 | 69 | 0.000 | 0.464 | 0.449 | 0.087 | | | | |
| Groove pattern LM2 | 79 | | 0.177 | 0.785 | 0.038 | | | | |
| Cusp number LM1 | 68 | 0.000 | 0.338 | 0.662 | | | | | |
| Cusp number LM2 | 77 | 0.078 | 0.922 | 0.000 | | | | | |
| Deflecting wrinkle LM1 | 52 | 0.365 | 0.115 | 0.154 | 0.365 | | | | |
| Trigonid crest LM1 | 74 | 0.986 | 0.014 | | | | | | |
| Protostylid LM1 | 109 | 0.679 | 0.321 | 0.000 | 0.000 | 0.000 | 0.000 | 0.000 | 0.000 |
| Cusp 7 LM1 | 116 | 0.922 | 0.026 | 0.034 | 0.017 | 0.000 | 0.000 | | |
| Tomes' root LP1 | 64 | 0.000 | 0.328 | 0.344 | 0.078 | 0.234 | 0.000 | 0.016 | 0.000 |
| Root number LC | 91 | | 1.000 | 0.000 | | | | | |
| Root number LM1 | 108 | | 0.000 | 0.963 | 0.037 | | | | |
| Root number LM2 | 86 | | 0.442 | 0.558 | 0.000 | | | | |
| Odontome UP1,2/LP1,2 | 73 | 0.904 | 0.096 | | | | | | |

## Corondo

| Trait | n | 0 | 1 | 2 | 3 | 4 | 5 | 6 | 7 |
|---|---|---|---|---|---|---|---|---|---|
| | | Grade | | | | | | | |
| Winging UI1 | 20 | | 0.500 | 0.000 | 0.450 | 0.050 | | | |
| Shoveling UI1 | 20 | 0.000 | 0.000 | 0.050 | 0.400 | 0.300 | 0.100 | 0.150 | 0.000 |
| Double-shoveling UI1 | 21 | 0.000 | 0.048 | 0.476 | 0.286 | 0.048 | 0.143 | 0.000 | |
| Interruption grooves UI2 | 16 | 0.438 | 0.563 | | | | | | |
| Tuberculum dentale UI2 | 14 | 0.714 | 0.000 | 0.000 | 0.000 | 0.000 | 0.000 | 0.000 | 0.286 |
| Bushman canine UC | 18 | 1.000 | 0.000 | 0.000 | 0.000 | | | | |
| Distal acc. ridge UC | 15 | 0.133 | 0.000 | 0.200 | 0.200 | 0.400 | 0.067 | | |
| Uto-Aztecan UP1 | 34 | 1.000 | 0.000 | | | | | | |
| Hypocone UM2 | 51 | 0.039 | 0.000 | 0.118 | 0.235 | 0.569 | 0.020 | 0.020 | |
| Cusp 5 UM1 | 38 | 0.632 | 0.184 | 0.105 | 0.053 | 0.026 | 0.000 | | |
| Carabelli's cusp UM1 | 39 | 0.128 | 0.231 | 0.231 | 0.179 | 0.128 | 0.051 | 0.051 | 0.000 |
| Parastyle UM3 | 29 | 0.897 | 0.000 | 0.034 | 0.000 | 0.034 | 0.034 | 0.000 | |
| Enamel extension UM1 | 61 | 0.328 | 0.311 | 0.016 | 0.344 | | | | |
| Root number UP1 | 42 | | 0.905 | 0.095 | 0.000 | | | | |
| Root number UM2 | 26 | | 0.500 | 0.231 | 0.269 | | | | |
| PRM UM3 | 52 | 0.731 | 0.077 | 0.019 | 0.173 | | | | |
| Lingual cusp no. LP2 | 47 | 0.000 | 0.638 | 0.319 | 0.043 | | | | |
| Groove pattern LM2 | 54 | | 0.167 | 0.833 | 0.000 | | | | |
| Cusp number LM1 | 39 | 0.000 | 0.359 | 0.641 | | | | | |
| Cusp number LM2 | 56 | 0.161 | 0.750 | 0.089 | | | | | |
| Deflecting wrinkle LM1 | 28 | 0.107 | 0.143 | 0.214 | 0.536 | | | | |
| Trigonid crest LM1 | 52 | 0.942 | 0.058 | | | | | | |
| Protostylid LM1 | 57 | 0.684 | 0.193 | 0.018 | 0.035 | 0.035 | 0.018 | 0.018 | 0.000 |
| Cusp 7 LM1 | 61 | 0.902 | 0.033 | 0.049 | 0.016 | 0.000 | 0.000 | | |
| Tomes' root LP1 | 13 | 0.308 | 0.077 | 0.462 | 0.077 | 0.000 | 0.000 | 0.077 | 0.000 |
| Root number LC | 38 | | 0.974 | 0.026 | | | | | |
| Root number LM1 | 38 | | 0.000 | 0.946 | 0.054 | | | | |
| Root number LM2 | 42 | | 0.714 | 0.286 | 0.000 | | | | |
| Odontome UP1,2/LP1,2 | 25 | 0.960 | 0.040 | | | | | | |

## Herradura and Teatinos

| | | Grade | | | | | | | |
|---|---|---|---|---|---|---|---|---|---|
| **Trait** | *n* | **0** | **1** | **2** | **3** | **4** | **5** | **6** | **7** |
| Winging UI1 | 52 | | 0.404 | 0.096 | 0.500 | 0.000 | | | |
| Shoveling UI1 | 48 | 0.000 | 0.000 | 0.104 | 0.542 | 0.208 | 0.083 | 0.063 | 0.000 |
| Double-shoveling UI1 | 45 | 0.067 | 0.133 | 0.267 | 0.378 | 0.089 | 0.067 | 0.000 | |
| Interruption grooves UI2 | 46 | 0.522 | 0.478 | | | | | | |
| Tuberculum dentale UI2 | 48 | 0.750 | 0.000 | 0.063 | 0.000 | 0.000 | 0.000 | 0.000 | 0.187 |
| Bushman canine UC | 43 | 0.953 | 0.023 | 0.023 | 0.000 | | | | |
| Distal acc. ridge UC | 15 | 0.200 | 0.067 | 0.133 | 0.400 | 0.200 | 0.000 | | |
| Uto-Aztecan UP1 | 50 | 1.000 | 0.000 | | | | | | |
| Hypocone UM2 | 64 | 0.109 | 0.016 | 0.156 | 0.328 | 0.391 | 0.000 | 0.000 | |
| Cusp 5 UM1 | 38 | 0.921 | 0.000 | 0.000 | 0.026 | 0.053 | 0.000 | | |
| Carabelli's cusp UM1 | 39 | 0.333 | 0.231 | 0.205 | 0.205 | 0.000 | 0.026 | 0.000 | 0.000 |
| Parastyle UM3 | 59 | 0.932 | 0.034 | 0.017 | 0.017 | 0.000 | 0.000 | 0.000 | |
| Enamel extension UM1 | 102 | 0.108 | 0.118 | 0.010 | 0.765 | | | | |
| Root number UP1 | 107 | | 0.963 | 0.037 | 0.000 | | | | |
| Root number UM2 | 71 | | 0.493 | 0.155 | 0.338 | 0.014 | | | |
| PRM UM3 | 115 | 0.670 | 0.026 | 0.000 | 0.304 | | | | |
| Lingual cusp no. LP2 | 26 | 0.000 | 0.731 | 0.231 | 0.038 | | | | |
| Groove pattern LM2 | 65 | | 0.031 | 0.862 | 0.108 | | | | |
| Cusp number LM1 | 37 | 0.000 | 0.595 | 0.405 | | | | | |
| Cusp number LM2 | 61 | 0.115 | 0.738 | 0.147 | | | | | |
| Deflecting wrinkle LM1 | 32 | 0.250 | 0.063 | 0.219 | 0.469 | | | | |
| Trigonid crest LM1 | 37 | 0.892 | 0.108 | | | | | | |
| Protostylid LM1 | 45 | 0.733 | 0.267 | 0.000 | 0.000 | 0.000 | 0.000 | 0.000 | 0.000 |
| Cusp 7 LM1 | 52 | 0.942 | 0.019 | 0.000 | 0.019 | 0.019 | 0.000 | | |
| Tomes' root LP1 | 60 | 0.000 | 0.483 | 0.367 | 0.133 | 0.017 | 0.000 | 0.000 | 0.000 |
| Root number LC | 94 | | 0.989 | 0.011 | | | | | |
| Root number LM1 | 123 | | 0.000 | 0.943 | 0.057 | | | | |
| Root number LM2 | 90 | | 0.467 | 0.533 | 0.000 | | | | |
| Odontome UP1,2/LP1,2 | 27 | 1.000 | 0.000 | | | | | | |

## Peru 1 and 2

| | | Grade | | | | | | | |
|---|---|---|---|---|---|---|---|---|---|
| **Trait** | *n* | **0** | **1** | **2** | **3** | **4** | **5** | **6** | **7** |
| Winging UI1 | 49 | | 0.857 | 0.000 | 0.143 | 0.000 | | | |
| Shoveling UI1 | 50 | 0.000 | 0.000 | 0.100 | 0.400 | 0.220 | 0.160 | 0.120 | 0.000 |
| Double-shoveling UI1 | 16 | 0.063 | 0.188 | 0.188 | 0.250 | 0.000 | 0.313 | 0.000 | |
| Interruption grooves UI2 | 34 | 0.618 | 0.382 | | | | | | |
| Tuberculum dentale UI2 | 39 | 0.615 | 0.000 | 0.128 | 0.077 | 0.000 | 0.000 | 0.000 | 0.179 |
| Bushman canine UC | 60 | 1.000 | 0.000 | 0.000 | 0.000 | | | | |
| Distal acc. ridge UC | 21 | 0.190 | 0.095 | 0.095 | 0.381 | 0.238 | 0.000 | | |
| Uto-Aztecan UP1 | 118 | 1.000 | 0.000 | | | | | | |
| Hypocone UM2 | 344 | 0.067 | 0.032 | 0.113 | 0.276 | 0.439 | 0.073 | 0.000 | |
| Cusp 5 UM1 | 162 | 0.846 | 0.019 | 0.074 | 0.056 | 0.006 | 0.000 | | |
| Carabelli's cusp UM1 | 226 | 0.376 | 0.230 | 0.235 | 0.106 | 0.035 | 0.009 | 0.009 | 0.000 |
| Parastyle UM3 | 128 | 0.953 | 0.023 | 0.023 | 0.000 | 0.000 | 0.000 | 0.000 | |
| Enamel extension UM1 | 310 | 0.335 | 0.181 | 0.019 | 0.465 | | | | |
| Root number UP1 | 450 | | 0.867 | 0.129 | 0.004 | | | | |
| Root number UM2 | 240 | | 0.379 | 0.175 | 0.446 | | | | |
| PRM UM3 | 333 | 0.718 | 0.009 | 0.015 | 0.258 | | | | |
| Lingual cusp no. LP2 | 82 | 0.000 | 0.756 | 0.207 | 0.037 | | | | |
| Groove pattern LM2 | 152 | | 0.211 | 0.711 | 0.079 | | | | |
| Cusp number LM1 | 119 | 0.000 | 0.387 | 0.613 | | | | | |
| Cusp number LM2 | 197 | 0.086 | 0.822 | 0.092 | | | | | |
| Deflecting wrinkle LM1 | 58 | 0.345 | 0.069 | 0.259 | 0.328 | | | | |
| Trigonid crest LM1 | 121 | 0.967 | 0.033 | | | | | | |
| Protostylid LM1 | 160 | 0.731 | 0.250 | 0.000 | 0.019 | 0.000 | 0.000 | 0.000 | 0.000 |
| Cusp 7 LM1 | 217 | 0.876 | 0.018 | 0.032 | 0.041 | 0.023 | 0.009 | | |
| Tomes' root LP1 | 86 | 0.035 | 0.326 | 0.221 | 0.128 | 0.267 | 0.000 | 0.023 | 0.000 |
| Root number LC | 199 | | 0.985 | 0.015 | | | | | |
| Root number LM1 | 217 | | 0.000 | 0.949 | 0.051 | | | | |
| Root number LM2 | 234 | | 0.308 | 0.692 | 0.000 | | | | |
| Odontome UP1,2/LP1,2 | 154 | 0.968 | 0.032 | | | | | | |

## *Preceramic Peru*

| Trait | n | Grade 0 | 1 | 2 | 3 | 4 | 5 | 6 | 7 |
|---|---|---|---|---|---|---|---|---|---|
| Winging UI1 | 7 | | 0.571 | 0.000 | 0.429 | 0.000 | | | |
| Shoveling UI1 | 34 | 0.000 | 0.000 | 0.147 | 0.559 | 0.088 | 0.147 | 0.059 | 0.000 |
| Double-shoveling UI1 | 32 | 0.188 | 0.094 | 0.344 | 0.094 | 0.125 | 0.156 | 0.000 | |
| Interruption grooves UI2 | 20 | 0.350 | 0.650 | | | | | | |
| Tuberculum dentale UI2 | 20 | 0.700 | 0.000 | 0.050 | 0.100 | 0.000 | 0.000 | 0.000 | 0.150 |
| Bushman canine UC | 15 | 1.000 | 0.000 | 0.000 | 0.000 | | | | |
| Distal acc. ridge UC | 12 | 0.000 | 0.250 | 0.250 | 0.167 | 0.250 | 0.083 | | |
| Uto-Aztecan UP1 | 18 | 1.000 | 0.000 | | | | | | |
| Hypocone UM2 | 44 | 0.136 | 0.045 | 0.091 | 0.341 | 0.318 | 0.068 | 0.000 | |
| Cusp 5 UM1 | 28 | 0.893 | 0.071 | 0.000 | 0.000 | 0.000 | 0.036 | | |
| Carabelli's cusp UM1 | 36 | 0.250 | 0.278 | 0.333 | 0.056 | 0.000 | 0.056 | 0.028 | 0.000 |
| Parastyle UM3 | 13 | 0.923 | 0.077 | 0.000 | 0.000 | 0.000 | 0.000 | 0.000 | |
| Enamel extension UM1 | 53 | 0.302 | 0.189 | 0.000 | 0.509 | | | | |
| Root number UP1 | 57 | | 0.895 | 0.105 | 0.000 | | | | |
| Root number UM2 | 28 | | 0.071 | 0.000 | 0.929 | | | | |
| PRM UM3 | 35 | 0.657 | 0.000 | 0.029 | 0.314 | | | | |
| Lingual cusp no. LP2 | 10 | 0.000 | 0.800 | 0.000 | 0.200 | | | | |
| Groove pattern LM2 | 44 | | 0.227 | 0.705 | 0.068 | | | | |
| Cusp number LM1 | 35 | 0.000 | 0.514 | 0.486 | | | | | |
| Cusp number LM2 | 49 | 0.020 | 0.837 | 0.143 | | | | | |
| Deflecting wrinkle LM1 | 21 | 0.333 | 0.190 | 0.333 | 0.143 | | | | |
| Trigonid crest LM1 | 26 | 0.923 | 0.077 | | | | | | |
| Protostylid LM1 | 30 | 0.667 | 0.267 | 0.000 | 0.000 | 0.033 | 0.033 | 0.000 | 0.000 |
| Cusp 7 LM1 | 31 | 0.968 | 0.032 | 0.000 | 0.000 | 0.000 | 0.000 | | |
| Tomes' root LP1 | 14 | 0.000 | 0.000 | 0.286 | 0.357 | 0.357 | 0.000 | 0.000 | 0.000 |
| Root number LC | 21 | | 1.000 | 0.000 | | | | | |
| Root number LM1 | 53 | | 0.000 | 0.981 | 0.019 | | | | |
| Root number LM2 | 41 | | 0.195 | 0.780 | 0.024 | | | | |
| Odontome UP1,2/LP1,2 | 23 | 0.913 | 0.087 | | | | | | |

| Trait | n | Grade 0 | 1 | 2 | 3 | 4 | 5 | 6 | 7 |
|---|---|---|---|---|---|---|---|---|---|
| Winging UI1 | 42 | | 0.619 | 0.024 | 0.333 | 0.024 | | | |
| Shoveling UI1 | 50 | 0.000 | 0.000 | 0.100 | 0.480 | 0.260 | 0.120 | 0.040 | 0.000 |
| Double-shoveling UI1 | 49 | 0.082 | 0.918 | | | | | | |
| Interruption grooves UI2 | | | | | | | | | |
| Tuberculum dentale UI2 | | | | | | | | | |
| Bushman canine UC | 61 | 0.951 | 0.016 | 0.000 | 0.033 | | | | |
| Distal acc. ridge UC | 23 | 0.261 | 0.000 | 0.174 | 0.304 | 0.217 | 0.043 | | |
| Uto-Aztecan UP1 | 55 | 1.000 | 0.000 | | | | | | |
| Hypocone UM2 | 66 | 0.045 | 0.015 | 0.061 | 0.258 | 0.561 | 0.061 | 0.000 | |
| Cusp 5 UM1 | 34 | 0.824 | 0.029 | 0.147 | 0.000 | 0.000 | 0.000 | | |
| Carabelli's cusp UM1 | 49 | 0.143 | 0.469 | 0.143 | 0.102 | 0.020 | 0.061 | 0.061 | 0.000 |
| Parastyle UM3 | 58 | 0.966 | 0.017 | 0.017 | 0.000 | 0.000 | 0.000 | 0.000 | |
| Enamel extension UM1 | 97 | 0.299 | 0.144 | 0.021 | 0.536 | | | | |
| Root number UP1 | 83 | | 0.867 | 0.120 | 0.012 | | | | |
| Root number UM2 | 79 | | 0.241 | 0.152 | 0.608 | | | | |
| PRM UM3 | 89 | 0.831 | | | 0.169 | | | | |
| Lingual cusp no. LP2 | 54 | 0.000 | 0.593 | 0.333 | 0.074 | | | | |
| Groove pattern LM2 | 81 | | 0.136 | 0.691 | 0.173 | | | | |
| Cusp number LM1 | 47 | 0.000 | 0.468 | 0.532 | | | | | |
| Cusp number LM2 | 73 | 0.096 | 0.904 | 0.000 | | | | | |
| Deflecting wrinkle LM1 | 31 | 0.065 | 0.032 | 0.355 | 0.548 | | | | |
| Trigonid crest LM1 | 67 | 0.970 | 0.030 | | | | | | |
| Protostylid LM1 | 61 | 0.820 | 0.148 | 0.016 | 0.016 | 0.000 | 0.000 | 0.000 | 0.000 |
| Cusp 7 LM1 | 82 | 0.915 | 0.012 | 0.012 | 0.049 | 0.012 | 0.000 | | |
| Tomes' root LP1 | 43 | 0.977 | 0.023 | | | | | | |
| Root number LC | 61 | | 0.984 | 0.016 | | | | | |
| Root number LM1 | 91 | | 0.000 | 0.956 | 0.044 | | | | |
| Root number LM2 | 79 | | 0.316 | 0.684 | 0.000 | | | | |
| Odontome UP1,2/LP1,2 | 39 | 0.923 | 0.077 | | | | | | |

## Santa Elena

| Trait | n | Grade 0 | 1 | 2 | 3 | 4 | 5 | 6 | 7 |
|---|---|---|---|---|---|---|---|---|---|
| Winging UI1 | 19 | | 0.684 | 0.000 | 0.316 | 0.000 | | | |
| Shoveling UI1 | 31 | 0.000 | 0.000 | 0.000 | 0.258 | 0.323 | 0.355 | 0.065 | 0.000 |
| Double-shoveling UI1 | 31 | 0.032 | 0.968 | | | | | | |
| Interruption grooves UI2 | | | | | | | | | |
| Tuberculum dentale UI2 | | | | | | | | | |
| Bushman canine UC | 35 | 1.000 | 0.000 | 0.000 | 0.000 | | | | |
| Distal acc. ridge UC | 27 | 0.111 | 0.074 | 0.074 | 0.222 | 0.481 | 0.037 | | |
| Uto-Aztecan UP1 | 25 | 1.000 | 0.000 | | | | | | |
| Hypocone UM2 | 34 | 0.059 | 0.029 | 0.088 | 0.235 | 0.500 | 0.088 | 0.000 | |
| Cusp 5 UM1 | 26 | 0.808 | 0.154 | 0.000 | 0.000 | 0.038 | 0.000 | | |
| Carabelli's cusp UM1 | 27 | 0.148 | 0.370 | 0.185 | 0.111 | 0.074 | 0.037 | 0.074 | 0.000 |
| Parastyle UM3 | 31 | 0.935 | 0.065 | 0.000 | 0.000 | 0.000 | 0.000 | 0.000 | |
| Enamel extension UM1 | 20 | 0.250 | 0.300 | 0.050 | 0.400 | | | | |
| Root number UP1 | 23 | | 0.696 | 0.304 | 0.000 | | | | |
| Root number UM2 | 21 | | 0.048 | 0.095 | 0.857 | | | | |
| PRM UM3 | 33 | 1.000 | | | 0.000 | | | | |
| Lingual cusp no. LP2 | 24 | 0.000 | 0.708 | 0.292 | 0.000 | | | | |
| Groove pattern LM2 | 30 | | 0.167 | 0.800 | 0.033 | | | | |
| Cusp number LM1 | 10 | 0.000 | 0.200 | 0.800 | | | | | |
| Cusp number LM2 | 35 | 0.057 | 0.943 | 0.000 | | | | | |
| Deflecting wrinkle LM1 | 5 | 0.000 | 0.200 | 0.200 | 0.600 | | | | |
| Trigonid crest LM1 | 12 | 1.000 | 0.000 | | | | | | |
| Protostylid LM1 | 32 | 0.750 | 0.250 | 0.000 | 0.000 | 0.000 | 0.000 | 0.000 | 0.000 |
| Cusp 7 LM1 | 25 | 1.000 | 0.000 | 0.000 | 0.000 | 0.000 | 0.000 | | |
| Tomes' root LP1 | 17 | 0.941 | 0.059 | | | | | | |
| Root number LC | 25 | | 1.000 | 0.000 | | | | | |
| Root number LM1 | 43 | | 0.000 | 0.953 | 0.047 | | | | |
| Root number LM2 | 36 | | 0.278 | 0.722 | 0.000 | | | | |
| Odontome UP1,2/LP1,2 | 32 | 0.938 | 0.063 | | | | | | |

## Southeast Asia

### *Borneo*

| Trait | *n* | Grade | | | | | | | |
|---|---|---|---|---|---|---|---|---|---|
| | | 0 | 1 | 2 | 3 | 4 | 5 | 6 | 7 |
| Winging UI1 | 14 | | 0.286 | 0.000 | 0.714 | 0.000 | | | |
| Shoveling UI1 | 17 | 0.118 | 0.235 | 0.353 | 0.176 | 0.059 | 0.059 | 0.000 | 0.000 |
| Double-shoveling UI1 | 17 | 0.882 | 0.000 | 0.059 | 0.000 | 0.000 | 0.059 | 0.000 | |
| Interruption grooves UI2 | 20 | 0.700 | 0.300 | | | | | | |
| Tuberculum dentale UI2 | 22 | 0.773 | 0.000 | 0.045 | 0.091 | 0.045 | 0.000 | 0.000 | 0.045 |
| Bushman canine UC | 27 | 0.889 | 0.074 | 0.037 | 0.000 | | | | |
| Distal acc. ridge UC | 6 | 0.833 | 0.167 | 0.000 | 0.000 | 0.000 | 0.000 | | |
| Uto-Aztecan UP1 | 4 | 0.500 | 0.000 | 0.250 | 0.000 | 0.250 | 0.000 | | |
| Hypocone UM2 | 79 | 0.063 | 0.013 | 0.025 | 0.177 | 0.582 | 0.139 | 0.000 | |
| Cusp 5 UM1 | 60 | 0.800 | 0.133 | 0.033 | 0.000 | 0.033 | 0.000 | | |
| Carabelli's cusp UM1 | 70 | 0.286 | 0.214 | 0.057 | 0.171 | 0.014 | 0.114 | 0.043 | 0.100 |
| Parastyle UM3 | 39 | 0.923 | 0.000 | 0.026 | 0.000 | 0.026 | 0.026 | 0.000 | |
| Enamel extension UM1 | 85 | 0.365 | 0.306 | 0.012 | 0.318 | | | | |
| Root number UP1 | 133 | | 0.534 | 0.451 | 0.015 | | | | |
| Root number UM2 | 105 | | 0.057 | 0.124 | 0.819 | | | | |
| PRM UM3 | 104 | 0.721 | 0.000 | 0.029 | 0.250 | | | | |
| Lingual cusp no. LP2 | 37 | 0.000 | 0.162 | 0.811 | 0.027 | | | | |
| Groove pattern LM2 | 55 | | 0.255 | 0.509 | 0.236 | | | | |
| Cusp number LM1 | 43 | 0.000 | 0.651 | 0.349 | | | | | |
| Cusp number LM2 | 52 | 0.250 | 0.577 | 0.173 | | | | | |
| Deflecting wrinkle LM1 | 10 | 0.800 | 0.000 | 0.200 | 0.000 | | | | |
| Trigonid crest LM1 | 13 | 0.923 | 0.077 | | | | | | |
| Protostylid LM1 | 49 | 0.714 | 0.204 | 0.000 | 0.061 | 0.020 | 0.000 | 0.000 | 0.000 |
| Cusp 7 LM1 | 60 | 0.850 | 0.017 | 0.017 | 0.050 | 0.033 | 0.033 | | |
| Tomes' root LP1 | 31 | 0.258 | 0.226 | 0.065 | 0.129 | 0.161 | 0.032 | 0.032 | 0.097 |
| Root number LC | 58 | | 0.983 | 0.017 | | | | | |
| Root number LM1 | 85 | | 0.000 | 0.859 | 0.141 | | | | |
| Root number LM2 | 73 | | 0.205 | 0.795 | 0.000 | | | | |
| Odontome UP1,2/LP1,2 | 37 | 0.973 | 0.027 | | | | | | |

## *Calatagan*

| | | Grade | | | | | | | |
|---|---|---|---|---|---|---|---|---|---|
| **Trait** | *n* | **0** | **1** | **2** | **3** | **4** | **5** | **6** | **7** |
| Winging UI1 | 7 | | 0.286 | 0.000 | 0.714 | 0.000 | | | |
| Shoveling UI1 | 14 | 0.000 | 0.000 | 0.500 | 0.214 | 0.214 | 0.071 | 0.000 | 0.000 |
| Double-shoveling UI1 | 6 | 0.833 | 0.000 | 0.167 | 0.000 | 0.000 | 0.000 | 0.000 | |
| Interruption grooves UI2 | 14 | 0.643 | 0.357 | | | | | | |
| Tuberculum dentale UI2 | 14 | 0.643 | 0.000 | 0.214 | 0.000 | 0.000 | 0.000 | 0.000 | 0.142 |
| Bushman canine UC | 16 | 1.000 | 0.000 | 0.000 | 0.000 | | | | |
| Distal acc. ridge UC | 10 | 0.400 | 0.100 | 0.200 | 0.300 | 0.000 | 0.000 | | |
| Uto-Aztecan UP1 | 21 | 1.000 | 0.000 | | | | | | |
| Hypocone UM2 | 27 | 0.111 | 0.074 | 0.000 | 0.259 | 0.519 | 0.037 | 0.000 | |
| Cusp 5 UM1 | 29 | 0.793 | 0.069 | 0.069 | 0.069 | 0.000 | 0.000 | | |
| Carabelli's cusp UM1 | 30 | 0.367 | 0.200 | 0.067 | 0.033 | 0.000 | 0.133 | 0.133 | 0.067 |
| Parastyle UM3 | 19 | 1.000 | 0.000 | 0.000 | 0.000 | 0.000 | 0.000 | 0.000 | |
| Enamel extension UM1 | 31 | 0.355 | 0.226 | 0.065 | 0.355 | | | | |
| Root number UP1 | 29 | | 0.655 | 0.276 | 0.069 | | | | |
| Root number UM2 | 30 | | 0.133 | 0.167 | 0.667 | 0.033 | | | |
| PRM UM3 | 28 | 0.929 | | | 0.071 | | | | |
| Lingual cusp no. LP2 | 22 | 0.000 | 0.182 | 0.773 | 0.045 | | | | |
| Groove pattern LM2 | 36 | | 0.333 | 0.444 | 0.222 | | | | |
| Cusp number LM1 | 32 | 0.000 | 0.688 | 0.313 | | | | | |
| Cusp number LM2 | 35 | 0.314 | 0.572 | 0.114 | | | | | |
| Deflecting wrinkle LM1 | 28 | 0.607 | 0.000 | 0.179 | 0.214 | | | | |
| Trigonid crest LM1 | 31 | 0.935 | 0.065 | | | | | | |
| Protostylid LM1 | 35 | 0.829 | 0.171 | 0.000 | 0.000 | 0.000 | 0.000 | 0.000 | 0.000 |
| Cusp 7 LM1 | 36 | 0.917 | 0.028 | 0.000 | 0.028 | 0.000 | 0.028 | | |
| Tomes' root LP1 | 17 | 0.000 | 0.176 | 0.353 | 0.176 | 0.176 | 0.118 | 0.000 | 0.000 |
| Root number LC | 35 | | 0.971 | 0.029 | | | | | |
| Root number LM1 | 42 | | 0.000 | 0.762 | 0.238 | | | | |
| Root number LM2 | 40 | | 0.250 | 0.750 | 0.000 | | | | |
| Odontome UP1,2/LP1,2 | 27 | 0.926 | 0.074 | | | | | | |

## *Malay*

| Trait | *n* | Grade 0 | 1 | 2 | 3 | 4 | 5 | 6 | 7 |
|---|---|---|---|---|---|---|---|---|---|
| Winging UI1 | 24 | | 0.042 | 0.000 | 0.958 | 0.000 | | | |
| Shoveling UI1 | 14 | 0.143 | 0.143 | 0.429 | 0.286 | 0.000 | 0.000 | 0.000 | 0.000 |
| Double-shoveling UI1 | 12 | 0.917 | 0.083 | | | | | | |
| Interruption grooves UI2 | 17 | 0.647 | 0.353 | | | | | | |
| Tuberculum dentale UI2 | 18 | 0.667 | 0.000 | 0.056 | 0.111 | 0.000 | 0.000 | 0.000 | 0.167 |
| Bushman canine UC | 23 | 0.957 | 0.043 | 0.000 | 0.000 | | | | |
| Distal acc. ridge UC | 12 | 0.167 | 0.167 | 0.250 | 0.167 | 0.167 | 0.083 | | |
| Uto-Aztecan UP1 | 13 | 1.000 | 0.000 | | | | | | |
| Hypocone UM2 | 42 | 0.048 | 0.048 | 0.095 | 0.214 | 0.381 | 0.214 | 0.000 | |
| Cusp 5 UM1 | 40 | 0.550 | 0.100 | 0.100 | 0.150 | 0.075 | 0.025 | | |
| Carabelli's cusp UM1 | 43 | 0.256 | 0.302 | 0.023 | 0.047 | 0.116 | 0.116 | 0.023 | 0.116 |
| Parastyle UM3 | 35 | 0.914 | 0.000 | 0.057 | 0.029 | 0.000 | 0.000 | 0.000 | |
| Enamel extension UM1 | 48 | 0.479 | 0.146 | 0.042 | 0.333 | | | | |
| Root number UP1 | 50 | | 0.380 | 0.620 | 0.000 | | | | |
| Root number UM2 | 44 | | 0.068 | 0.227 | 0.705 | | | | |
| PRM UM3 | 49 | 0.816 | | | 0.184 | | | | |
| Lingual cusp no. LP2 | 33 | 0.000 | 0.182 | 0.727 | 0.091 | | | | |
| Groove pattern LM2 | 39 | | 0.385 | 0.436 | 0.179 | | | | |
| Cusp number LM1 | 23 | 0.000 | 0.652 | 0.348 | | | | | |
| Cusp number LM2 | 39 | 0.333 | 0.641 | 0.026 | | | | | |
| Deflecting wrinkle LM1 | 13 | 0.538 | 0.000 | 0.231 | 0.231 | | | | |
| Trigonid crest LM1 | 23 | 0.826 | 0.174 | | | | | | |
| Protostylid LM1 | 31 | 0.710 | 0.258 | 0.032 | 0.000 | 0.000 | 0.000 | 0.000 | 0.000 |
| Cusp 7 LM1 | 32 | 0.938 | 0.000 | 0.031 | 0.000 | 0.000 | 0.031 | | |
| Tomes' root LP1 | 31 | 0.710 | 0.032 | 0.000 | 0.065 | 0.032 | 0.000 | 0.000 | 0.161 |
| Root number LC | 30 | | 1.000 | 0.000 | | | | | |
| Root number LM1 | 40 | | 0.000 | 0.900 | 0.100 | | | | |
| Root number LM2 | 41 | | 0.268 | 0.683 | 0.049 | | | | |
| Odontome UP1,2/LP1,2 | 45 | 1.000 | 0.000 | | | | | | |

*Philippines*

| Trait | n | Grade 0 | 1 | 2 | 3 | 4 | 5 | 6 | 7 |
|---|---|---|---|---|---|---|---|---|---|
| Winging UI1 | 52 | | 0.135 | 0.000 | 0.865 | 0.000 | | | |
| Shoveling UI1 | 54 | 0.000 | 0.074 | 0.500 | 0.278 | 0.130 | 0.019 | 0.000 | 0.000 |
| Double-shoveling UI1 | 29 | 0.828 | 0.000 | 0.103 | 0.000 | 0.000 | 0.069 | 0.000 | |
| Interruption grooves UI2 | 54 | 0.722 | 0.278 | | | | | | |
| Tuberculum dentale UI2 | 58 | 0.776 | 0.000 | 0.052 | 0.034 | 0.000 | 0.017 | 0.000 | 0.120 |
| Bushman canine UC | 76 | 0.974 | 0.013 | 0.013 | 0.000 | | | | |
| Distal acc. ridge UC | 14 | 0.286 | 0.071 | 0.357 | 0.143 | 0.143 | 0.000 | | |
| Uto-Aztecan UP1 | 4 | 1.000 | 0.000 | | | | | | |
| Hypocone UM2 | 149 | 0.121 | 0.040 | 0.067 | 0.215 | 0.503 | 0.054 | 0.000 | |
| Cusp 5 UM1 | 124 | 0.895 | 0.073 | 0.016 | 0.000 | 0.016 | 0.000 | | |
| Carabelli's cusp UM1 | 147 | 0.401 | 0.231 | 0.082 | 0.095 | 0.027 | 0.095 | 0.034 | 0.034 |
| Parastyle UM3 | 101 | 0.990 | 0.000 | 0.000 | 0.010 | 0.000 | 0.000 | 0.000 | |
| Enamel extension UM1 | 132 | 0.379 | 0.220 | 0.076 | 0.326 | | | | |
| Root number UP1 | 172 | | 0.674 | 0.291 | 0.035 | | | | |
| Root number UM2 | 148 | | 0.135 | 0.135 | 0.716 | 0.014 | | | |
| PRM UM3 | 172 | 0.773 | 0.012 | 0.023 | 0.192 | | | | |
| Lingual cusp no. LP2 | 90 | 0.000 | 0.333 | 0.644 | 0.022 | | | | |
| Groove pattern LM2 | 124 | | 0.323 | 0.540 | 0.137 | | | | |
| Cusp number LM1 | 98 | 0.020 | 0.592 | 0.388 | | | | | |
| Cusp number LM2 | 123 | 0.285 | 0.577 | 0.138 | | | | | |
| Deflecting wrinkle LM1 | 13 | 0.462 | 0.077 | 0.385 | 0.077 | | | | |
| Trigonid crest LM1 | 20 | 1.000 | 0.000 | | | | | | |
| Protostylid LM1 | 122 | 0.811 | 0.148 | 0.000 | 0.033 | 0.000 | 0.008 | 0.000 | 0.000 |
| Cusp 7 LM1 | 129 | 0.938 | 0.016 | 0.008 | 0.008 | 0.016 | 0.016 | | |
| Tomes' root LP1 | 63 | 0.444 | 0.143 | 0.143 | 0.063 | 0.143 | 0.032 | 0.016 | 0.016 |
| Root number LC | 37 | | 1.000 | 0.000 | | | | | |
| Root number LM1 | 127 | | 0.000 | 0.827 | 0.173 | | | | |
| Root number LM2 | 122 | | 0.279 | 0.721 | 0.000 | | | | |
| Odontome UP1,2/LP1,2 | 117 | 0.974 | 0.026 | | | | | | |

*Taiwan*

| Trait | n | Grade 0 | 1 | 2 | 3 | 4 | 5 | 6 | 7 |
|---|---|---|---|---|---|---|---|---|---|
| Winging UI1 | 38 | | 0.237 | 0.053 | 0.684 | 0.026 | | | |
| Shoveling UI1 | 45 | 0.000 | 0.044 | 0.244 | 0.467 | 0.133 | 0.089 | 0.022 | 0.000 |
| Double-shoveling UI1 | 43 | 0.535 | 0.186 | 0.023 | 0.000 | 0.000 | 0.256 | 0.000 | |
| Interruption grooves UI2 | 28 | 0.464 | 0.536 | | | | | | |
| Tuberculum dentale UI2 | 27 | 0.519 | 0.000 | 0.111 | 0.000 | 0.000 | 0.037 | 0.000 | 0.333 |
| Bushman canine UC | 44 | 0.977 | 0.023 | 0.000 | 0.000 | | | | |
| Distal acc. ridge UC | | | | | | | | | |
| Uto-Aztecan UP1 | | | | | | | | | |
| Hypocone UM2 | 57 | 0.211 | 0.088 | 0.105 | 0.158 | 0.421 | 0.018 | 0.000 | |
| Cusp 5 UM1 | 38 | 0.947 | 0.000 | 0.000 | 0.026 | 0.000 | 0.026 | | |
| Carabelli's cusp UM1 | 53 | 0.434 | 0.283 | 0.075 | 0.019 | 0.038 | 0.113 | 0.038 | 0.000 |
| Parastyle UM3 | 18 | 1.000 | 0.000 | 0.000 | 0.000 | 0.000 | 0.000 | 0.000 | |
| Enamel extension UM1 | 61 | 0.164 | 0.148 | 0.066 | 0.623 | | | | |
| Root number UP1 | 56 | | 0.643 | 0.321 | 0.036 | | | | |
| Root number UM2 | 43 | | 0.047 | 0.140 | 0.814 | | | | |
| PRM UM3 | 48 | 0.500 | 0.000 | 0.042 | 0.458 | | | | |
| Lingual cusp no. LP2 | 24 | 0.000 | 0.208 | 0.750 | 0.042 | | | | |
| Groove pattern LM2 | 19 | | 0.526 | 0.368 | 0.105 | | | | |
| Cusp number LM1 | 15 | 0.000 | 0.533 | 0.467 | | | | | |
| Cusp number LM2 | 21 | 0.190 | 0.762 | 0.048 | | | | | |
| Deflecting wrinkle LM1 | | | | | | | | | |
| Trigonid crest LM1 | | | | | | | | | |
| Protostylid LM1 | 29 | 0.931 | 0.000 | 0.000 | 0.034 | 0.000 | 0.034 | 0.000 | 0.000 |
| Cusp 7 LM1 | 33 | 0.939 | 0.030 | 0.030 | 0.000 | 0.000 | 0.000 | | |
| Tomes' root LP1 | 13 | 0.000 | 0.692 | 0.154 | 0.077 | 0.077 | 0.000 | 0.000 | 0.000 |
| Root number LC | 14 | | 1.000 | 0.000 | | | | | |
| Root number LM1 | 25 | | 0.000 | 0.960 | 0.040 | | | | |
| Root number LM2 | 21 | | 0.381 | 0.619 | 0.000 | | | | |
| Odontome UP1,2/LP1,2 | 18 | 0.944 | 0.056 | | | | | | |

Printed in the USA
CPSIA information can be obtained
at www.ICGtesting.com
LVHW080624131223
766279LV00006B/482

9 781107 480735